INTRODUCTION TO SUPRAMOLECULAR CHEMISTRY

Introduction to Supramolecular Chemistry

by

Helena Dodziuk

Institute of Physical Chemistry,
Polish Academy of Sciences,
Warsaw, Poland

ISBN 1-4020-0214-9

Dodziuk: Introduction to Supramolecular Chemistry

First Indian Reprint 2007

ISBN 978-81-8128-734-2

This edition is manufactured in India for sale only in India, Pakistan, Bangladesh, Nepal and Sri Lanka and any other country as authorized by the publisher.

This edition is published by Springer (India) Private Limited,
A part of Springer Science+Business Media, Registered Office: 906-907,
Akash Deep Building, Barakhamba Road, New Delhi – 110 001, India.

Printed in India by Rajkamal Electric Press, Delhi.

Contents

PREFACE

Supramolecular chemistry emerged only a few decades ago but it is developing rapidly despite the lack of a precise definition of this domain. Interacting with chemistry, physics, biology, and technology it is gaining its strength from fruitful collaborations of researchers representing these diverse fields. It promises, on the one hand, a better understanding of processes in living organisms on the molecular level and, on the other, numerous applications which will change our everyday life. A supermolecule, the subject of study in supramolecular chemistry, is composed of molecules and/or ions held together by weak nonbonding interactions. Weak, but numerous, these interactions may dramatically change the properties of constituent parts of the association. Anions of alkaline metals created owing to a high affinity of cryptands to these metals, nitrogen atoms and He_2 and Ne_2 molecules isolated in fullerene cages, and stable, otherwise short-lived, species obtained in 'molecular flasks' are probably the most spectacular examples of nontrivial effects resulting from the supermolecule creation. The aim of this book is an introductory presentation of this fascinating field to research scientists working in related areas and to Ph.D. students. It will be useful to specialists as well since it gives a comprehensive, fully referenced, concise and balanced view of the subject. The book is divided into two parts.

General ideas constituting the basis of supramolecular chemistry, its interdisciplinary character, present and future potential applications are presented in the first part. The second part gives a brief but complete overview of important groups of compounds and systems involved. I have been fascinated by their variety and by prospects of industrial applications and hope to transmit my fascination to the reader.

While working on the book I received generous help from many people. Dr. O. Lukin and Mr. G. Dolgonos took an integral part in the process from stimulating comments in the beginning to formatting formulae, preparing drawings and the camera ready copy required by the publisher in the end. Comments and critical remarks by Professors Z. R. Grabowski, B. Korybut-Daszkiewicz, J. Lipkowski, W. Kutner, W. Pasik-Bronikowska, M. Geller, J. F. Biernat, A. Poniewierski and R. Nowakowski lead to numerous improvements of the presentation and are gratefully acknowledged. Thanks are due to Professors A. Harada, J. Lipkowski and J. A. Ripmeester for supplying me with drawings.

Finally, I would like to express my hope that readers' pleasure while reading this book will not be less than that I have experienced in writing it.

Chapter 1

SUPRAMOLECULAR CHEMISTRY - WHAT IS THIS?

Supramolecular chemistry [1] is a new emerging domain lying amidst chemistry, biochemistry, physics, and material science (or technology). Its foundations were laid down less than 50 years ago and in 1987 its founding fathers, Pedersen. Cram and Lehn, were awarded the Nobel Prize in Chemistry [2] for their works on molecular recognition. According to one definition proposed by Lehn [1b], supramolecular chemistry is chemistry beyond the molecule. A concept of supermolecule was coined much earlier in the thirties [3] and was later applied to describing objects studied in this research area. Lehn's definition is not very specific. For instance, in accordance with it a monocrystal and a solution of sodium chloride in water are gigantic supermolecules. This situation could result in claims that supramolecular chemistry does not exist at all because it simply encompasses all chemistry and a great deal of physics.

Another Lehn's definition stresses the role of nonbonded interactions in supramolecular chemistry as opposed to that played by covalent interactions in classical organic chemistry. Nonbonded interactions forcing the association of molecules are characterized by much smaller energies than those of 200-400 kJ/mol typical for covalent chemical bonds. In addition to relatively strong ion-ion electrostatic interactions of ca. 4-40 kJ/mol and hydrogen bonding of ca. 1-80 kJ/mol, they include much smaller London dispersion forces, ion-induced dipole and dipole-dipole interactions that are less than 4 kJ/mol strong. Hydrophobic effects are also of this order of magnitude. The definition of supramolecular chemistry on the basis of noncovalent interactions seems a little more specific.

Unfortunately, it also covers too vast an area. It does not exclude crystals and solutions mentioned above. Moreover, it also includes polymers, in which nonbonded interactions play such an important role, into the realm of supramolecular chemistry.

In spite of the lack of a precise definition, the domain of supramolecular chemistry is blooming. It has diversified enormously and includes charge-transfer complexes [4], inclusion complexes (incorporating e.g. Cram's hemicarcerands [1e, 5] and cyclodextrins [6]), mono- and polylayers, micelles (see examples 2, 5-8 below), vesicles (Figure 1.4) [1d], liquid crystals [7] and cocrystals consisting of at least two different kinds of molecules [8] which form highly specific domains differing in the objects studied and research techniques. The specificity and separateness of the first group, i.e., charge-transfer complexes, and those of liquid crystals seem generally recognized. On the other hand, as concerns inclusion complexes or other molecular aggregates consisting of only few molecules, higher molecular aggregates, and cocrystals formed by at least two types of molecules the situation is not that clear. The objects studied in these areas differ essentially as concerns the number of molecules which are formed of and the typical methods of research used.

Inclusion, that is host-guest, complexes and small aggregates typically consist of a few (usually two) molecules and the physicochemical methods applied in their studies are very close to those used in classical organic chemistry. Contrary to such aggregates, larger molecular assemblies (micelles, vesicles, mono- and polylayers) are characterized by much larger, ill-defined number of objects forming them. In this respect they are similar to polymers of which the molecular weight is also only approximately given. The assemblies have found numerous applications but their internal structure and the mechanism in which such structures are built from isolated molecules are not fully understood. Studying such complicated structures requires novel experimental techniques other than those used to analyze single molecules. On the other hand, to study the last group of supermolecules involving crystals the standard X-ray technique is used. This group is of practical importance for the new research area bearing the name crystal engineering. The aim of this domain consists in obtaining crystals with predefined desirable properties.

Science is a complicated matter and any definition of a research area is an oversimplification. This is especially true for a new domain *in statu nascendi* such as supramolecular chemistry [9]. However, a recent development in

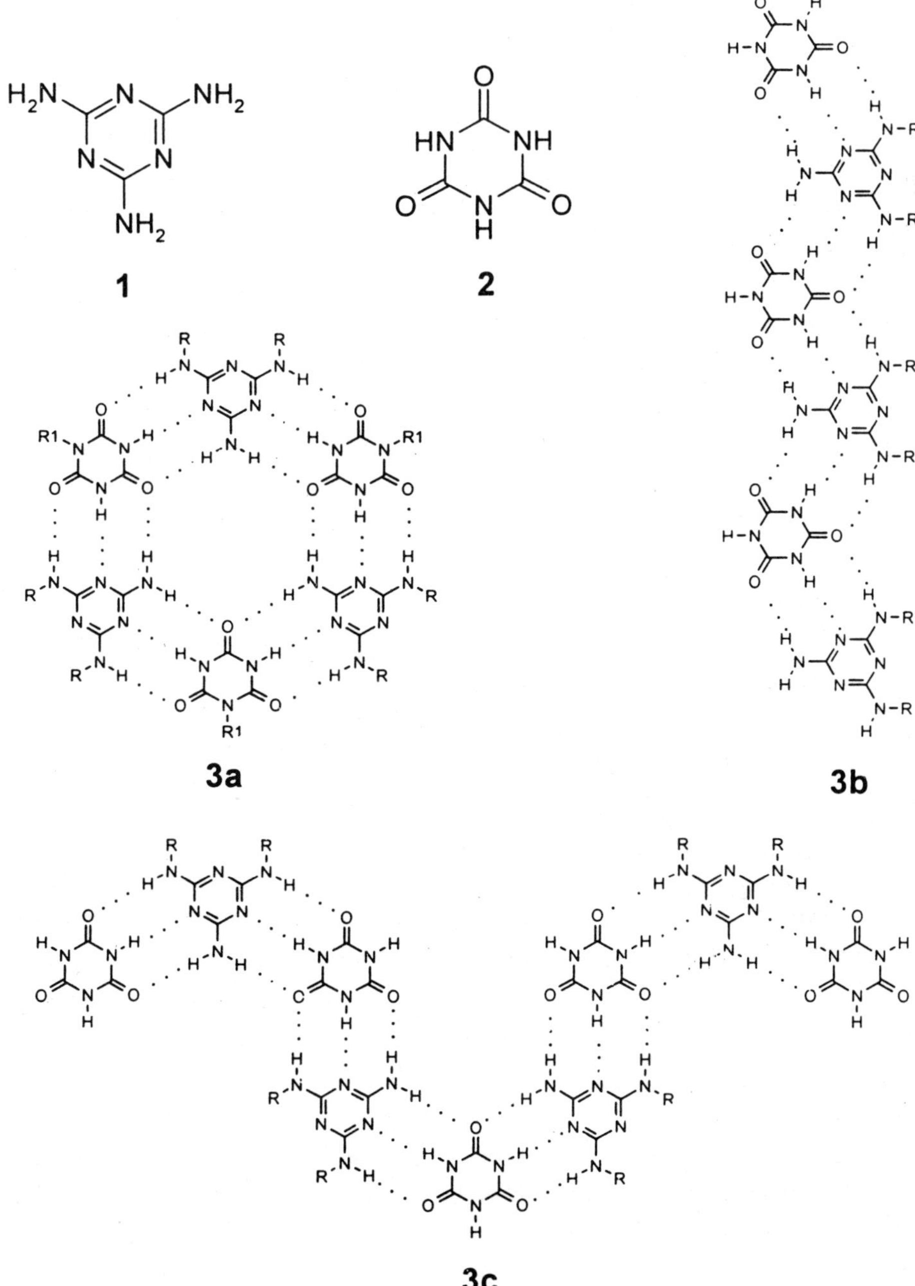

Figure 1.1. Some supramolecular architectures of hydrogen bonded Whitesides complexes.

supramolecular chemistry is so innovative, involving both novel concepts and ideas as well as specific experimental techniques, that it justifies the establishment of this new field even if at present it lacks any precise definition. Let us look at a few examples showing what makes supramolecular chemistry different from the classical organic chemistry.

1. Melamine **1** and cyanuric acid **2** derivatives can form various types of stable aggregates characterized by different hydrogen bonding patterns such as those presented in Figure 1.1 [10]. The structure of these aggregates influences their properties as reflected, amongst others, by their NMR spectra. The energy of a single hydrogen bond is much smaller than that of a covalent bond. However, one of the most complicated systems of this kind created by the Whitesides group contains as many as 54 hydrogen bonds. Even assuming a moderate value of 16 kJ/mol for the energy of one of such bonds, one arrives at more than 800 kJ/mol for the energy of the whole H-bonded system. Interestingly, the energy of these bonds is much higher than that of a standard covalent C-C bond influencing the properties of the whole system.

2. Cyclobutadiene **4** is extremely unstable under normal conditions. However, it was obtained and kept at room temperature for several months inside **5** by Cram and coworkers [5], who called the latter molecule a molecular flask.

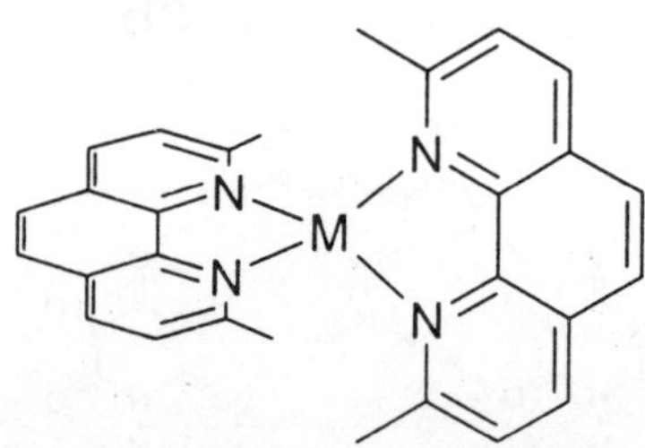

Figure 1.2. Perpendicular orientation of phenanthroline fragments complexed with metal.

3. The synthesis of a molecular knot **6** [11], olympiadane **7** [12], and many other topological molecules discussed in Sections 2.3 and 8.1 would not be possible without preorganization of substrates forcing their appropriate orientation. In this case the preorganization is accomplished by the complexation of phenanthroline fragments with a metal ion (Figure 1.2).

Thus there is an essential difference between classical homogeneous reactions in organic chemistry and reactions such as those in which catenanes and knots are formed. In the latter, there are heterogeneities on the micro scale. Thus supramolecular chemistry lies also in the border area between classical organic chemistry and surface chemistry.

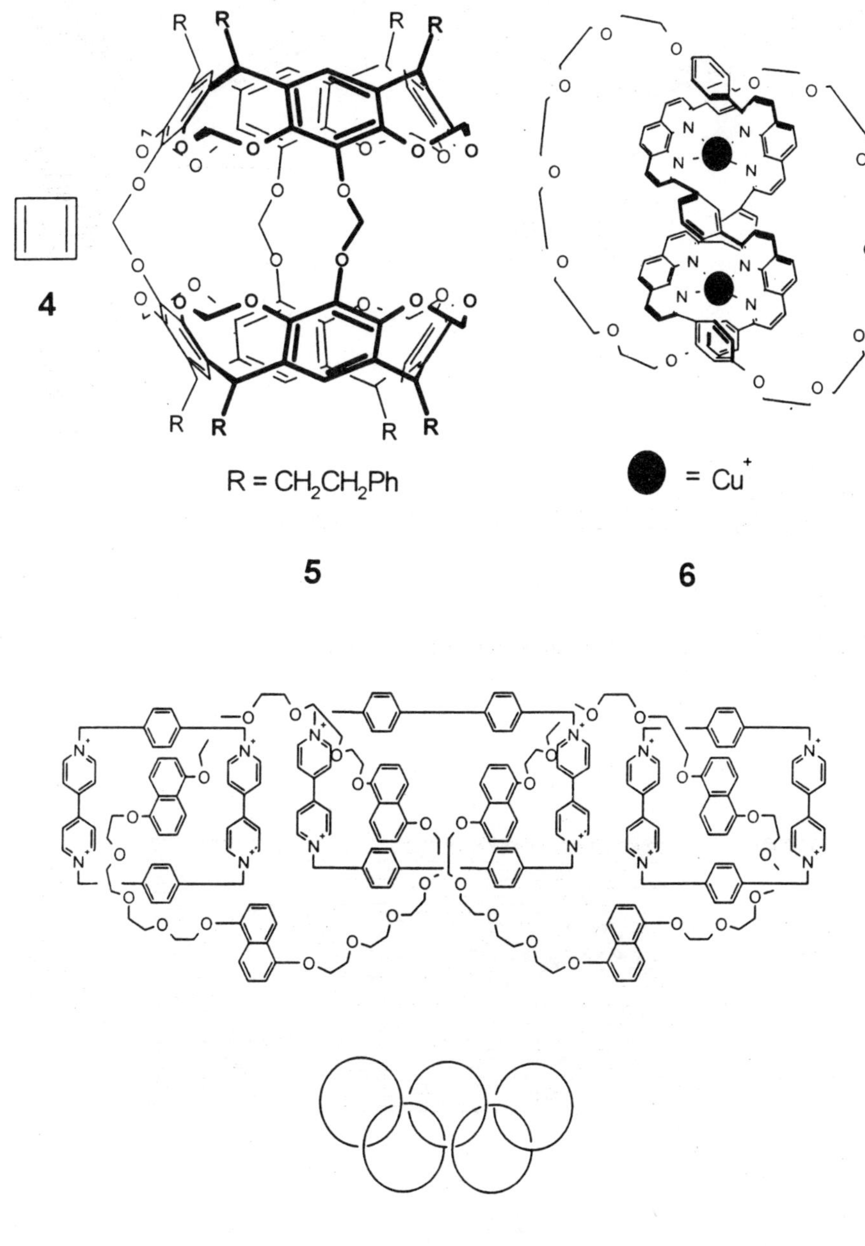
4
R = CH2CH2Ph
5
= Cu+
6
7

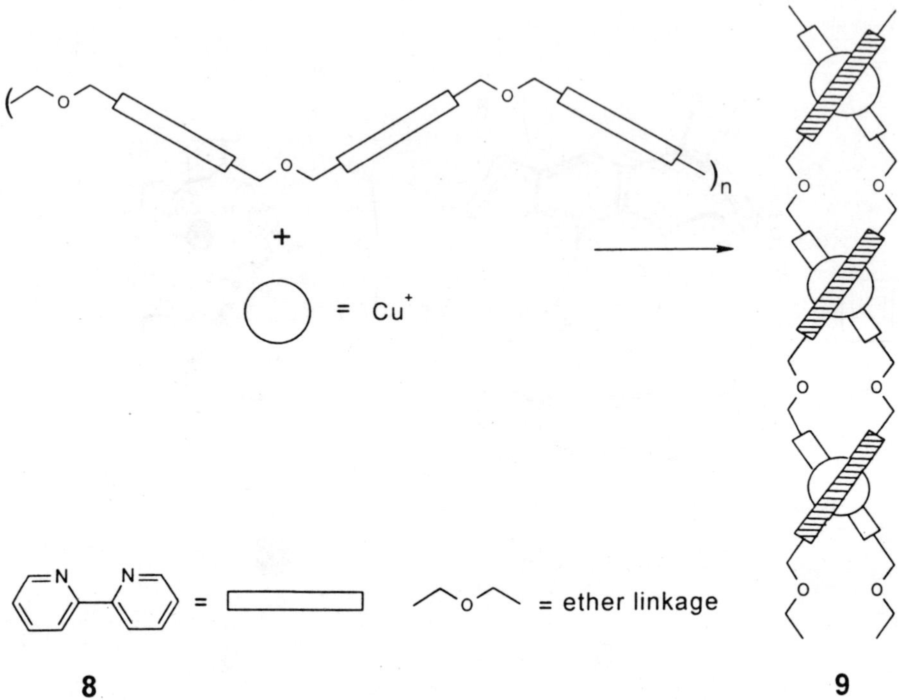

Figure 1.3. Self-entwining strands forming a helicate.

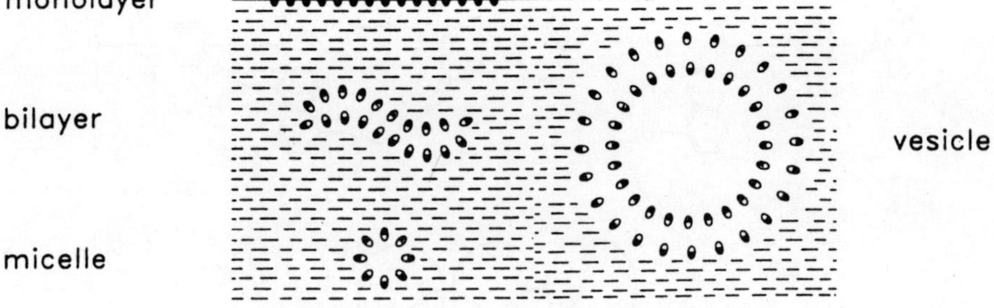

Figure 1.4. Schematic structures of aggregates of amphiphilic molecules in a polar solvent (the hydrophilic regions of each molecule are shaded).

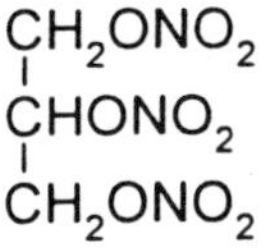

CH_2ONO_2
$CHONO_2$
CH_2ONO_2

10a 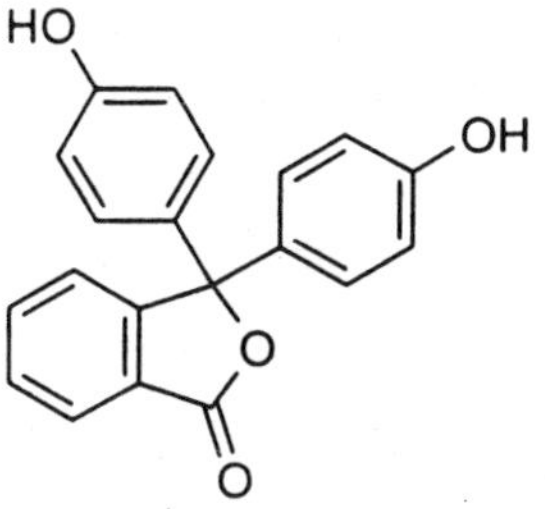 **10b** 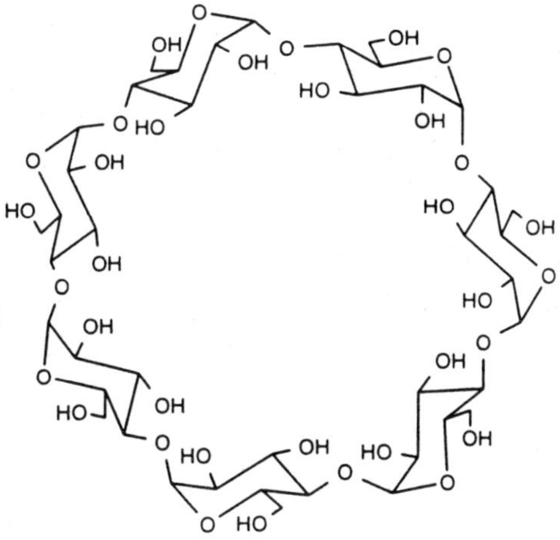 **11**

4. Polyether containing 2,2'-bipyridil units **8** spontaneously forms a double helicate **9** by multiple coordination with Cu^+ ions [13]. This process of self-organization is enforced by the proper orientation of coordinated bipyridyl units analogous to that shown in Figure 1.2. It is characterized by a positive cooperativity yielding no partly assembled species.

5. Nitroglycerine **10a** is both a drug and an explosive. Its inclusion into the cavity of β-cyclodextrin, β-CD, **11** prevents its decomposition and enhances its bioavailability [14]. The complex of **10a** with **11** is marketed under the name Nitropen as a coronary dilator sublingual tablets by Nippon Kayaku company in Japan.

6. In polar solvents amphiphilic molecules, that is molecules with a polar 'head' and hydrophobic 'tail', tend to form various aggregates. The structure of micelles is usually much more complicated than that schematically shown in Figure 1.4 (see the pertaining discussion in Section 2.3). Nevertheless, in water they can include nonpolar molecules into their voids acting like surfactants applied in toiletry [15]. Similarly to cyclodextrins such as **11** [6, 16] and liquid crystals [7] discussed in Section 2.6, surfactants are examples of few supramolecular systems which have found numerous practical applications.

7. The 'molecular necklace' **12** of α-cyclodextrin **13** 'beads' threaded on a polyether chain (Figure 1.5) forms spontaneously in solution [17]. This is an example of a so-called 'one-pot reaction' in which complicated structures are

obtained in one step as opposed to multistep reactions typical for chemistry of natural products.

8. The formation of supramolecular complexes catalyzes numerous reactions. In case of autocatalytic reaction one can speak about a self-replicating system crudely mimicking reproduction. An interesting example of this kind was provided by Luisi and coworkers [18]. The authors created a system of reverse micelles consisting of water droplets stabilized in organic solvent by a layer of surfactant, which promoting a reaction inside these micelles is capable of forming the new micelles. The system under consideration consists of 50 mM octanoid acid sodium salt acting as a surfactant, aqueous LiOH and 9:1 (v/v) mixture of isooctane with 1-octanol. The alcohol that serves as cosurfactant is

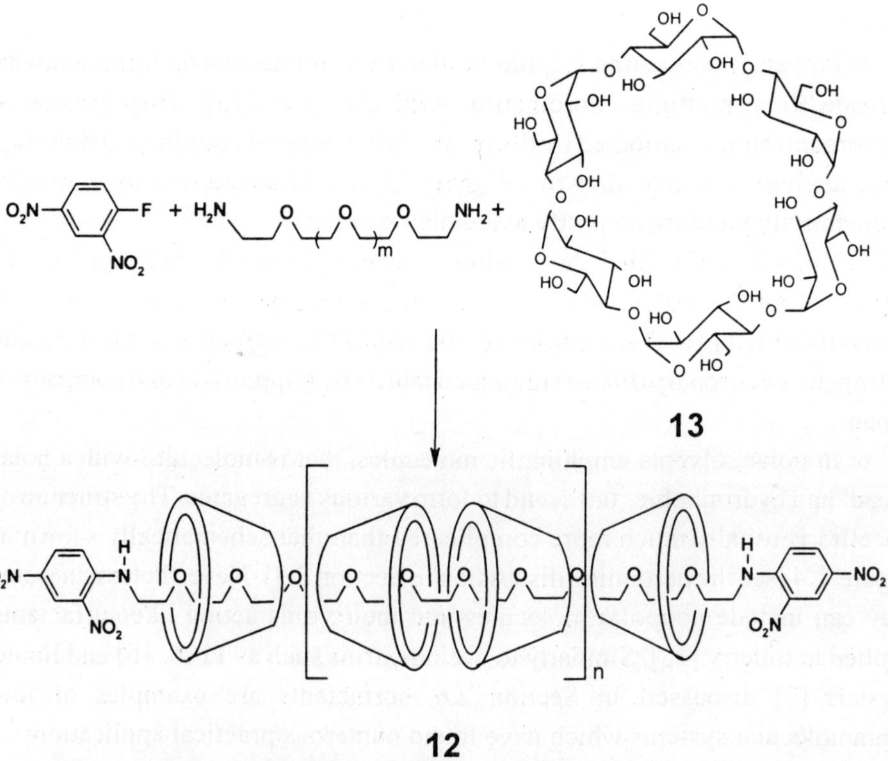

Figure 1.5. The 'one-pot' formation of a 'molecular necklace' involving 20-22 α-cyclodextrin 'beads' represented schematically by buckets.

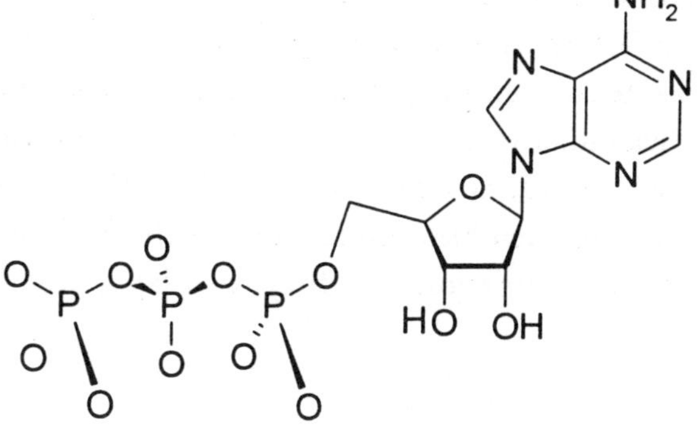

14

15

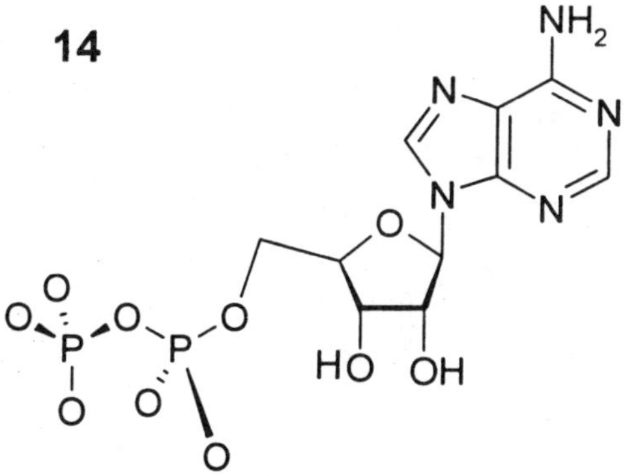

16

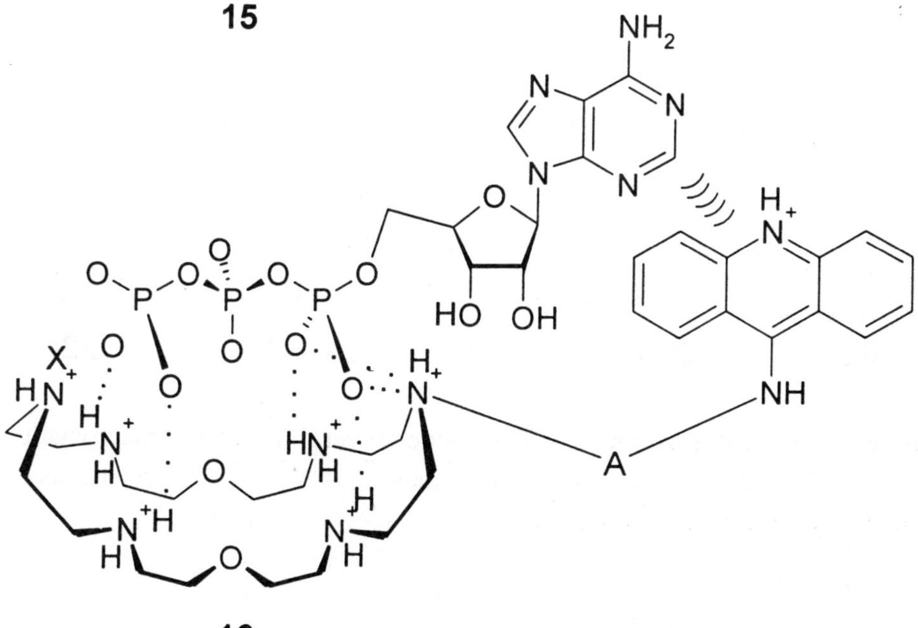

essential for the creation of stable reverse micelles partitions between the micelle layer and the bulk solvent. The reaction used was the hydrolysis of octanoic acid octyl ester catalyzed by LiOH. In control experiments the reaction producing new micelles was shown to depend critically on the presence of reverse micelles.

9. The hydrolysis of adenosine triphosphate **14**, ATP, to adenosine diphosphate **15**, ADP, is of considerable chemical and biochemical importance since such processes catalyzed by numerous enzymes play a crucial role in

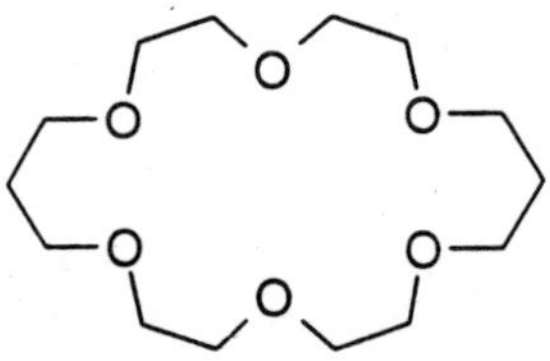

17 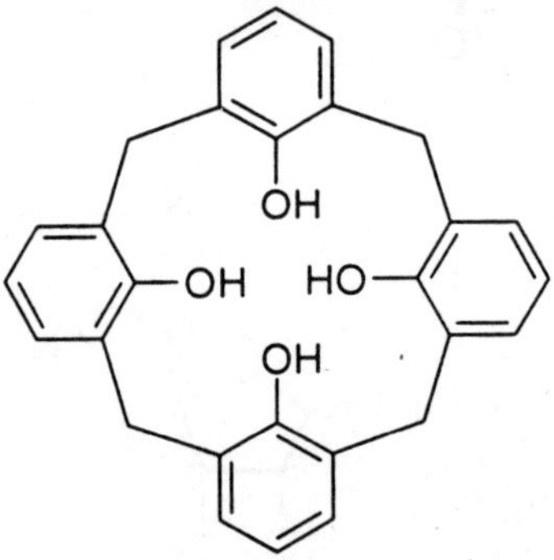 **18** 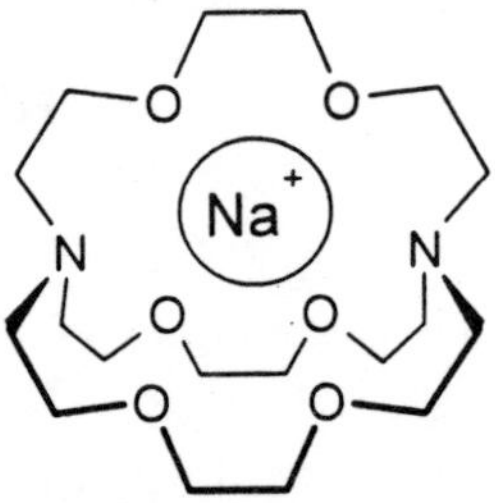 **19**

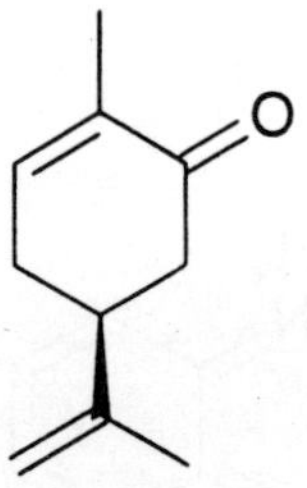

20a 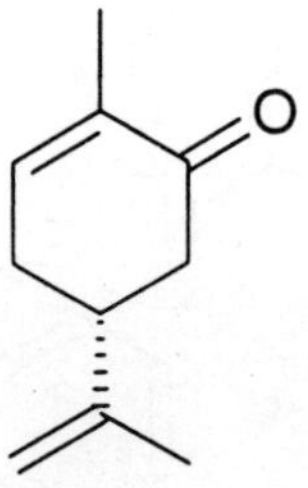 **20b** 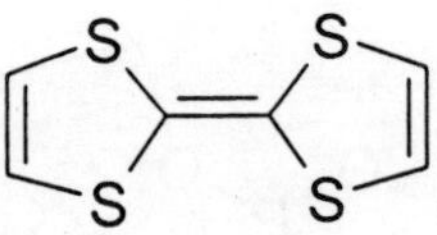 **21** 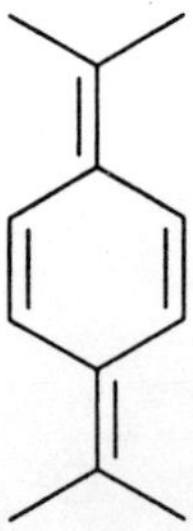 **22**

biology. Lehn with coworkers [19] developed several substituted macrocycles which catalyze among others the transformation of ATP to ADP by means of formation of intermediate complex **16**.

10. Selective complexation of cations by crown ethers **17** [1b, 1g] and calixarenes **18** [1f] depending on the rings size was proposed to be used in sensors.

11. Sodium and other alkali metals are known to easily form cations. Surprisingly, they can also form anions, which are the strongest known reducing agents. One of the most stable of such salts consisting of a Na^+ cation trapped in cryptand **19** and Na^- is relatively easy to obtain and does not decomposes in vacuum at room temperature. Its X-ray analysis and NMR spectra prove the existence of such highly untypical anions [20].

12. Wonderful colours of butterfly and bird wings emerge as a result of diffraction or scattering of light by thin-film nanostructures. [21]

In all examples presented above the systems have changed their properties upon association. Cyclobutadiene **4** has become stable after being complexed with **5**, in spite of it being a highly reactive species under normal conditions [5]. Somewhat similarly, the possibility of nitroglycerine **10a** explosion is considerably diminished after complexation with β-cyclodextrin **11** [14]. Micelles and vesicles allow one to introduce nonsoluble agents into a solution. The spatial reorientation of reaction substrates, i.e. their preorganization, owed to the complexation with metals allowed Dietrich-Buchecker and Sauvage with collaborators to obtain a molecule twisted into a knot **6** [11]. Similarly, the synthesis of olympiadane **7** by Stoddart's group [12] would not be possible without the preorganization forced by π-stacking interactions. All these examples and many other discussed in this book show that a system of interacting molecules or ions is different from the sum of its separated parts thus pointing to the most essential specificity of supramolecular chemistry. The above examples point to a basic property of the complexation processes under consideration and of supramolecular chemistry in general, namely, molecular recognition. According to Lehn [22] it "is defined by the energy and the information involved in the binding and selection of substrate(s) by a given receptor molecule; it may also involve a specific function". This translates into the selectivity of intermolecular binding making possible by "pattern recognition process through a structurally well-defined set of intermolecular interactions". The formation of Whitesides' hydrogen bonded aggregates **3a-c** [4] shown in Figure 1.1 is so efficient because: (1) there are favourable spatial relationships between melamine and cyanuric acid molecules; and (2) the electrostatic fields of both molecules complement each other. Thus suitable conditions for efficient intermolecular attractions are created and the molecules recognize each other. Similarly, one-pot synthesis of the 'necklace' **12** [7] would not be so effective (or even possible) with a larger cyclodextrin. Thus, also in this case the substrates recognize each other. The recognition phenomena in nature and host-guest chemistry are mostly analyzed using the concepts of receptor and substrate and that of 'key and lock' mechanism of the recognition process introduced by Emil Fischer more than 100

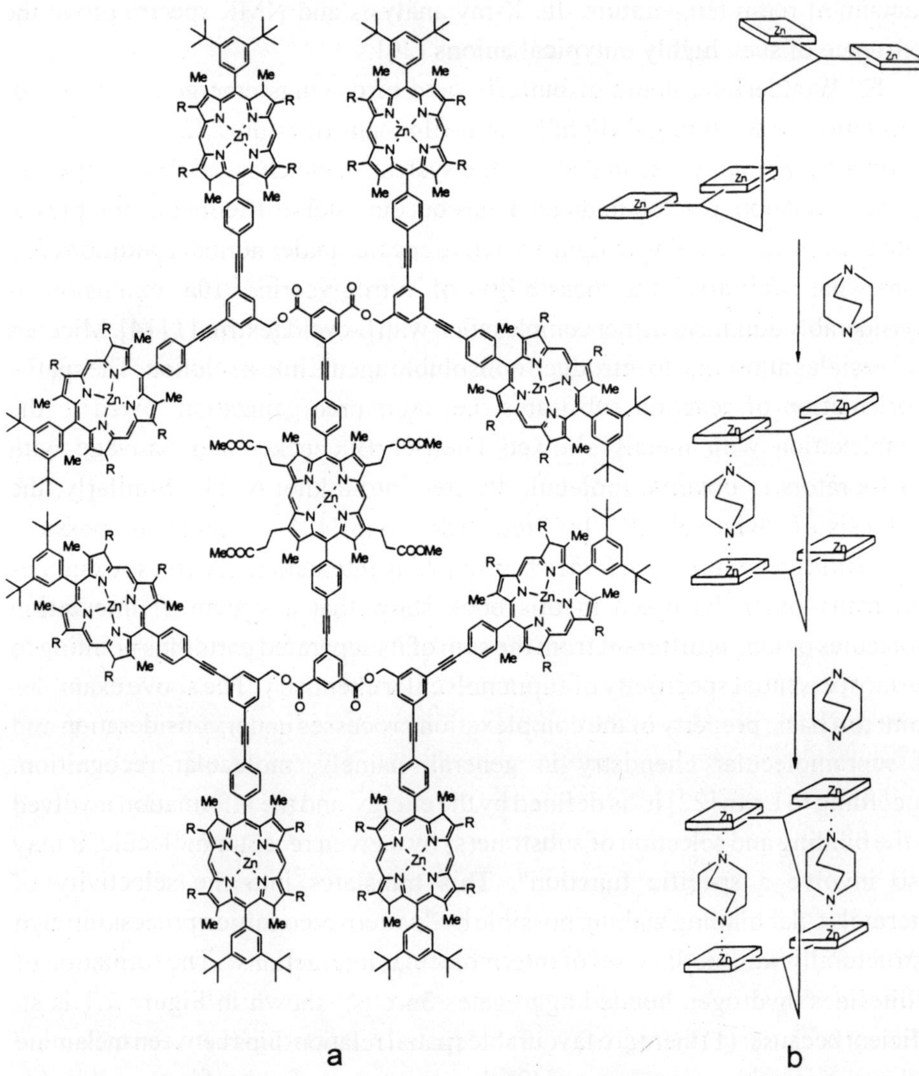

Figure 1.6. Example of induced fit complexation.

years ago [23]. They usually involve a larger molecule with a kind of cavity called receptor and a smaller one that fits into this cavity bearing the name substrate. According to this model, these two parts of the system fit as a key into a lock. Today we know that this a is somewhat oversimplified picture of the recognition phenomenon, and a more subtle model involving induced fit [24a] will be presented in Chapter 3. An impressive example of the dendrimer host adaptation to the complexed guest presented by the Sanders group [24b] is schematically visualized in Figure 1.6b although it is not clear why the dendrimer molecule depicted in Figure 1.6a does not complex four bicyclic amines.

Supramolecular chemistry owes its importance to a great extent to the abundance of recognition and assembling processes in living Nature. To name but a few:

1. Enzymes recognize substrates highly specifically and carry out reactions in very efficient way. Thus *L*-, not *D*-, amino acids are predominantly synthesized in living organisms. However, contrary to common opinion, they are not exclusive [25].

2. The sensitivity of our (or better dogs') noses to fragrances is based on the ability of the smell receptors to discriminate between sometimes very small differences in molecular shape and charge distribution. Noses recognize fragrances at molecular level very precisely. For instance, by smelling one can easily differentiate between (+)- and (−)-carvone **20a,b** which differ only in the configuration on one carbon atom [26]. The carvone isomers are mirror images, and this type of recognition bears the name chiral recognition.

3. The central part of cell walls is a membrane consisting of complex self-assembled structures with built-in channels that execute complicated functions, e.g., the transport of ions briefly discussed in Section 5.3.4). Creating artificial membranes mimicking the functioning of biological membranes is one of the important tasks of supramolecular chemistry.

4. As discussed in detail in Section 5.2.1, a living creature, tobacco mosaic virus, is built of a helical strand of RNA enclosed by a sheath composed of 2130 proteins. Amazingly, by changing the experimental conditions one can decompose the virus into its constituents parts and then reassemble it by switching to the former conditions [27a]. This means that a kind of living organism [27b] could be obtained from the fragments which, at least in principle, can be synthesized in a test tube. Such observations further complicate the answer to the fundamental question 'What is life?'. To understand the structure and behaviour

of supramolecular assemblies in Nature one can model them by simpler systems called biomimetic structures. This is one of the most important tasks of supramolecular chemistry.

In spite of its importance, the significance of supramolecular chemistry cannot be limited to the understanding of molecular foundations of life. Present or prospective practical applications of molecular assemblies are another driving force for the rapid development of this domain. The use of the complex of nitroglycerine **10a** with β-cyclodextrin **11** in the pharmaceutical industry was mentioned above. Such a mode of drug administration not only prevents the decomposition but also enhances its solubility resulting in its increased bioavailability [16]. Similarly, the complexation of fragrances or spices with cyclodextrins allows one to store them without loss for a long time [28]. Adding cyclodextrins to waste water enables its more effective purification [29]. Another field of practical applications of supramolecular assemblies provides liquid crystals [7] widely used, amongst others, as displays (see, however, the discussion in Section 4.2.6).

The prospective applications of molecular assemblies seem so wide that their limits are difficult to set. The sizes of electronic devices in the computer industry are close to their lower limits. One simply cannot fit many more electronic elements into a cell since the 'walls' between the elements in the cell would become too thin to insulate them effectively. Thus further miniaturization of today's devices will soon be virtually impossible. Therefore, another approach 'from bottom up' was proposed. It consists in the creation of electronic devices of the size of a single molecule or of a well-defined molecular aggregate. This is an enormous technological task and only the first steps in this direction have been taken. In the future, organic compounds and supramolecular complexes will serve as conductors, as well as semi- and superconductors, since they can be easily obtained with sufficient, controllable purity and their properties can be fine tuned by minor adjustments of their structures. For instance, the charge-transfer complex of tetrathiafulvalene **21** with tetramethylquinodimethane **22** exhibits room- temperature conductivity [30] close to that of metals. Therefore it could be called an organic metal. Several systems which could serve as molecular devices have been proposed. One example of such a system which can also act as a sensor consists of a basic solution of phenolophthalein dye **10b** with β-cyclodextrin **11**. The purple solution of the dye not only loses its colour upon the complexation but the colour comes back when the solution is heated [31].

Therefore after scaling it could serve as thermometer. The complicated processes involved in de- and re-colouration are not fully understood, but the latter is undoubtedly associated with the complex decomposition triggered by thermal motion of the cyclodextrin involved. Thus it reflects the dynamic character of the phenolophthalein complex with **11** (see Section 3.4 for a short discussion of dynamic character of supramolecular complexes). Optoelectronics making use of nonlinear optical phenomena is yet another field of prospective applications of molecular assemblies [32].

Another aspect of future applications of supramolecular chemistry, as opposed to classical organic chemistry, is that it opens the possibility for much cleaner technological processes on the one hand, and provides means for the removal of toxic wastes from the environment on the other (see Section 6.3.4).

It should be noted that the word 'complex', often used in supramolecular chemistry, is not very specific. It is applied to charge-transfer complexes like the one formed by **21** with **22** [30] as well as to coordination complexes consisting of one or more atoms or ions with *n* ligands like $K_2[Pt(NO_2)_4]$. The same name complex also covers the Whitesides' hydrogen bonded systems [10] shown in Figure 1.1 and inclusion complexes of **4** embedded in **5**. Thus the term complex without any adjective has no specificity and can be applied to any type of molecular associates.

According to Lehn [33] "A receptor-substrate supermolecule (*i.e.* supramolecular complex) is characterized by its geometric (structure, conformation), its thermodynamic (stability, enthalpy and entropy of formation) and its kinetic (rates of formation and of dissociation) features." It should be stressed that due to its smaller energy the 'intermolecular bonding' in supramolecular systems is much softer than a covalent chemical bond. Therefore, (1) in solution some of these complexes, e.g. cyclodextrin or donor-acceptor complexes, exist as mixtures of rapidly interconverting free and complexed species. The processes of overall and local molecular motions can be studied by means of NMR relaxation experiments [34], which in certain cases indicate very short lifetimes of the complexes, comparable with the overall reorientation rates [35], raising the question about the criterion of existence of the complexes under study. Moreover: (2) as discussed in Section 3.3, the complex structure in the solid state can be different from that in solution in analogy with a famous biphenyl case [36]. Also, as the result of a weak 'bonding' in supramolecular systems the dynamics of the motion of molecules constituting the complex under investigation may be, and usually is, different from those of its free constituent parts. Also in this case the investigation of nuclear relaxation is a method of choice.

To summarize, supramolecular chemistry is a rapidly developing, but ill-defined, field encompassing at least three highly specific domains mostly characterized by different objects and research techniques. As discussed in some detail in Section 4.1, shortly after its establishment supramolecular chemistry has ripened into being divided into small aggregate chemistry which encompasses host-guest (or inclusion) chemistry, the chemistry of higher aggregates which at present lacks a proper name (aggregate chemistry?) and crystal engineering.

Numerous supramolecular systems have found practical applications but their internal structure and the mechanism of their formation from isolated molecules are not fully understood. Their study requires the application of new experimental techniques. Thus, in addition to the classical physicochemical methods (IR, UV, NMR and ESR), novel specific experimental techniques evolve. They include Scanning Probe Microscopy, SPM [37a], (in particular, Atomic Force Microscopy, AFM) [37b], Small Angle X-ray Scattering SAXS [38], Extended X-ray Absorption Fine Structure EXAF [39], Brewster Angle Light Microscopy [40], Langmuir Balance [41], electrochemical techniques [42], Thermogravimetric Analysis and Differential Scanning Calorimetry [43], to name but a few. The complex structure of supramolecular assemblies and their dynamic character call for a wide, but cautious (see Section 7.4.3), use of molecular modelling for investigation of the structure and behaviour of supramolecular assemblies [44].

REFERENCES

1. (a) *Comprehensive Supramolecular Chemistry*, J.-M. Lehn, J. L. Atwood, J. E. D. Davies, D. D. MacNicol, F. Vögtle, Eds., Pergamon, Oxford, 1996; (b) J.-M. Lehn, *Supramolecular Chemistry: Concepts and Perspectives.* VCH, Weinheim, 1995; (c) F. Vögtle, *Supramolecular Chemistry,* J. Wiley, New York, 1991; (d) J.-H. Fuhrhop, J. Köning, *Membranes and Molecular Assemblies: The Synkinetic Approach,* Monographs in Supramolecular Chemistry, J. F. Stoddart, Ed., The Royal Society of Chemistry, Cambridge, United Kingdom, 1994; (e) D. J. Cram, J. M. Cram, *Container Molecules and Their Guests,* Monographs in Supramolecular Chemistry, J. F. Stoddart, Ed., The Royal Society of Chemistry, Cambridge, United Kingdom, 1994; (f) C. D. Gutsche, *Calixarenes,* Monographs in Supramolecular Chemistry, J. F. Stoddart, Ed., The Royal Society of

Chemistry. Cambridge, United Kingdom, 1989; (g) G. Gokel, *Crown Ethers and Cryptands*, Monographs in Supramolecular Chemistry, J. F. Stoddart, Ed., The Royal Society of Chemistry. Cambridge. United Kingdom. 1989; (h) F. Diederich. *Cyclophanes*. Monographs in Supramolecular Chemistry. J. F. Stoddart. Ed.. The Royal Society of Chemistry. Cambridge. United Kingdom. 1989; (i) H. Dodziuk. *Modern Conformational Analysis. Elucidating Novel Exciting Molecular Structures.* Chapter 10. VCH Publishers. New York. 1995.

2. It is interesting to note that Pedersen is one of the few (if not the only) Nobel Prize-Winners in sciences without a Ph.D.
3. R. Pfeffer. *Organische Molekülverbindungen*. Enke. Stuttgart. 1927.
4. *Molecular Association, Including Molecular Complexes*, R. Foster. Ed.. Academic Press. New York. 1979; R. Foster, *Charge-Transfer Complexes*. Academic Press. New York. 1969.
5. D. J. Cram, M. E. Tanner, R. Thomas, Angew. Chem. Int. Ed. Engl., 1991. 30. 1024.
6. *New Trends in Cyclodextrins and Derivatives*, D. Duchene, Ed., Edition de Sante, Paris, France, 1991.
7. H. Kelker, R. Hatz. *Handbook of Liquid Crystals*, VCH, Weinheim, 1980; *Phase Transitions in Liquid Crystals*, NATO ASI Series B, V. 290. S. Martelluci. A. N. Chester. Plenum. New York, 1992.
8. *Crystallography of Supramolecular Compounds*. G. Tsoucaris, J. L. Atwood, J. Lipkowski. Eds.. Kluwer Academic Publishers, Dordrecht, 1996; V. G. Videnova-Adrabinska, *The Hydrogen Bond as a Design Element of the Crystal Architecture. Crystal Engineering from Biology to Materials*, Oficyna Wydawnicza Politechniki Wrocławskiej, Wrocław, Poland.
9. *In statu nascendi* means in the state of emerging.
10. (a) J. A. Zerkowski. C. T. Seto. G. M. Whitesides. J. Am. Chem. Soc.. 1992. 114, 5473; (b) E. E. Simanek, M. Mammen, D. M. Gordon, D. Chin, J. P. Mathias. C. T. Seto. G. M. Whitesides. Tetrahedron, 1995, 51, 607, and references cited therein.
11. C.-O. Dietrich-Buchecker, J.-P. Sauvage, Angew. Chem. Int. Ed. Engl., 1989. 28. 189.
12. D. B. Amabilino, P. R. Ashton, A. S. Reder, N. Spencer, J. F. Stoddart, Angew. Chem. Int. Ed. Engl.. 1994, 33. 1286.
13. J.-M. Lehn, A. Rigault, J. Siegel. J. Harrowfield, B. Chevrier, D. Moras, Proc. Natl. Acad. Sci. USA. 1987, 84, 2565.
14. A. Stadler-Szöke, J. Szejtli, Acta Pharm. Hung., 1979, 49, 30.
15. J. H. Clint, *Surfactant Aggregation*, Blackie, Glasgow, 1992.
16. J. Szejtli. *Cyclodextrin Technology*, Kluwer Academic Publishers. Dordrecht. 1988.
17. A. Harada, J. Kamachi, Nature, 1992, 356, 325; J. F. Stoddart. Angew. Chem. Int. Ed. Engl.. 1992. 31. 846.

18. P. A. Bachmann, P. Walde, P. L. Luisi, J. Lang, J. Am. Chem. Soc., 1990, 112, 8200.

19. W. M. Hosseini, A. J. Blacker, J.-M. Lehn, J. Am. Chem. Soc., 1990, 112, 3896.

20. F. J. Tehan, B. L. Barnett, J. L. Dye, J. Am. Chem. Soc., 1974, 96, 7203.

21. M. Srinivasarao, Chem. Rev., 1999, 99, 1935.

22. Ref. 1b, p. 11.

23. E. Fischer, Ber., 1894, 27, 2985.

24a. D. E. Koshland, Jr., Angew. Chem. Int. Ed. Engl., 1994, 33, 2475; (b) C. C. Mak, N. Bampos, J. K. M. Sanders, Angew. Chem. Int. Ed. Engl., 1998, 37, 3020.

25. Ref. 1e, p. 119. Moreover, special enzymes for D-amino acids exist.

26. K. Bauer, D. Garbe, H. Surburg, *Common Fragrances and Flavor Materials,* VCH, Weinheim, 1990, p. 51.

27. (a) H. Fraenkel-Conrat, R. C. Williams, Proc. Natl. Acad. Sci. USA, 1955, 41, 690; (b) Viruses replicate only in other, higher organisms. Thus they actually occupy an intermediate position between the living and non-living Nature.

28. Extract from garlic is marketed in form of a cyclodextrin complex.

29. K. Gruiz, E. Fenyvesi, E. Kriston, M. Molnar, B. Horvath, in Proceedings of the Eigth International Symposium on Cyclodextrins, J. Szejtli, L. Szente, Eds., Kluwer Academic Publishers, Dordrecht, 1996, p. 609.

30. S. S. Shaik, M.-H. Whangbo, Inorg. Chem., 1986, 25, 1201.

31. K. Taguchi, J. Am. Chem. Soc., 1986, 108, 2705.

32. (a) G. H. Wagniere, *Linear and Nonlinear Properties of Molecules,* VCH, Weinheim, 1993; (b) J.-M. Andre, J. Delhalle, Chem. Rev., 1991, 91, 843.

33. Ref. 1b, p. 51.

34. (a) A. Abragam, *The Principles of Nuclear Magnetism,* Clarendon Press, Oxford, 1961; (b) H. Friebolin, *Basic One- and Two-Dimentional NMR Spectroscopy,* VCH, Weinheim, 1993, Chapter 7.

35. C. Brevard, J.-M. Lehn, J. Am. Chem. Soc., 1970, 92, 4987.

36. The barrier to internal rotation in biphenyl is smaller than crystalline forces, thus the considerable nonplanarity of the molecule disappears in the solid state. G. Bastiansen, Acta Chem. Scand., 1952, 6, 205; C. P. Brock, K. L. Haller, J. Phys. Chem., 1984, 88, 3570; G. P. Charbonneau, Y. Delugeard, Acta Crystallogr. Sect. B, 1976, 32, 1420.

37. (a) R. Wiesendanger, Ed., *Scanning Probe Microscopy,* Springer, Berlin, 1998; R. Wiesendanger, H.-J. Güntherodt, Eds., Springer, Berlin, 1996; (b) G. Kaupp, in *Comprehensive Supramolecular Chemistry,* v. 8, p. 381; J. Frommer, Angew. Chem. Int. Ed. Engl., 1992, 31, 1298.

38. *Neutron, X-Ray and Light Scattering : Introduction to an Investigative Tool for Colloidal and Polymeric Systems,* P. Lindner. T. Zemb, North-Holland. Amsterdam. 1991; *Small Angle X-Ray Scattering,* O. Glatter, O. Kratky, Eds., Academic Press. New York. 1982.
39. R. M. White. T. H. Geballe, *Long Range Order in Solids. Solid State Physics, Supplement 15,* Academic Press. New York, 1979, p. 359.
40. D. Wollhardt. Adv. Colloid Interface Sci., 1996. 64, 143.
41. B. S. Murrey. P. V. Nelson. Langmuir. 1996. 12. 5973.
42. A. E. Kaifer. in *Comprehensive Supramolecular Chemistry,* v. 8. p. 499.
43. M. A. White. in *Comprehensive Supramolecular Chemistry,* v. 8. p. 179.
44. *Computational Approaches in Supramolecular Chemistry,* G. Wipff. Ed., NATO ASI Series C. vol. 426. Kluwer. Dordrecht, 1994.

Chapter 2

MOLECULAR AND CHIRAL RECOGNITION. SELF-ORGANIZATION, SELF-ASSEMBLY AND PREORGANIZATION

2.1 Molecular and Chiral Recognition

Molecular recognition, *self-organization* and *self-assembly* are the central concepts in supramolecular chemistry. The recognition consists in selective binding of a substrate molecule, called a guest in supramolecular chemistry, by a receptor bearing the host name. As mentioned in Chapter 1, according to Lehn [1] a supramolecular complex is characterized by the energy and the information involved in its binding, by the selection of substrate(s) by a given receptor molecule, and sometimes by a specific function [2]. Strong bonding need not necessarily be accompanied by selectivity, thus, it is different from molecular recognition. The macrocyclic tetraphenolate **23** is a strong binder of neurotransmitter cholin $(CH_3)_2N^+CH_2CH_2OH\,OH$ **24** (the association constant $K = 50\,000 M^{-1}$ [3]). However, such a large value is characteristic of not only this but of all guest molecules possessing a $N^+(CH_3)_3$ group that lacks considerable steric hindrance. Thus the complexation of **24** by **23** is very selective for the latter group but does not recognize the rest of the molecule. An illustration of higher affinity but lower selectivity in chiral recognition by cyclodextrins is presented below. Some examples of the recognition were briefly presented in Chapter 1. For instance, the highly selective and diversified aggregation of melamine **1** with

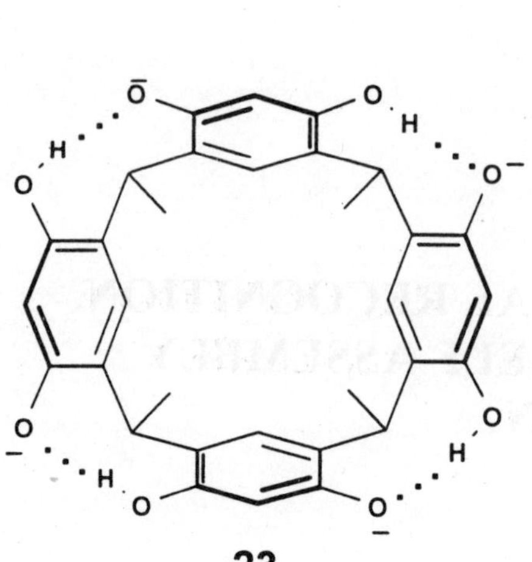

23

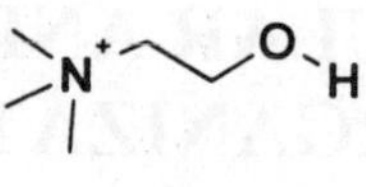

24

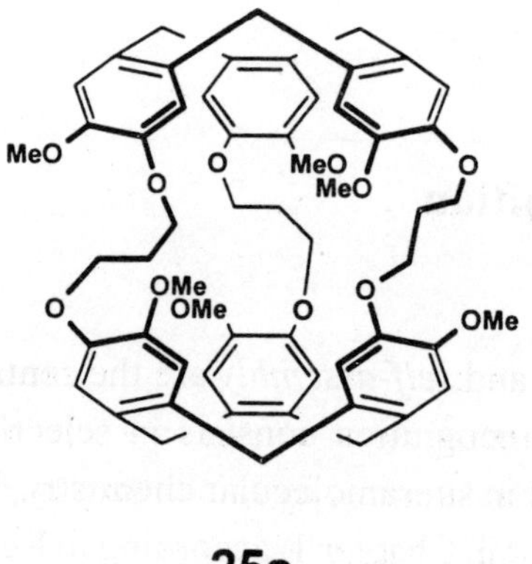

25a

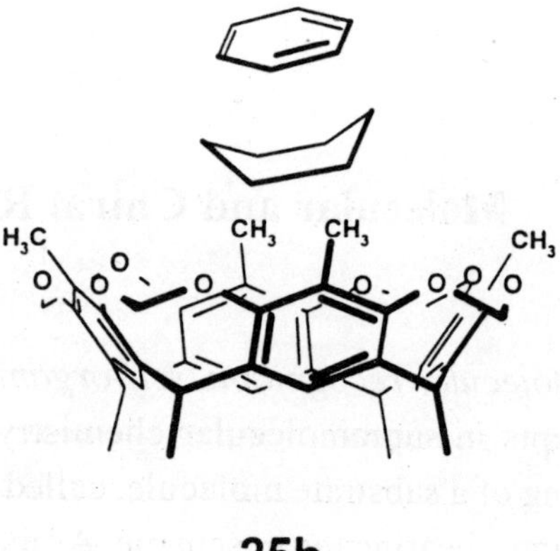

25b

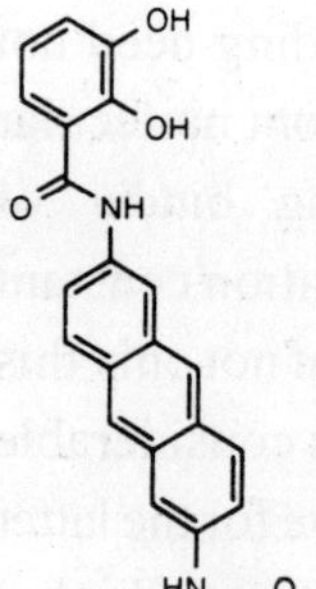

26a

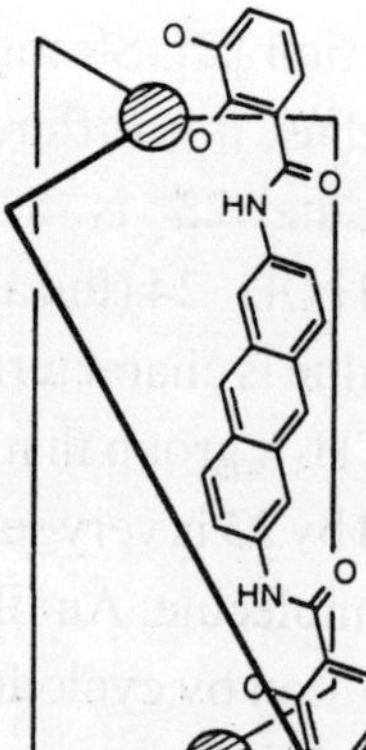

26b

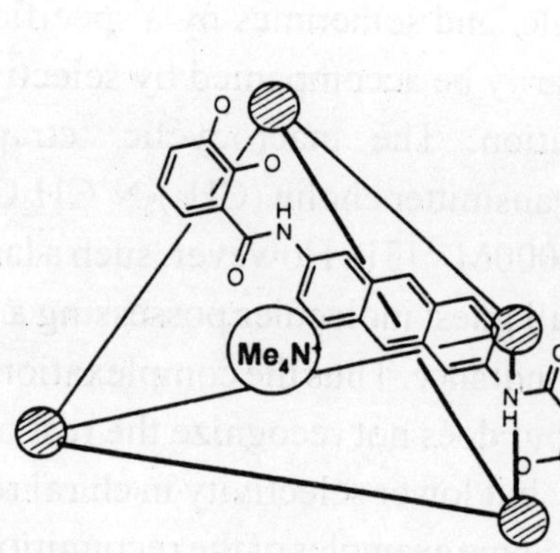

26c

uronic acid **2** and/or of their derivatives is made possible by complementarity of their donor and acceptor sites enabling multiple hydrogen bond formation [4]. Similarly, the favourable orientation of 2,2'-bipyridine units **8** coordinated with Cu^+ ions forces the formation of the double helicate **9** [5a] and knot **6** [5b]. On the other hand, weak but numerous dispersive interactions are one of the main driving forces for the cyclodextrin complexation (such as that of nitroglycerine **10** with β-cyclodextrin **11** [6] and the 'molecular necklace' of **12** and **13** [7]). Molecular and chiral recognition in nature (exemplified, amongst others, by enzymatic reactions, the formation of the DNA double helix ,and the reassembling of the decomposed tobacco mosaic virus [8] discussed in some detail in Chapter 5) is much more efficient, enabling unrivaled specificity of reaction chains in living organisms. As discussed in brief in Chapters 1 and 5, the 'lock and key' [9a] and 'induced fit' models [9b] have been proposed for describing recognition processes. In agreement with the latter model, some enzymes were found to undergo conformational changes promoting their action [10]. Another example showing that the host is not rigid and adapts itself to the anionic guests of varying size is provided by cryptophane **25a** [11a]. This host includes not only molecules the van der Waals radii of which perfectly match the size of its cavity [11b] but also a relatively large chloroform guest. In agreement with the 'induced fit' model, this indicates the host ability to undergo changes to adapt itself to the guest. On the other hand, the ternary complex involving cavitand **25b**, benzene and cyclohexane in the highly unusual boat conformation in the solid state represents a fascinating example of the accomodation [12]. One of the most spectacular changes upon complexation was reported by the Raymond group [13a]. The latter authors have shown that the ligand **26a** forms complexes **26b** and **26c** not only of different spatial structure but also of different stoichiometry with X = $[Ti(acac)_2]$ or $[Ga(acac)_2]$ depending on the presence of the Me_4N^+ guest.

The building of a cavity around the guest is an extension of 'induced fit' concept. This is the case with hexokinase enzyme [13b] and foldamers [13c] that wrap themselves around the guest.

By analogy with molecular recognition, *chiral recognition* consists in the selective binding of enantiomers, that is, of the molecules that are mirror images of each other, such as **27a** and **27b**. A small child trying to put his left foot into the right shoe is probably the best visualization of this phenomenon. As discussed in Chapter 5, chiral recognition is especially important in living organisms.

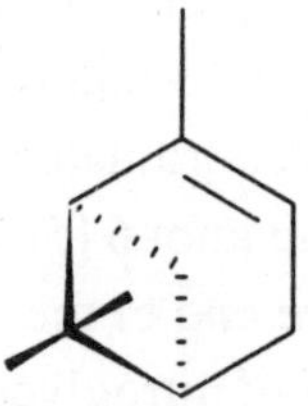

27a 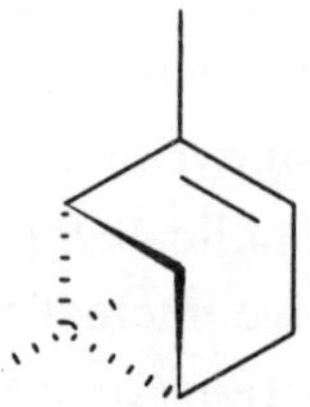 **27b** 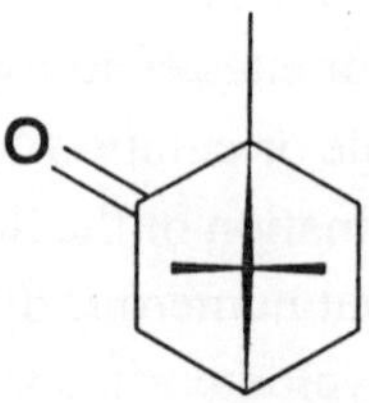 **28a** 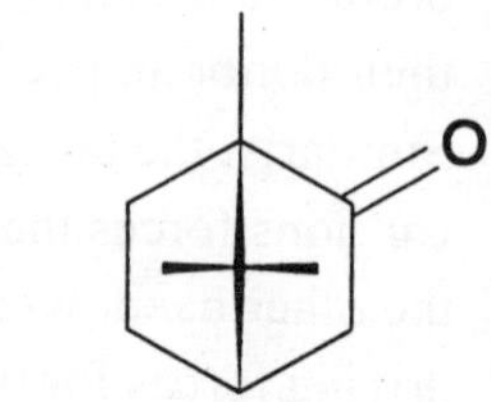 **28b**

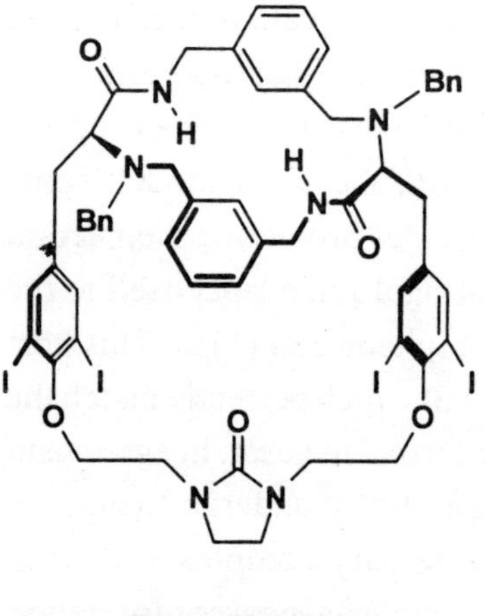

Bn = CH_2Ph

29a **29b**

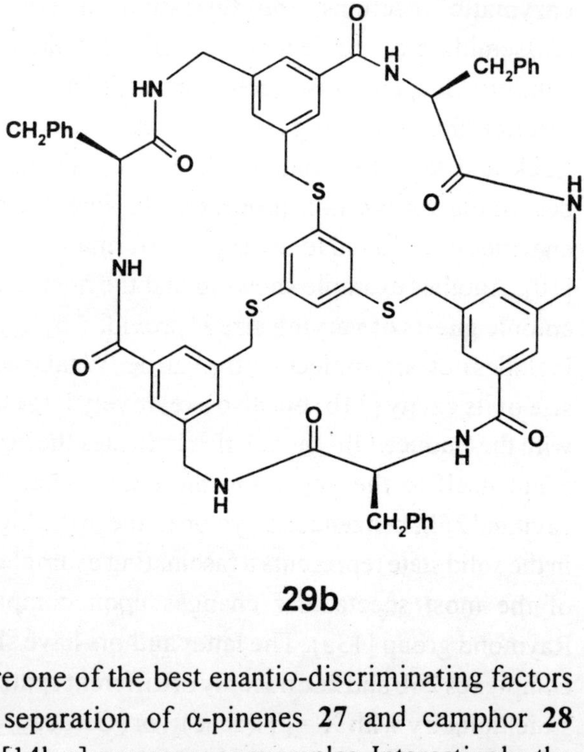

Cyclodextrins (Section 7.4) are one of the best enantio-discriminating factors [14a]. The chromatographic separation of α-pinenes **27** and camphor **28** enantiomers by α-cyclodextrin [14b,c] may serve as examples. Interestingly, the latter host **13** recognizes the enantiomers of **27** although the stability constants of the complexes are smaller than those with **11** that does not recognize them [14d]. Specific hosts such as **29** for very effective enantio-selective binding of aminoacid derivatives have been synthesized by Still group [15]. The free energy difference between diastereomeric complexes formed by a host with enantiomeric guests are usually less than 0.3 kcal/mol. However, for the complex of **29a** with enantiomers of an alanine dipeptide this difference is equal to 1.3 kcal/mol [15b],

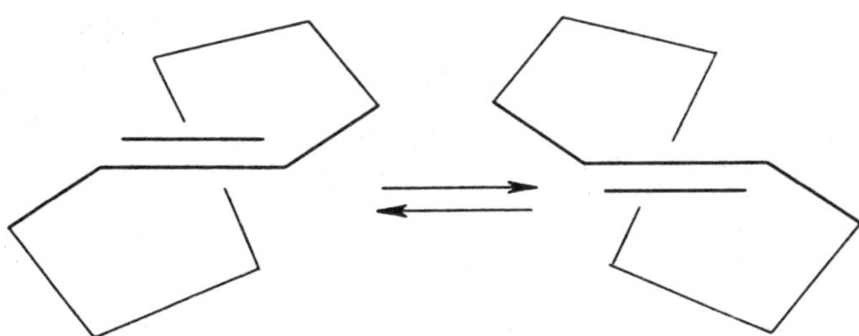

Figure 2.1. Enantiomerization of cyclohexene.

and it reaches the unusually high a value of 3 kcal/mol for **29b** complexed with enantiomers of a simple peptide [15c]. Interesting example of solvent (e.g., diethyl ether, pentane) polarity affecting the product chirality was reported by Inoue and Wada [16]. The photochemical isomerisation of *cis*-cyclooctene carried out by the authors yielded *M* or *P* enantiomer (Figure 2.1). The effect should be cleared up since, contrary to other factors used by Inoue and Wada to influence the outcome of the reaction, an achiral solvent should not, in principle, generate such effects.

2.2 Self-Assembly and Self-Organization

The spontaneous formation of complicated well-defined architectures such as hydrogen bonded Whitesides systems (Figure 1.1), those of intertwined helicates **9** (Figure 1.3) and 'molecular necklaces' presented in Figure 1.5, as well as those of the aggregates shown in Figure 1.4, illustrates self-organization of molecular components leading to the self-assembly of complicated supramolecular systems. One can distinguish between chemical (i.e., covalent) self-assembly and supramolecular one induced by intermolecular interactions such as hydrogen bonding, ion-ion, ion-induced dipole, dipole-dipole, and van der Waals interactions. A few examples of covalent self-assembly are given in Section 2.4, while those of supramolecular self-assembly will be amply discussed in several chapters of this book. Self-assembly is based on the *template effect* (see below) often involving not one but several steps taking place spontaneously in a single cooperative operation. The formation of the double helix of model nucleic acids, the all or nothing process discussed in Section 2.2, exemplifies such *cooperativity*.

A spontaneous arrangement of molecules with respect to each other facilitating chemical reactions is called *preorganization* [17a]. Some examples of the latter phenomenon in the domain of topological chemistry are given in Sections 2.3 and 8.2. A factor that forces preorganization by appropriate spatial arrangement of reagents, thus assisting self-assembling processes, is called a *template* [17b,c,d].

Molecular imprinting is a special polymerization technique making use of molecular recognition [18] consisting in the formation of a cross-linked polymer around an organic molecule which serves as a template. An imprinted active site capable of binding is created after removal of the template. This process can be applied to create effective chromatographic stationary phases for enantiomers separation. An example of such a sensor is presented in Section 6.3.2.3.

Allosteric effect operates in a system exhibiting conformational mobility when inclusion of one guest creates an additional cavity for a second guest (Figure 2.2). A similar example with two identical guests was presented in Figure 1.6.

Intermolecular forces can induce creation of larger polymolecular assemblies. For instance, *amphiphilic* molecules (see Chapter 4) having a polar 'head' and

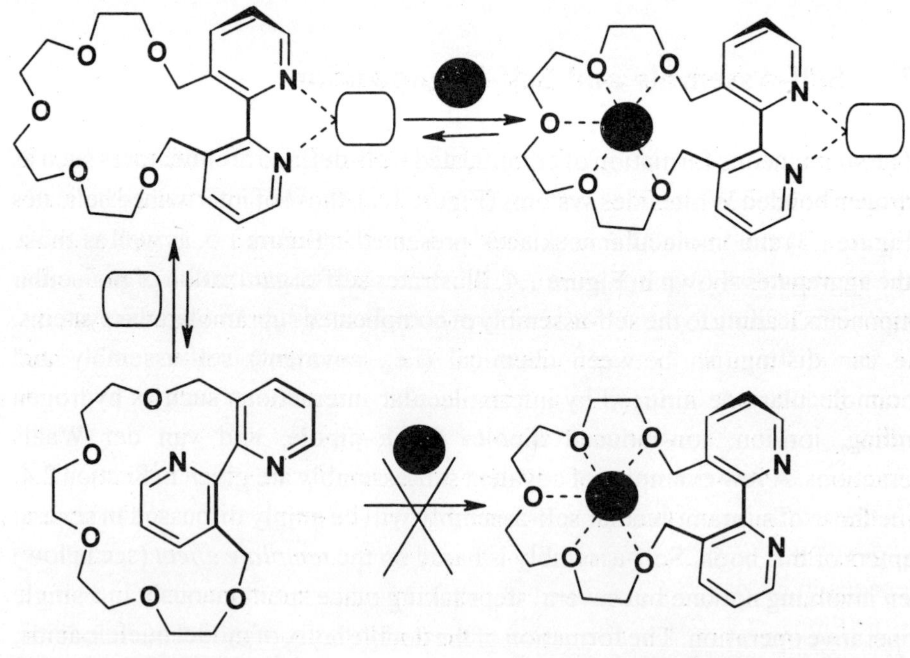

Figure 2.2. Owing to the allosteric effect the inclusion of an alkaline ion into the crown part of the ligand is favoured by the first complexation involving bipyridyl moiety.

apolar 'tail' can form layers, micelles, or vesicles (held together by weak noncovalent interactions) which were shown schematically in Figure 1.4. The central part of cell walls is a membrane consisting of a phospolipid bilayer. Thus studies of natural and model artificial membranes are of basic importance, enabling the understanding of the membranes' operation in living organisms. In particular, the membranes with inserted pores [19a] serve as models for the transport of ions through the cell walls. These problems will be discussed shortly in Chapter 4.

Supramolecular chemistry is a rapidly developing domain creating its own language, e.g., recently one even started to speak about the synthesis of a noncovalent molecular assembly. In analogy with the concepts of synthesis and synthons in organic chemistry, Fuhrhop and König [19b] have introduced the word '*synkinesis*' for the supramolecular assembly process, and the word '*synkinon*' for the building blocks of such assemblies. Tecton is another word proposed for these blocks [20].

2.3 The Role of Preorganization in the Synthesis of Topological Molecules. Template Reactions

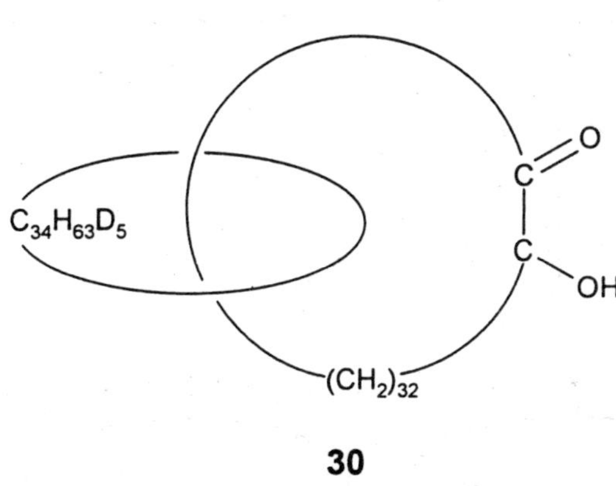

30

Since Möbius works in the 1820s [21a] mathematicians' studies of the relationships between sets and topology have evolved as a branch of mathematics dealing with such relationships. If a set can be transformed into another by a continuous transformation then these sets are topologically equivalent. For instance (Figure 2.3), two circles of different diameters or a circle and a triangle are topologically equivalent, whilst a circle and an interval or knot are not. Links bearing the name catenanes in chemistry, such as **30** [22], the knot **6** [23], and the Möbius strip **31a, b** (Figure 2.4) [21b], all have distinct topological properties. The latter molecule is obtained by glueing the ends of an interval after one of them is turned

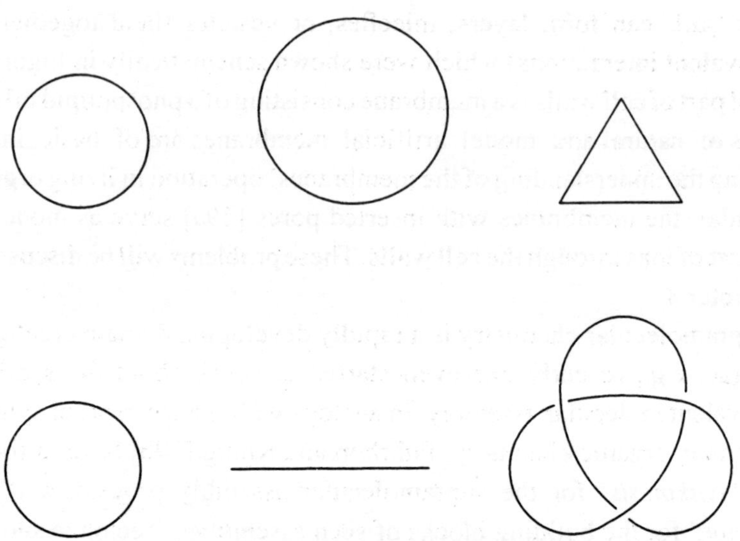

Figure 2.3. Equivalent and nonequivalent topological objects.

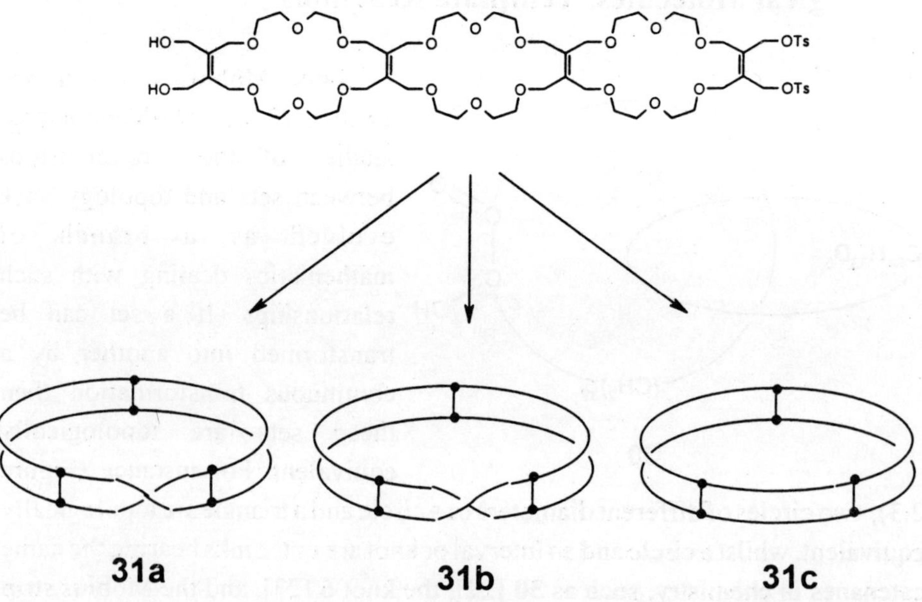

Figure 2.4. The scheme of Möbius strip formation.

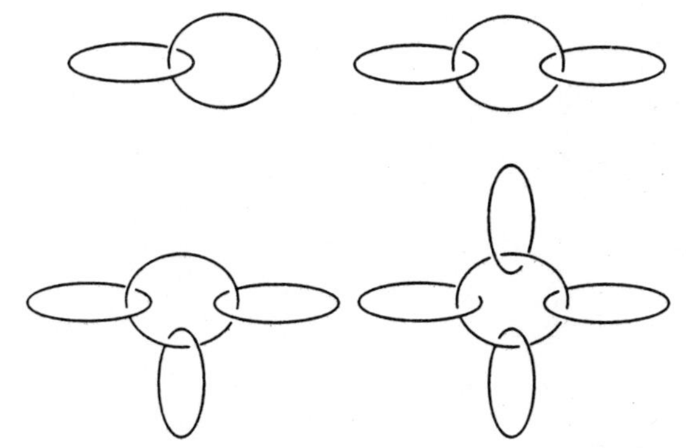

32

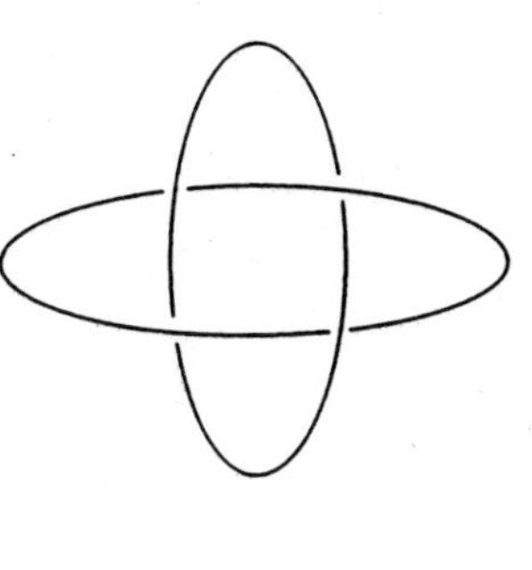

33

34

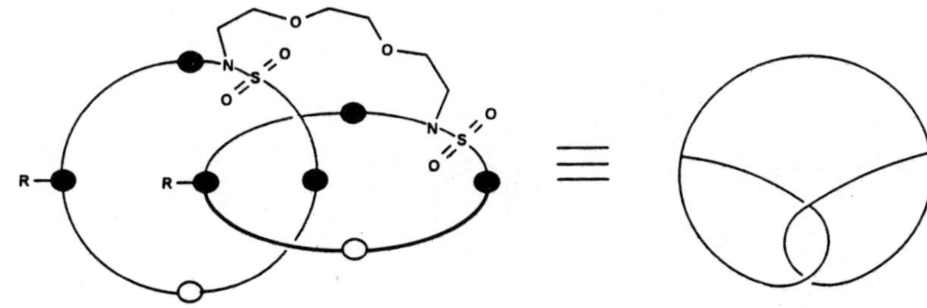

35

36

37

38

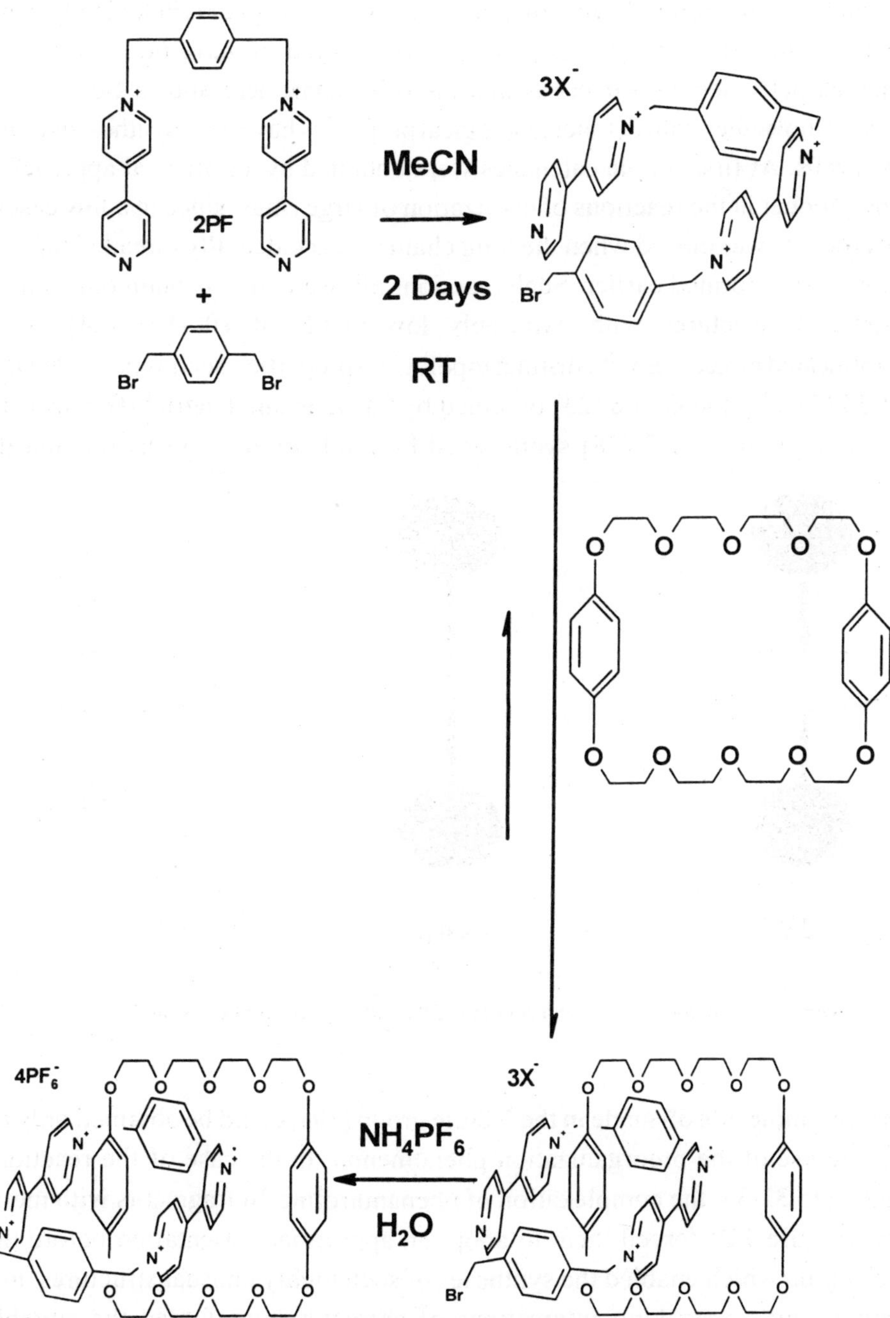

Figure 2.5. The highly-efficient (70%) formation of [2]catenane making use of π-stacking interaction [30].

around by 180⁰ (before connecting the ends) while the other is held firmly. Until recently the study of such objects was considered to have no relation with chemist activities. However, as discussed in detail here and in Section 8.1, several molecules with distinct topological properties have been synthesized since the 1960s. At first simple catenanes were obtained by a statistical approach as a by-product in the reactions of cyclization of large rings, since in a few cases a macrocycle was closed when the long chain was accidentally threaded through another ring formed earlier. Such reactions allowed one to obtain only simple catenated structures with extremely low yields of 10^{-3}-10^{-4} [24]. More complicated molecules with distinct topological properties, such as polycatenanes **32-34** [25-27], the knot **6** [23] obtained by Sauvage and Dietrich-Buchecker's group, olympiadane **7** [28] synthesized by Stoddart and coworkers, and the

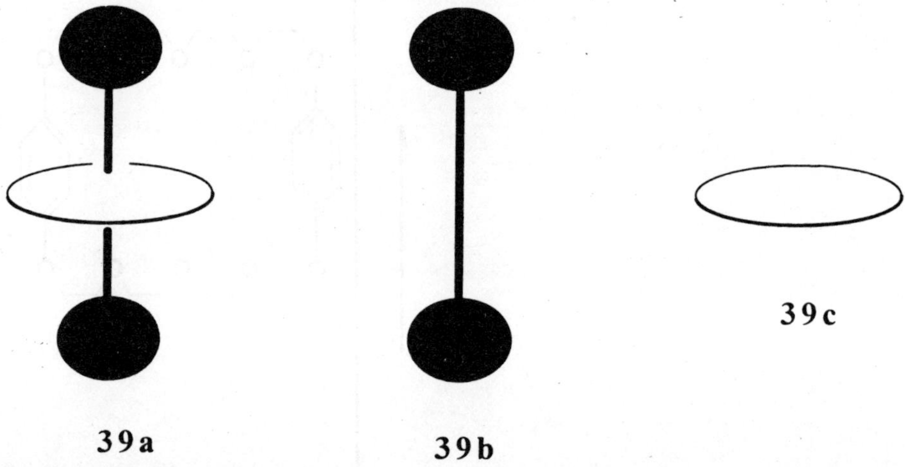

Figure 2.6. Schematic view of a rotaxane **39a** and its separated constituent parts.

'pretzel' molecule **35** made in the Vögtle group [29a] could be obtained only by making use of the preorganization phenomenon. In the case of the reactions leading to **32-34**, the complexation of phenanthroline **36** fragments with metal ions (Figure 1.2) forced them to adopt an appropriate orientation before the cyclization which enabled the syntheses of such highly unusual structures. In a similar way, π-stacking interactions of aromatic rings forced the suitable orientation of molecular fragments yielding catenane [30] (Figure 2.5) and olympiadane structure **7**. Hydrogen bonding is believed to enforce the formation

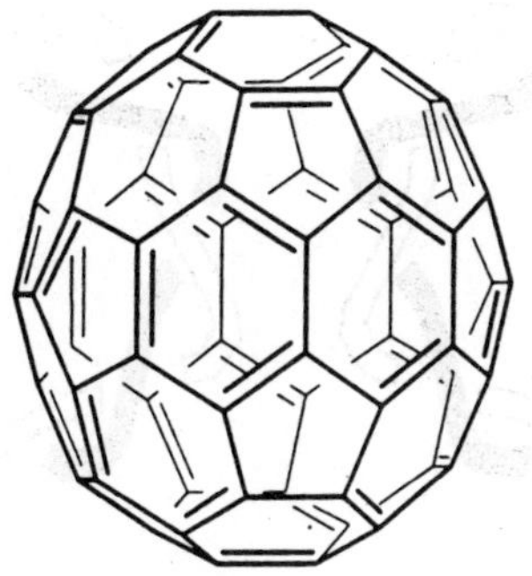

40

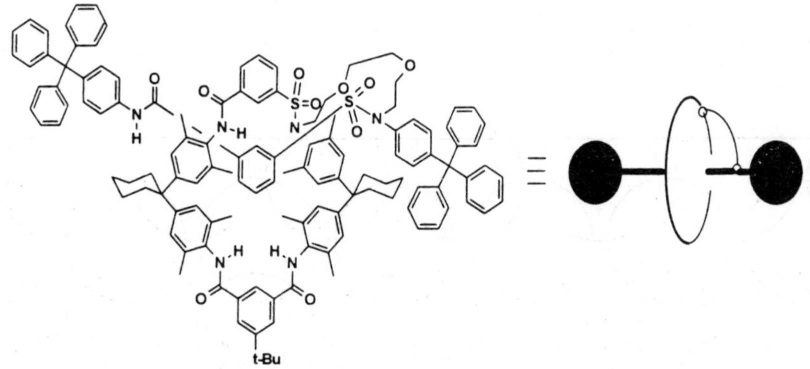

41a

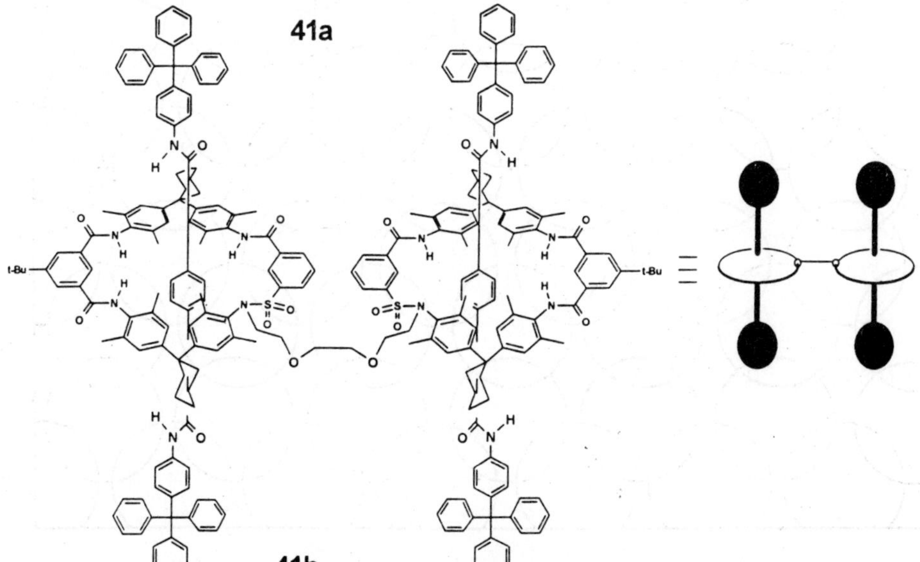

41b

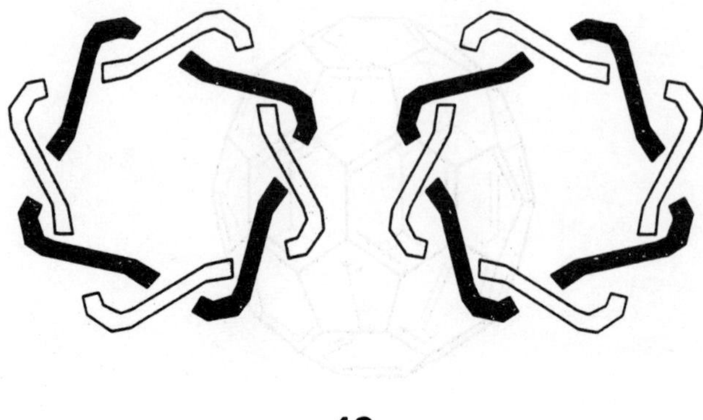

42

Figure 2.7. An 'idealized' net of the interwined circular DNAs.

of the 'pretzel' molecule **35** and catenane **37** [29b], whilst weak but numerous van der Waals interactions are responsible for the formation of the 'molecular necklace' **12** [31].The above reactions leading to the formation of topologically distinct molecules are examples of so called template reactions [18b, c, 32] making use of preorganization phenomenon in which one of the substrates forces appropriate spatial arrangement of the others leading to a spectacular increase in reaction yields. The template effect is also operating in the synthesis of crown ethers like **17** discussed in some detail in Chapter 3 and Section 7.1 [33], hemicarcerands like **5** [34], *p-t*-Bu-calix[6]arene **38** [35] and in many other reactions. Rotaxanes **39a** (Figure 2.6) have until recently not been included in the realm of topological chemistry, since they can be decomposed into separate fragments **39b** and **39c** by a sufficiently large distortion of the central ring. However, they were often discussed together with the catenanes they resemble. We have recently shown [36a] that one can define topological isomers in such a way that rotaxanes and endohedral fullerene complexes with one or more ions inside the C_{60} cage **40** and 'in' and 'out' isomers of hydrogenated fullerenes having CH bonds directed inside the cage [36b] will also constitute molecules with distinct topological properties. Some interesting rotaxanes **41** have been recently synthesized in Vögtle's group [37].

The molecules with distinct topological properties are not a mere curiosity, since they can be found in Nature. Circular DNA schematically presented as **42** are sometimes found in living organisms in the form of catenanes and knots [38], and special enzymes topoisomerases take part in their formation and transformations [39]. Circular DNA molecules can even form nets of catenated structures like that schematically presented in Figure 2.7 [40]. A discussion of biological topological structures falls outside the scope of this monograph; it should be stressed, however, that their role in Nature is not understood and warrants an explanation.

2.4 'One-Pot' Reactions. Covalent Self-Assembly Based on Preorganization

Both in nature and in test tubes many complicated reactions take place with apparent ease. Highly specific enzymatic reactions involving complementarity of substrate and receptor both with respect to their spatial structure and electrostatic fields lie outside the scope of this book. Here we would like only to mention a few

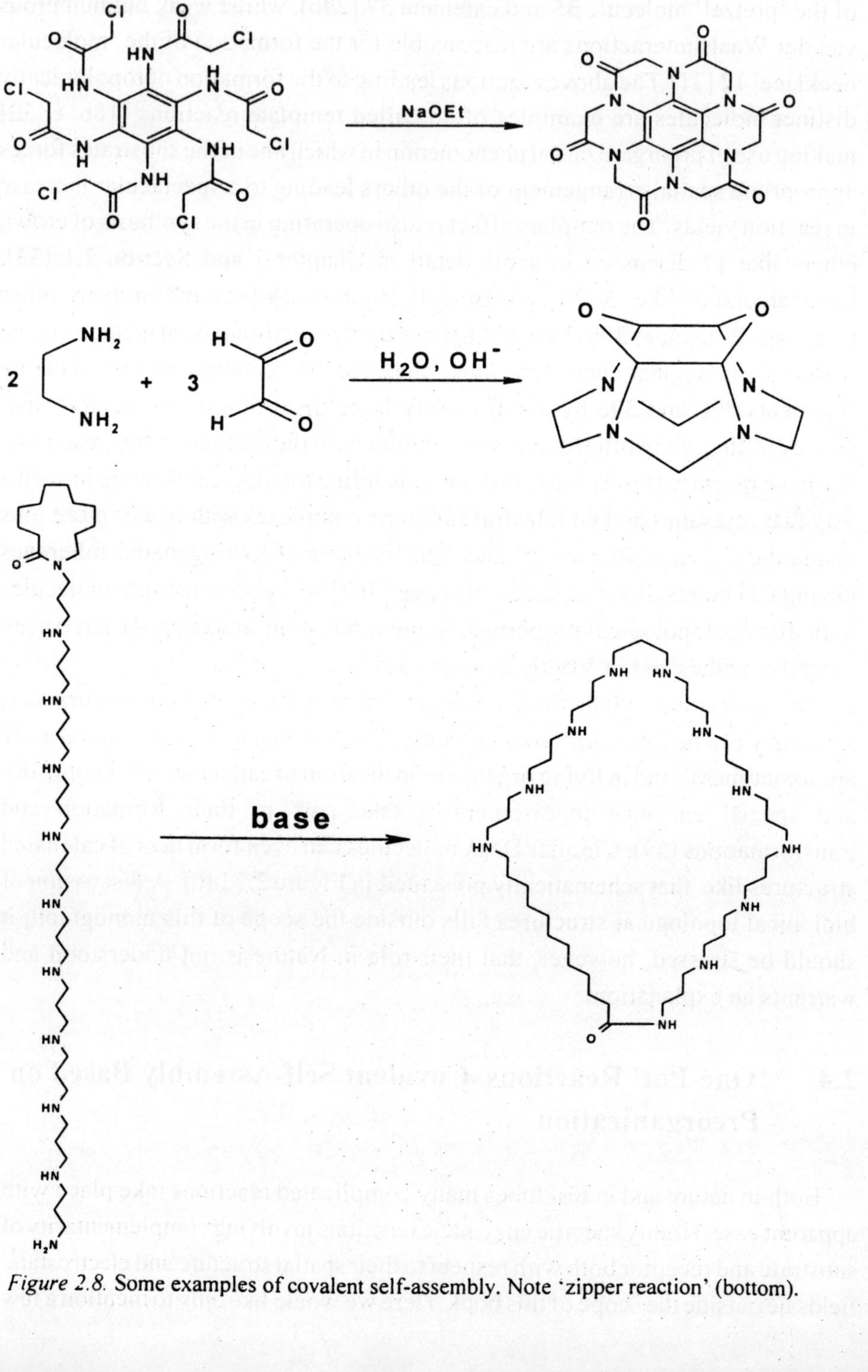

Figure 2.8. Some examples of covalent self-assembly. Note 'zipper reaction' (bottom).

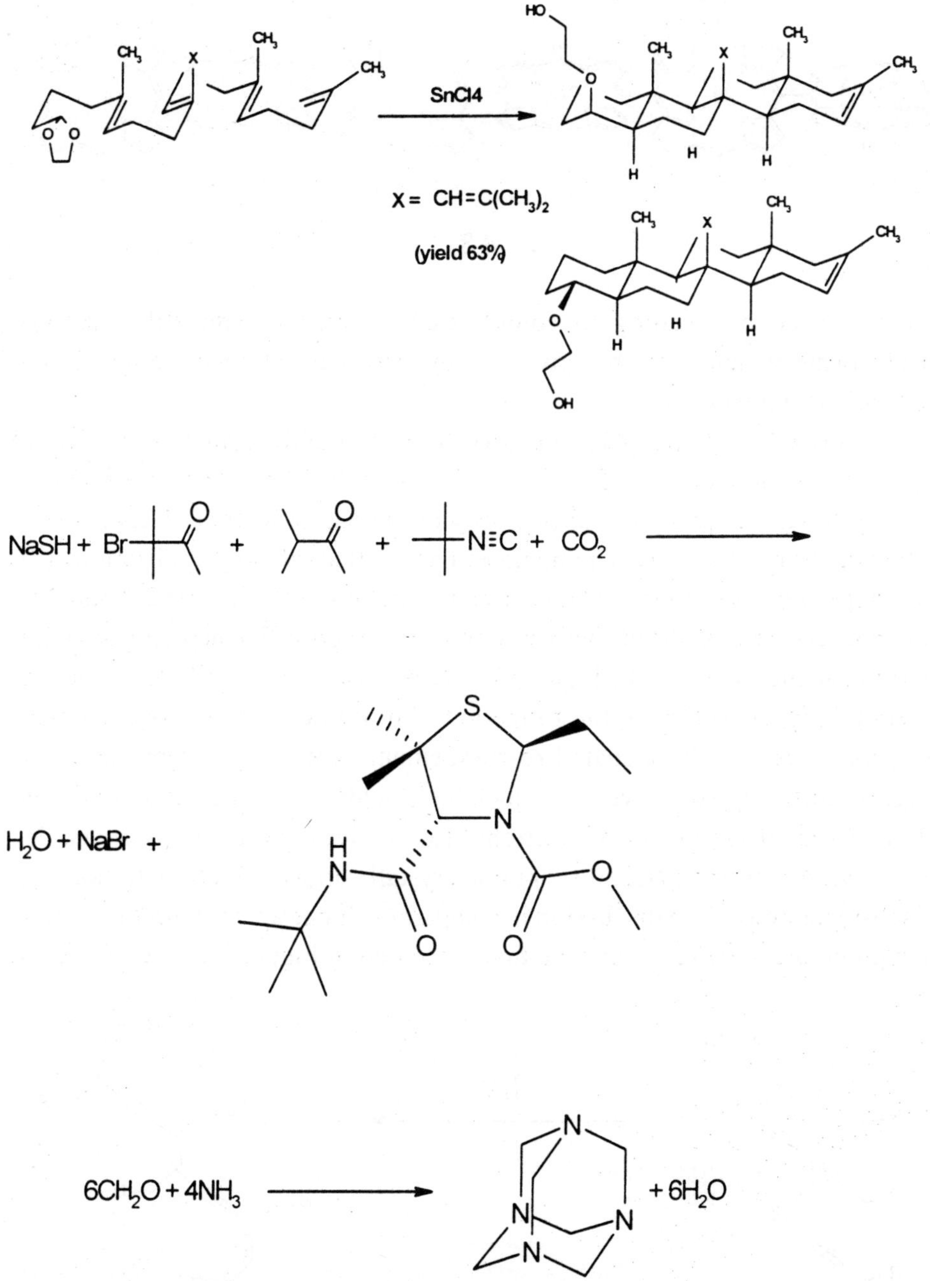

SnCl4

X = $CH{=}C(CH_3)_2$

(yield 63%)

NaSH + Br ... + ... + ... N≡C + CO_2 →

H_2O + NaBr +

$6CH_2O + 4NH_3$ → ... $+ 6H_2O$

Figure 2.9. Further examples of covalent self-assembly: of a tetracyclic hydrocarbon (top); the Ugi reaction (middle); atropine synthesis (bottom).

43

one-pot reactions, leading to complicated structures, with still unknown mechanisms which must be governed by preorganization of molecules or molecular fragments.

As noticed by Lehn [41], the formation of highly symmetrical C_{60} **40** represents an ultimate case of covalent self-assembly. The same is true for those of other fullerenes and nanotubes briefly discussed in Sections 4.2.4 and 7.5. Although the mechanisms of monomolecular reactions of (poly)cyclization [42], the 'zipper reaction' [43], and the cascade reaction [44] (Figures 2.8 and 2.9) have not yet been elucidated, they certainly involve very efficient preorganization rearrangements enabling the high yield one-pot reactions. Similarly, in the Ugi reaction [45], as well as in the formation of atropine cited in most chemical textbooks (Figure 2.9), a spatial rearrangement, that is a preorganization, of reagents must play a decisive role in their high yields. Obtaining oligoporphyrins **43** in one-pot electrochemical reactions also involves self-assembly of smaller units [46]. According to Ghadiri opinion crystallization is an ultimate example of self-recognition (see discussion in Chapter 6). If one agrees with this notion then photochemical reactions like that presented in Figure 2.10 are forced by

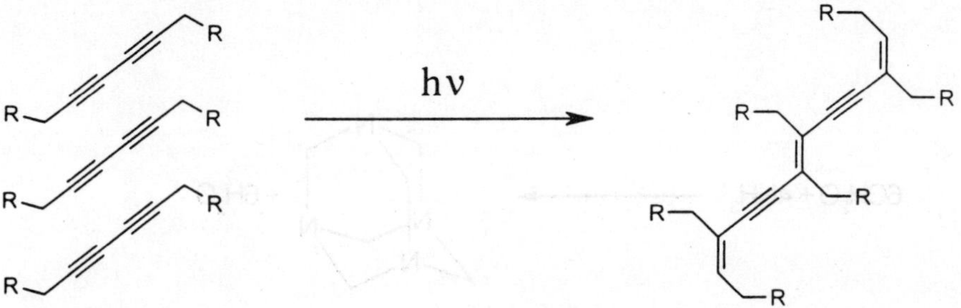

Figure 2.10. Photochemical solid-state trimerization.

preorganization of diyn molecules in the crystalline state [47a].

Monolayers discussed in Section 4.2 also introduce order preorganizing molecules. An analogous photochemical reaction making use of this phenomenon was carried out in diacetylene monolayers on gold [47b]. At present most 'one-pot reactions' are of no practical importance. However their occurence in prostaglandin synthesis [48] indicates that they will find their applications leading to more environmentally clean technologies.

REFERENCES

1. J.-M. Lehn. *Supramolecular Chemistry: Concepts and Perspectives.* VCH. Weinheim, 1995, p. 11.
2. We believe that one can, eventually, speak about the function of a molecular assembly in a living organism but the term seems inappropriate in supramolecular chemistry. Later there is a notion of 'molecular recognition as a binding with purpose'. Similarly, such an anthropomorphization seems justifiable only in biochemistry.
3. H.-J. Schneider. T. Blatter. U. Cuber, R. Juneja. T. Schiestel. U. Schneider. I. Theis. P. Zimmermann. *Shapes, Selectivity, and Complementarity in Molecular Recognition.* in Frontiers in Supramolecular Chemistry, H.-J. Schneider, H. Dürr. Eds., VCH, Weinheim. 1991, p. 29.
4. J. A. Zerkowski, C. T. Seto, G. M. Whitesides, J. Am. Chem. Soc., 1992. 114, 5473; E. E. Simanek. M. Mammen. D. M. Gordon. D. Chin, J. P. Mathias. C. T. Seto. G. M. Whitesides. Tetrahedron. 1995, 51, 607, and references cited therein.
5. (a) J.-M. Lehn, A. Rigault, J. Siegel, J. Harrowfield, B. Chevrier, D. Moras. Proc. Natl. Acad. Sci. USA. 1987, 84, 2565; (b) C. O. Dietrich-Buchecker, J.-P. Sauvage. Angew. Chem. Int. Ed. Engl., 1989, 28, 189.
6. A. Stadler-Szöke, J. Szejtli, Acta Pharm. Hung., 1979, 49, 30.
7. (a) A. Harada, J. Kamachi, Nature, 1992, 356, 325; (b) J. F. Stoddart, Angew. Chem. Int. Ed. Engl., 1992, 31, 846.
8. H. Fraenkel-Conrat. R. C. Williams, Proc. Natl. Acad. Sci. USA. 1955. 41, 690.
9. (a) E. Fischer, Ber., 1894, 27, 2985; (b) D. E. Koshland, Jr., Angew. Chem. Int. Ed. Engl., 1994. 33. 2475.
10. See, for instance, W. L. Jorgenson, Nature, 1991, 254, 954; B. E. Bernstein, P. A. M. Michels, W. G. J. Hol, Nature, 1997, 385, 275.
11. (a) A. Renault, D. Talham, J. Canceill, P. Batail, A. Collet, J. Lajzerowicz. Angew. Chem. Int. Ed. Engl., 1989. 28, 1249; (b) As discussed in H. Dodziuk. *Modern Conformational Analysis. Elucidating Novel Exciting Molecular Structures.* VCH Publishers. New York.

1995. van der Waals radii are ill-defined quantities which can be used only in semi-quantitative way.

12. D. J. Cram, S. Karbach, H.-E. Kim, C. B. Knobler, E. F. Maverick, J. L. Ericson, R. C. Helgeson, J. Am. Chem. Soc., 1988, 110, 2229.
13. (a) M. Scherer, D. L. Caulder, D. W. Johnson, K. N. Raymond, Angew. Chem. Int. Ed. Engl., 1999, 38, 1588; (b) W. H. Elliott, D. C. Elliott, *Biochemistry and Molecular Biology*, Oxford University Press, Oxford, 1997, p. 6; (c) R. B. Prince, S. A. Barnes, J. S. Moore., J. Am. Chem. Soc., 2000, 122, 2758.
14. (a) Cyclodextrins are one of the best enantiodiscriminating factors. D. Sybilska, J. Zukowski, in *Chiral Separations by HPLC*, A. M. Krstulovic, Ed., Wiley, New York, 1989, Chapter 7; (b) C. Moeder, T. O'Brian, R. Thompson, G. Bicker, J. Chromatography A, 1996, 736, 1; (c) H. Dodziuk, A. Ejchart, O. Lukin, M. O. Vysotsky, J. Org. Chem., 1999, 64, 1503; (d) C. Moeder, T. O'Brien, R. Thompson, G. Bicker, J. Chromatogr. A., 1996, 736, 1.
15. (a) W. C. Still, Acc. Chem. Res., 1996, 29, 155; (b) R. Liu, P. E. J. Sanderson, W. C. Still, J. Org. Chem., 1990, 55, 5184; (c) J.-I. Hong, S. K. Namgoong, A. Bernardi, W. C. Still, J. Am. Chem. Soc., 1991, 113, 5111.
16. Y. Inoue, T. Wada, J. Am. Chem. Soc., 2000, 122, 406.
17. (a) D. J. Cram, Angew. Chem. Int. Ed. Engl., 1988, 27, 1009; (b) D. H. Busch, J. Incl. Phenom. Mol. Recogn. Chem., 1992, 12, 389; (c) N. V. Gerbeleu, V. A. Arion, J. Burgess, *Template Synthesis of Macrocyclic Compounds*, Wiley-VCH, Weinheim, 1999; (d) R. Hoss, F. Vögtle, Angew. Chem. Int. Ed. Engl., 1994, 33, 375.
18. G. Wulff, Angew. Chem. Int. Ed. Engl., 1995, 34, 1812.
19. (a) J.-H. Fuhrhop, J. König, *Membranes and Molecular Assemblies: The Synkinetic Approach*, Monographs in Supramolecular Chemistry, J. F. Stoddart, Ed., The Royal Society of Chemistry, Cambridge, United Kingdom, 1994, p. 2; (b) Ref. 19a, Chapter 1.
20. M. Simard, D. Su, J. D. Wuest, J. Am. Chem. Soc., 1991, 113, 4696.
21. (a) A. F. Möbius, *Gesammelte Werke* (*Collected Works*, in German), Leipzig, 1885; (b) D. M. Walba, Tetrahedron, 1985, 41, 3161.
22. E. Wasserman, J. Am. Chem. Soc., 1960, 82, 4433.
23. C. O. Dietrich-Buchecker, J.-P. Sauvage, Angew. Chem. Int. Ed. Engl., 1989, 28, 189.
24. G. Schill, *Catenanes, Rotaxanes and Knots*, Academic Press, New York, 1971.
25. F. Bitsch, C. O. Dietrich-Buchecker, A.-K. Khemiss, J.-P. Sauvage, A. van Dorsselaer, J. Am. Chem. Soc., 1991, 113, 4023.
26. J.-F. Nierengarten, C. O. Dietrich-Buchecker, J.-P. Sauvage, J. Am. Chem. Soc., 1994, 116, 375.
27. C. O. Dietrich-Buchecker, B. Frommberger, I. Löer, J.-P. Sauvage, F. Vögtle, Angew. Chem. Int. Ed. Engl., 1993, 32, 1434.

28. D. B. Amabilino, P. R. Ashton, A. S. Reder, N. Spencer, J. F. Stoddart, Angew. Chem. Int. Ed. Engl., 1994, 33, 1286.

29. (a) R. Jäger, T. Schmidt, D. Karbach, F. Vögtle, Synlett, 1996, 726; (b) A. G. Johnston, D. A. Leigh, R. J. Pritchard, M. D. Deegan, Angew. Chem. Int. Ed. Engl., 1995, 34, 1209.

30. T. T. Goodnow, A. E. Kaifer, M. V. Reddington, A. M. Z. Slawin, N. Spencer, J. F. Stoddart, C. Vicent, D. J. Williams, Angew. Chem. Int. Ed. Engl., 1989, 28, 1396.

31. A. Harada, J. Kamachi, Nature, 1992, 356, 325; J. F. Stoddart, Angew. Chem. Int. Ed. Engl., 1992, 31, 846.

32. F. Vögtle, Top. Curr. Chem., 1982, 101, 1.

33. G. W. Gokel, *Crown Ethers and Cryptands*, The Royal Society of Chemistry, Cambridge, 1994, p. 27.

34. R. G. Chapman, J. C. Sherman, J. Am. Chem. Soc., 1995, 117, 9081.

35. C. D. Gutsche, unpublished results, cited in Ref. 9b of V. Böhmer, Angew. Chem. Int. Ed. Engl., 1995, 34, 713.

36. (a) H. Dodziuk, K. S. Nowiński, Tetrahedron, 1998, 54, 2917; (b) H. Dodziuk, K. Nowiński, Chem. Phys. Lett., 1996, 249, 406.

37. R. Jäger, F. Vögtle, Angew. Chem. Int. Ed. Engl., 1997, 36, 930.

38. E. M. Shekhtman, S. A. Wasserman, N. Cozzarelli, M. J. Solomon, New J. Chem., 1993, 17, 757.

39. J. M. Berger, S. J. Gamblin, S. C. Harrison, J. C. Wang, Nature, 1996, 379, 225.

40. J. Chen, C. A. Rauch, J. H. White, P. T. Englung, N. R. Cozarelli, Cell, 1995, 80, 61.

41. Ref. 1, p. 140.

42. (a) J. M. Edwards, U. Weiss, R. D. Gilardi, I. L. Karle, J. Chem. Soc. Chem. Commun., 1968, 1649; (b) W. S. Johnson, S. J. Telfer, S. Cheng, Schubert, J. Am. Chem. Soc., 1987, 109, 2517.

43. U. Kramer, A. Guggisberg, M. Hesse, H. Schmid, Angew. Chem. Int. Ed. Engl., 1978, 17, 200.

44. J. Thomaides, P. Maslak, R. Breslow, J. Am. Chem. Soc., 1988, 110, 3970.

45. A. Dömling, I. Ugi, Angew. Chem. Int. Ed. Engl., 1993, 32, 563.

46. T. Ogawa, Y. Nishimoto, N. Yoshida, N. Ono, A. Osukua, J. Chem. Soc. Chem. Commun., 1998, 337.

47. (a) V. Enkelmann, Adv. Polym. Sci., 1984, 63, 9; (b) M. Cai, M. D. Mowery, H. Menzel, C. E. Evans, Langmuir, 1999, 15, 1215.

48. C. R. Johnson, M. P. Brown, J. Am. Chem. Soc., 1993, 115, 11014.

Chapter 3

INCLUSION COMPLEXES: HOST-GUEST CHEMISTRY

3.1 Early Development of Host-Guest Chemistry. Pedersen's Works on Crown Ethers

"In a preparation of bis[2-(o-hydroxyphenoxy)ethyl] ether by reacting in aqueous 1-butanol bis(2-chloroethyl) ether with the sodium salt of 2-(o-hydroxyphenoxy)tetrahydropyran contaminated with some catechol, a very small amount of a white, fibrous, crystalline by-product was obtained. It was found to be 2,3,11,12-dibenzo-1,4,7,10,13,16-hexaoxacyclooctadeca-2,11-diene **44** (another number is given in the original paper), a cyclic polyether capable of forming stable complexes with many salts of the alkali and alkaline earth metals." [1] This modest observation on the by-product obtained owing to the presence of impurities in the reaction mixture, correctly assessed by Pedersen, marked the beginning of Nobel Prize winning research in host-guest (or inclusion) chemistry. As discussed earlier, in full analogy with biochemistry, it involves a larger host molecule with a cavity in which a guest molecule can be embedded (see, however, the discussion of the complications involving the choice of host and guest in the solid state supramolecular chemistry presented in Section 6.2 and the hexokinase enzyme [2a] and the foldamers [2b] which wrap themselves around the guest). The hosts are usually macrocyclic organic molecules, whilst neutral organic molecules, their ions, as well as metal ions, can serve as guests. Typical coordination complexes formed by transition metal ions with nitrogen-containing macrocycles such as cyclam **45** and porphyrins **46**, representing a border case, are not always included in this domain. Hemicarcerand **5** complex with cyclobutadiene **4** [3a], that of β-cyclodextrin **11** with nitroglycerine **10a** [3b] and the complexes of metal cations with crown ethers **17** [4] and calixarenes **18** [5] briefly presented in Chapter 1 are typical examples of inclusion complexes.

It should be stressed that there is not always justice in reseach evaluation. The selective formation of inclusion complexes by cyclodextrins (such as **11**) was established by Cramer [6] at least 15 years earlier than that by crown ethers. However, cyclodextrin studies forming an independent branch of host-guest chemistry seem underestimated in spite of their considerably greater practical importance at present than that of other host macrocycles (crown ethers **17**, calixarenes **18**, etc.). Sometimes they are even totally neglected by discussing inclusion phenomena [7].

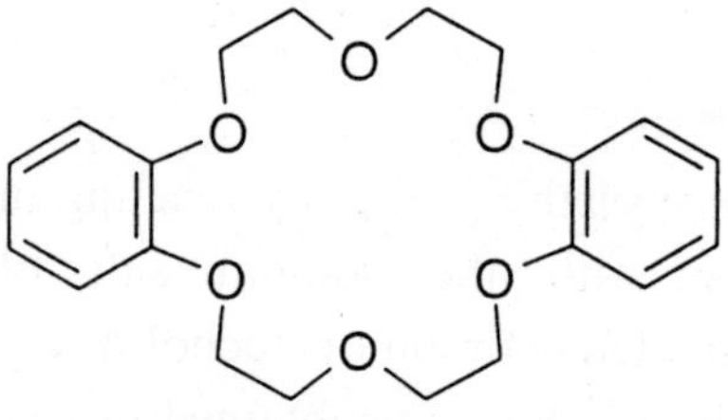

44 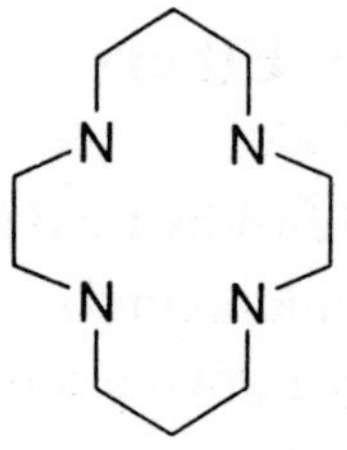 **45** 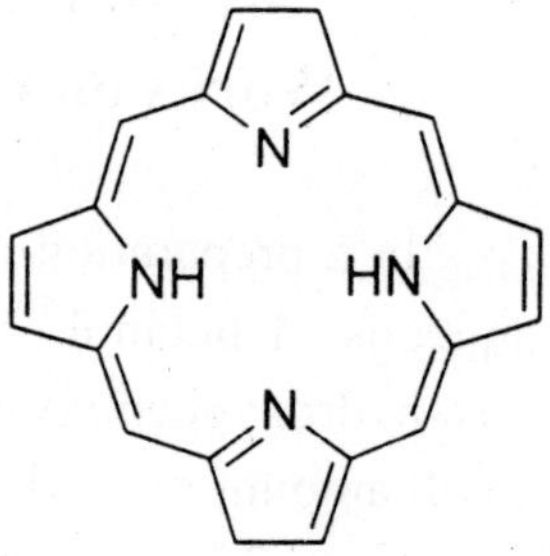 **46**

The Pedersen work on polyethers [1] was carried out about 30 years ago. Since then enormous progress has been made in this area of chemistry in general and in physical methods allowing one to determine the structure of a complex and its strength in particular. Therefore it is probably worth showing briefly how the crown ethers complexation was proved by Pedersen in order to compare his method of research with those used today. Three criteria have been used with this purpose in mind: (a) changes in the solubilities of the polyethers and salts upon complexation in different solvents; (b) analogous characteristic changes in the UV spectra of the aromatic polyethers and those in the IR spectra, and, naturally; (c) isolation of the complexes in form of pure compounds and their X-ray study. It was not always possible to use the last criterion, which could provide direct evidence since some complexes, stable in solution in certain solvents, could not be isolated.

Let us compare the methods applied by Pedersen for establishing the complex' formation with a modern approach. Today tedious solubility studies are carried out almost exclusively with practical applications in mind, but they are not performed to prove the complex' formation. For instance, one of the main reasons for the use of cyclodextrin complexes in the pharmaceutical industry is their solubilizing effect on drugs [8]. There, and almost only there, solubility studies are a must. As concerns spectroscopic methods, at present the NMR technique is one of the main tools enabling one to prove the formation of inclusion complex, carry out structural studies (for instance, making use of the NOE effect [9a]),. determine the complex stability [9b, c] and mobility of its constituent parts [9d]. However, at the time when Pedersen performed his work, the NMR method was in the early stage of development, and thus inaccurate, and its results proved inconclusive. UV spectra retained their significance in supramolecular chemistry, whilst at present the IR method is used to prove the complex formation only in very special cases.

Applying the above criteria to almost 50 crown ethers Pedersen was able to formulate conditions influencing the stability of the complexes he studied. The following were considered important:

1. The size of the ion must be smaller than that of the hole in the polyether ring. Thus, a stable complex was not formed if the ion was too large to fit into the polyether ring.

2. The larger the number of oxygen atoms, the greater the stability of the complex provided, that,

3. The oxygen atoms are coplanar (they lie in the same plane and the apex of C-O-C angle is centrally directed in the same plane) and they are symmetrically evenly spaced in a circle).

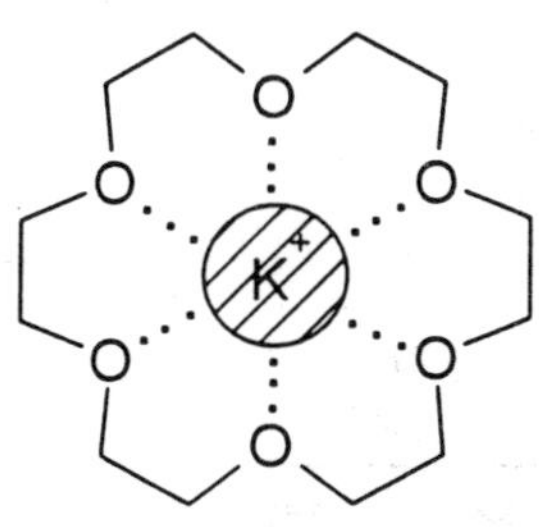

Figure 3.1. 18-crown-6 complex with K^+.

4. The stability of the complex is higher with higher basicity of the oxygen atoms. (The atoms attached to aromatic rings are less basic than the ones attached to aliphatic carbon atoms).

5. Steric hindrance in the polyether ring prevents the formation of the complex.

6. The strong tendency of the ion to associate with the solvent also hinders or even precludes the formation of the complex.

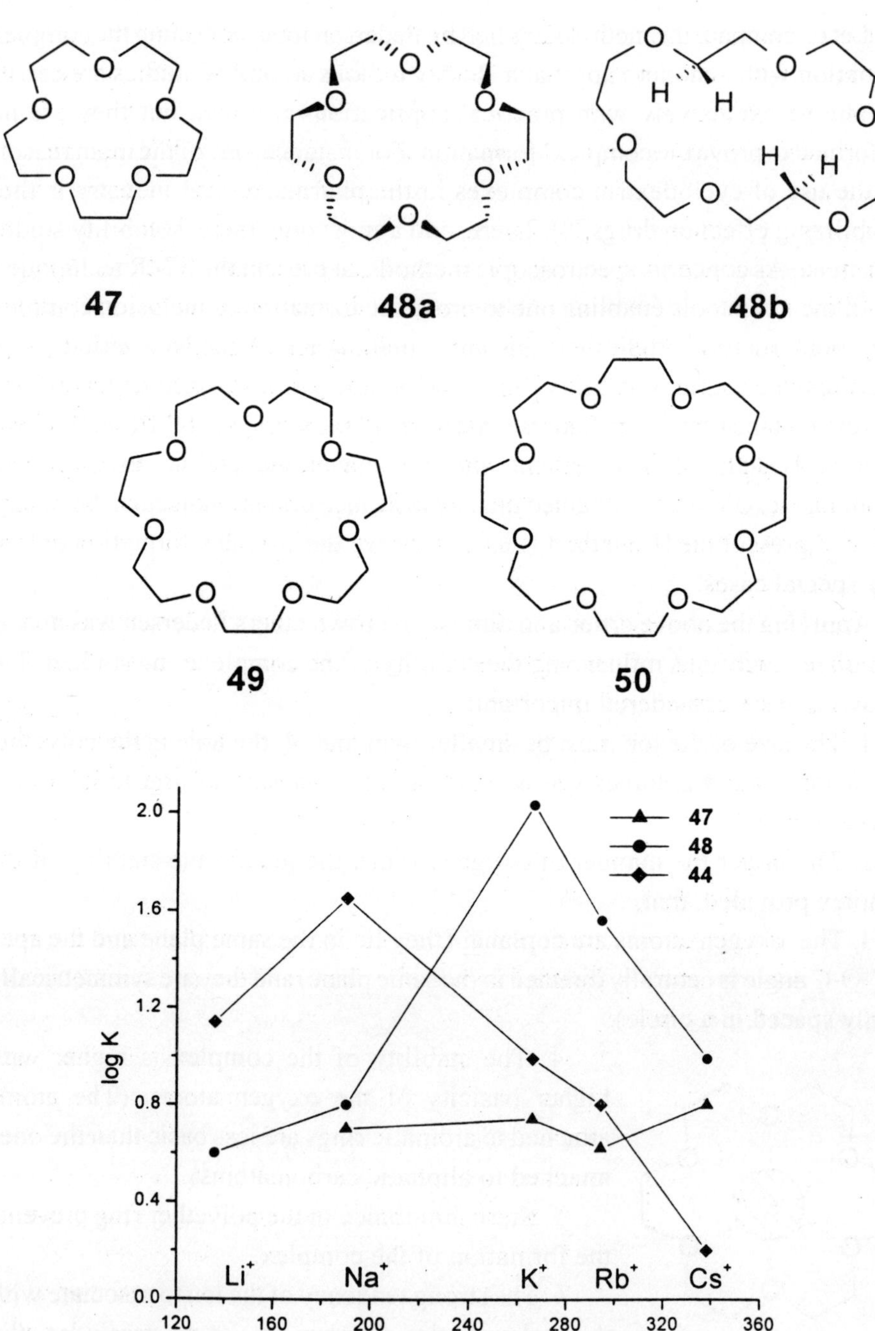

Figure 3.2. Dependence of the stability constant K on the cation diameters for alkali metal complexes with three crown ethers.

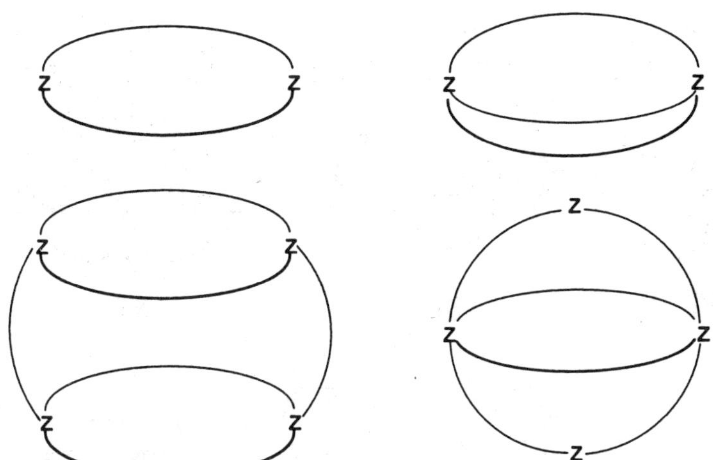

Figure 3.3. The Lehn classification of the recognition patterns.

7. The electrical charge on the ion should influence the effect under study. However, Pedersen has not studied it. He has only observed the 1 : 1 stoichiometry of the complexes (corresponding to one molecule of the crown to one of the guest) regardless of the valence.

51

52

As concerns the spatial fit of host and guest, **44** forms the most stable complex with K^+ (Figure 3.1) [10], since its radius of ca. 138 pm is approximately equal to the ionic radius of the guest. The dependence of stability constants of the complexes of **47**, **48** and **44** with alkali metal cations on the ion diameters is shown in Figure 3.2. The complicated character of the depicted relations indicates that more factors (*e.g.,* solvent effect) are at play in the ions recognition.

The studies started by Pedersen [1] have been extended by Lehn [11a] and Cram [2b] and their coworkers and, later, by many others. To be

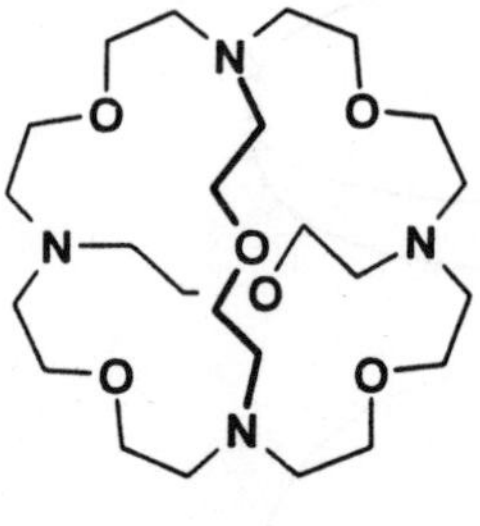

53

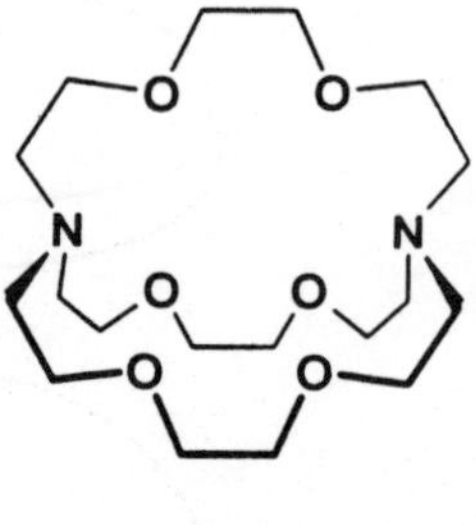

54

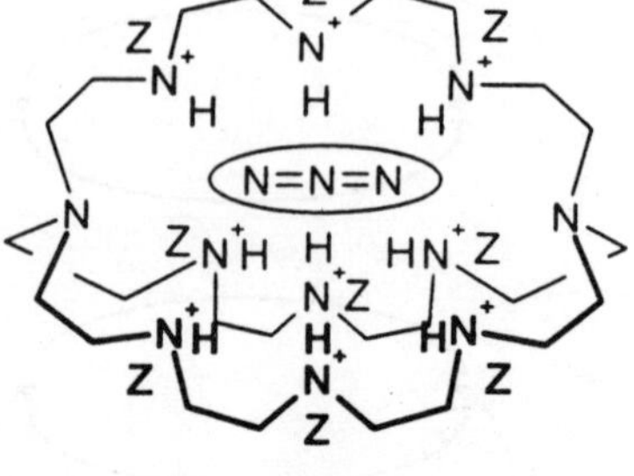

55

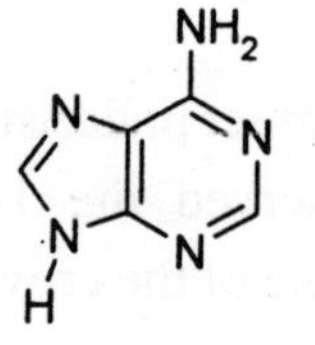

56

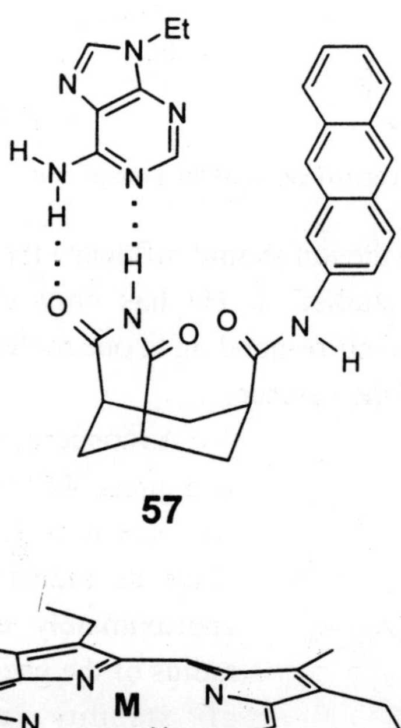

57

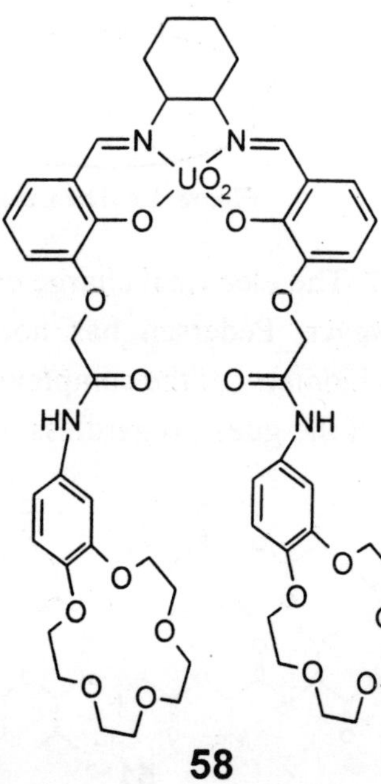

58

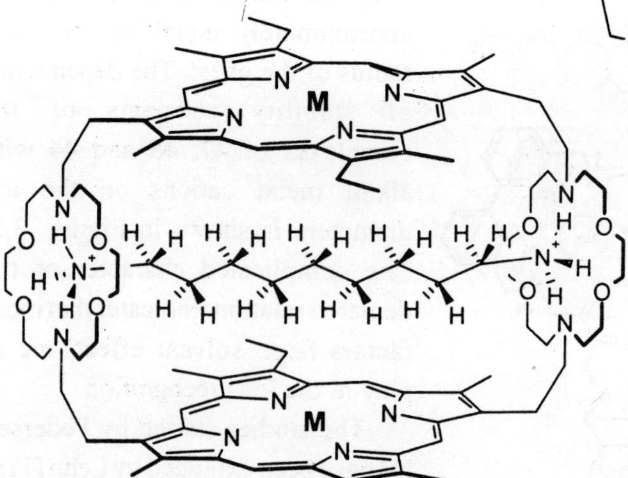

59

able to rationally design a host macrocycle for a specific guest Lehn introduced the concepts of spherical recognition of cations and anions, tetrahedral recognition of ammonium cation and water, linear recognition, and recognition of neutral molecules. Valinomycin **51** selectively binding K^+ ion [12], dibenzo-18-crown-6 **44** forming a complex with Rb^+ ion [1] and trinuclear sodium cryptate **52** [13] are examples of spherical recognition. The complexes of **53** [14] and **54** [15] with NH_4^+ ion represent tetrahedral recognition while complex **55** [16] corresponds to the linear recognition. Lehn's classification [11b] involving the recognition patterns shown in Figure 3.3 played an important role in the development of

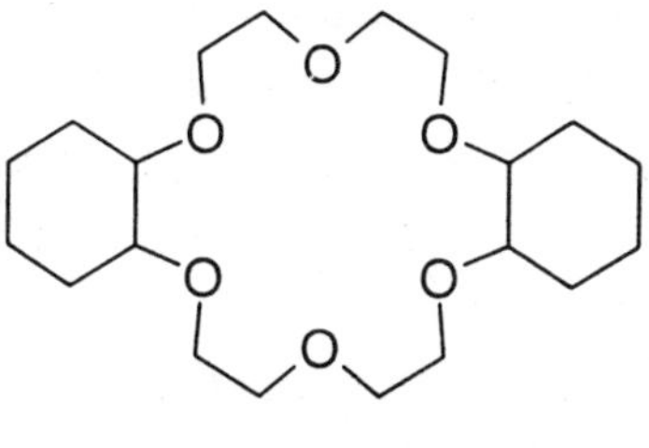

60 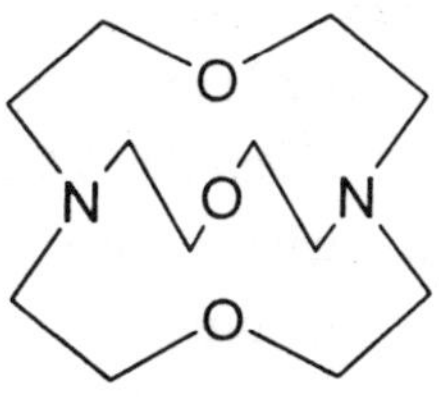 **61** 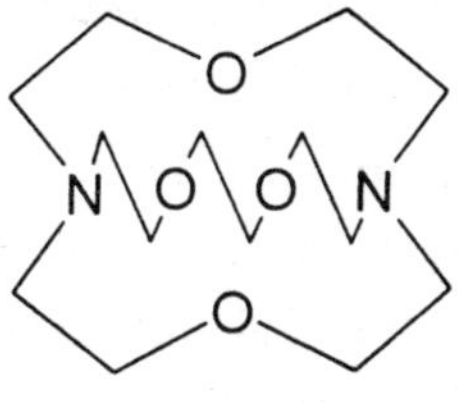 **62**

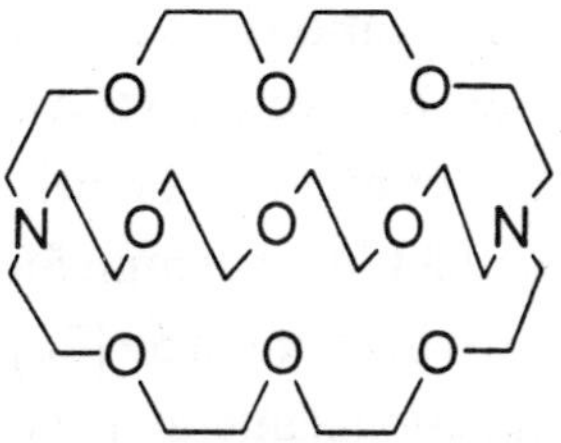

63 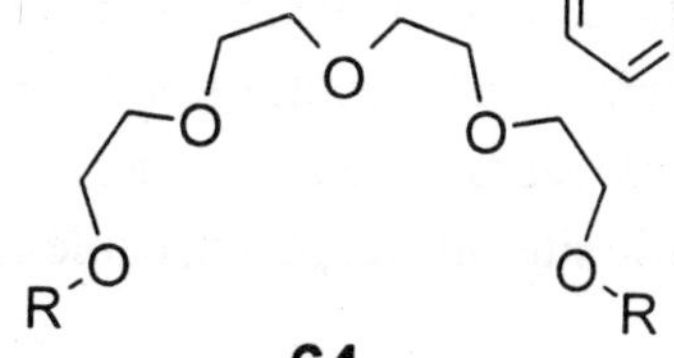 **64** 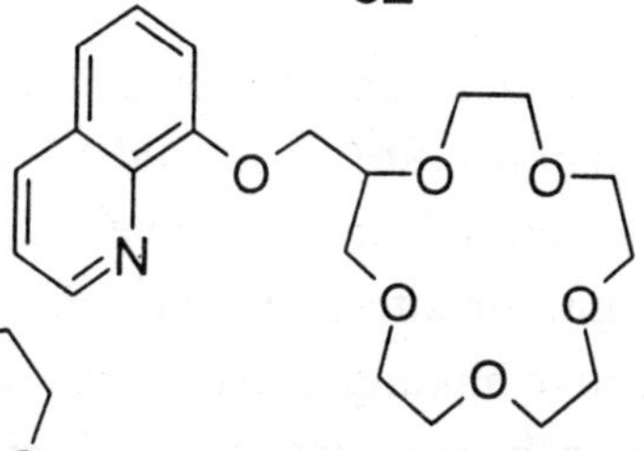 **65**

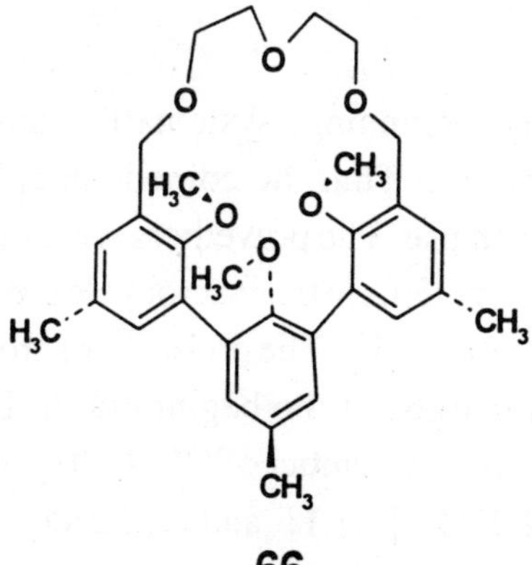

66 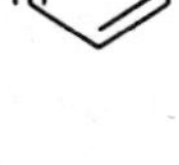 **67a** 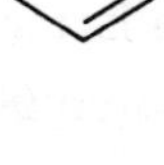 **67b**

host-guest chemistry paving the way to rational design of hosts for particular guests. Recently, this type of reasoning involving hydrogen bonding patterns has been frequently applied in crystal engineering and in host design discussed in Section 6.2 and Chapter 8. However, it is not general and does not cover, amongst others, the complexation by calixarenes **18**, hemicarcerands such as **5**, and cyclodextrins in which saturated or aromatic rings form the 'walls' of the cavities. What is more important, the recognition by these and many other macromolecules is based on less specific nonbonded interactions: binding of adenine **56** by Rebek's cleft **57** [17] through hydrogen bondings and π-stacking interactions: the beautiful Whitesides structures presented in Figure 1.1 (as well as other examples shown in Section 6.2) exemplify the recognition of neutral molecules; similarly, discrimination between α-pinene **27** [18a] and camphor **28** [18b] enantiomers by α-cyclodextrin **13** on the basis of van der Waals interactions (mentioned in the former Chapter), as well as most cases of recognition by cyclodextrins (discussed in Section 7.4) represent the same type.

The complexation studies involving anions have been started later than those involving cations but at present then constitute a rapidly growing domain [16, 19a].

Development of ditopic receptors **58** for both an anion and a cation is a logical extension [19b]. It should be stressed that the recognition by cyclodextrins, calixarenes, hemicarcerands, and many other systems does not involve specific recognition sites, whereas tetrahedral recognition by **53** and **54**, the adenine **56** selective binding, and multiple recognition like the one shown in formula **59** [20] require the presence of specific sites in receptor capable of directional interactions.

3.2 Nomenclature

As stated by Pedersen [1] at the very beginning, systematic names of polyethers are too cumbersome for repeated use. Thus, he coined an efficient system of their trivial names that is in common use. The polyethers were dubbed 'crown ethers' or 'crowns'. The total number of atoms in the macrocycle preceded this term whilst the number of oxygen atoms followed it. The presence of aromatic **44** or saturated **60** rings in the systems is marked at the beginning. In Lehn's system for bicyclic molecules called cryptands the number of OCH_2CH_2 units in each branch of the bicyclic hosts **61-63** ([1.1.1]-, [2.1.1]- and [2.2.2]cryptands

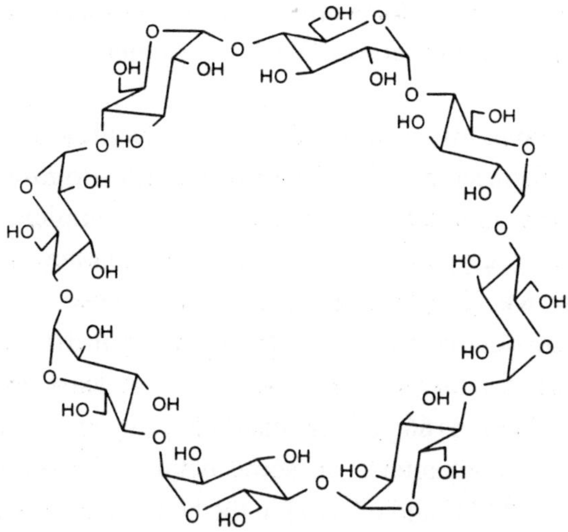

68

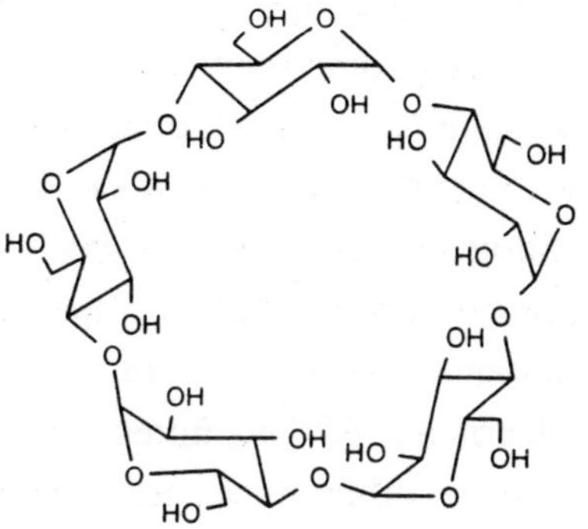

69

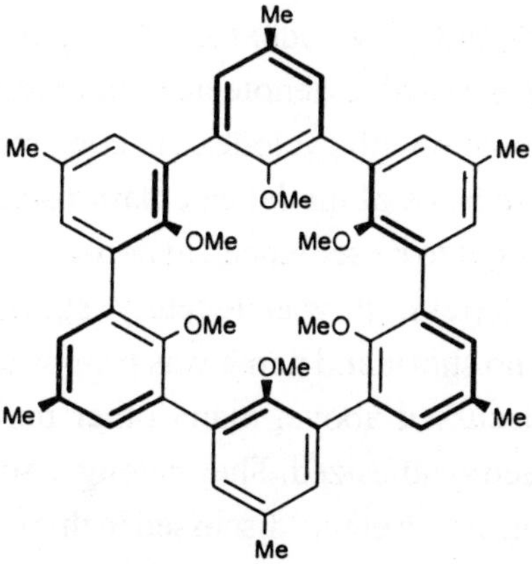

70

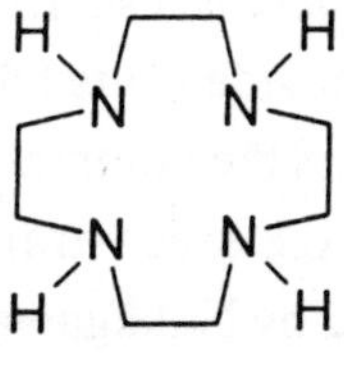

71

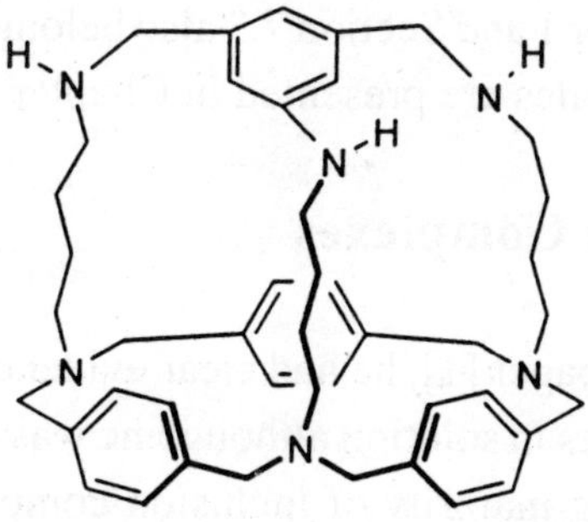

72

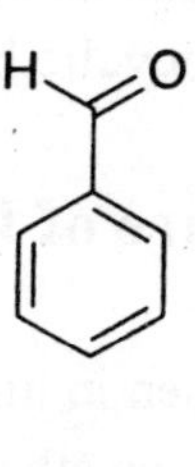

73

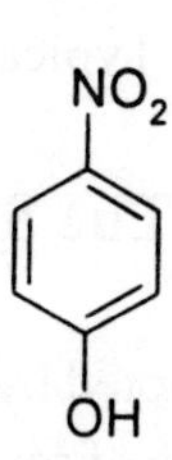

74

[21] is given. Thus **48** is denoted 18-crown-6, **44** as dibenzo-18-crown-6, and **60** as dicyclohexyl-18-crown-6, for which specification of the stereoisomer involved should be given.

As stated above, systematic names of macrocyclic host molecules were "absurdly complicated for routine discussions" [22]. Therefore Vögtle proposed the name 'coronand' for crown ethers, and that of coronates for their complexes while cryptand complexes were called 'cryptates'. The corresponding noncyclic analogues are podands such as **64** [23] and podates, respectively. The cumbersome name 'podando-coronands' (and correspondingly 'podando-coronates') was proposed for lariat ethers [24] having at least one sidearm like **65**. Examples of hemispherands **66** [25], cavitands **25** [26] and those of some other hosts are discussed in Chapter 7 in some detail, whilst the exceptional stability of fragile guests **4** [2a] and **67** [27] in the hemicarcerand **5** cavity are discussed in Chapters 1 and Section 7.3.

The most common cyclodextrins (discussed in Section 7.4) formed of 6, 7 or 8 *gluco*pyranoside units are called α-, β- and γ-cyclodextrins **11**, **13** and **68** [8, 9a]. Further letters of the Greek alphabet are used to denote next members of the series but there is a small designation problem with cyclodextrin consisting of 5 *gluco*pyranoside rings **69**. Namely, on the basis of model calculations for more than 20 years the latter molecule was thought to be sentenced to non-existence in view of excessive strain [28]. Nevertheless, it was recently successfully synthesized by Nakagawa et al. [29] but no shorthand name was proposed for it.

In addition to macrocyclic hosts discussed above, many other molecules capable of selective complexation have been synthesized. They belong to so-called macrocyclic chemistry [30] encompassing crown ethers discussed in this Chapter, cryptands **61-63** [21], spherands **70** [31], cyclic polyamines **71** [32], calixarenes **18** [5], and other cyclophane cages such as **72** [33] to name but a few. Hemicarcerand **5** [2b] discussed in Chapter 1 and Section 7.3 also belongs to this domain. Typical macrocyclic host molecules are presented in Chapter 7.

3.3 The Structure of Inclusion Complexes

As noted by Pedersen in his seminal paper [1], he had clear evidence of the existence of several crown ether complexes in solution although he was not able to crystallize them. In solution prevailing majority of inclusion complexes is present in the equilibrium with their free hosts and guests [9c]. Exceptions are

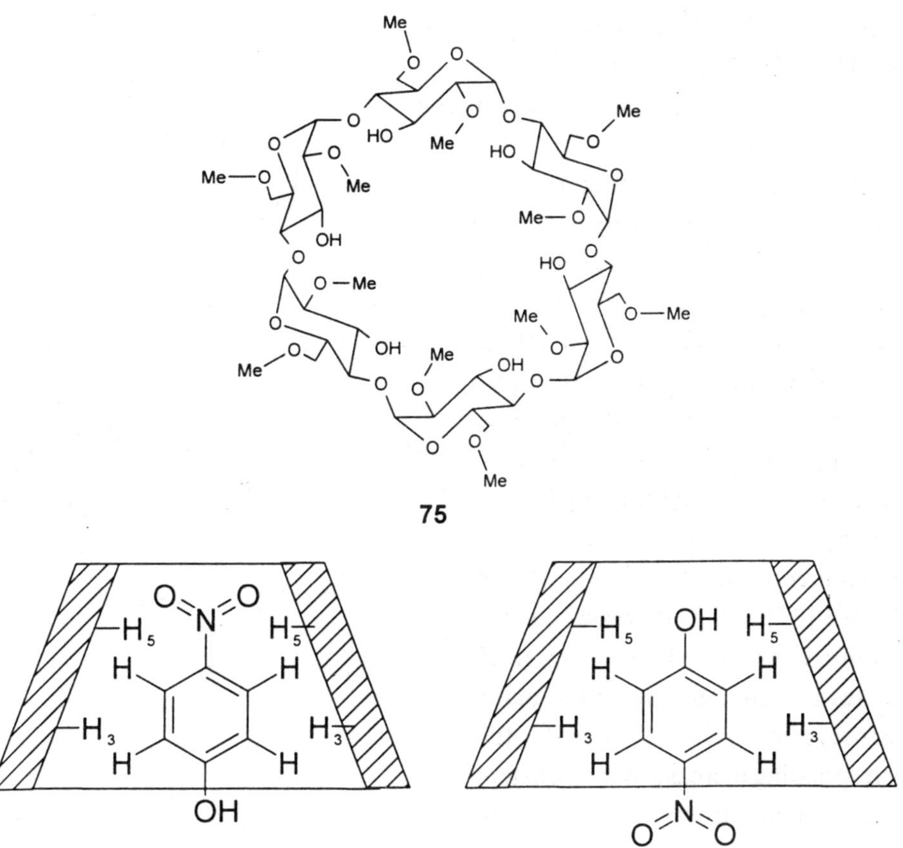

Figure 3.4. Schematic presentation of the mode of entrance of **74** into the **75** cavity in the solid state (left) and in the solution (right).

complexes involving hemicarcerands like **5** or C_{60} hosts which prevent their once entrapped guests **4** [2a] and **67** [27] escaping from the cavity. Understandably, in solution the complexed host and guest are less mobile than their free constituent parts. On the other hand, in the complexed form they are considerably more mobile in solution than in the solid state. As stated by Lehn [34], the delicate balance between the rigidity and flexibility of supramolecular complexes plays an important role in their stability and functioning. Comparison of X-ray results with solid state NMR spectra reveals that even in the solid state the complexes may be less rigid than they appear in X-ray studies [35]. For instance, NMR NOE results reveal that the aromatic ring of benzaldehyde **73** in α- and β-cyclodextrin (**13** and

11, respectively) complexes executes fast 180^0 flips around C1-C4 axis which are undetected by the former technique [36].

There is an interesting relation between ligand flexibility and the stability of the corresponding complexes. Podands like **64** are considerably more flexible than crown ethers. The latter, in turn, are more flexible than the corresponding cryptands. Their complexes reveal an opposite order of stability with their respective complex formation constants equal to 10^2-10^4, 10^4-10^6 and 10^6-10^8 [37]. The increase in the stability constant when going from podates to coronates bears the name *macrocyclic effect*, whilst the corresponding increase of the constant when going from coronates to cryptates is called the *macrobicyclic* or *cryptate effect* [38]. Thermodynamic behaviour of the complexes should be differentiated from their kinetic behaviour. For instance, conformationally most rigid cryptands form their complexes very slowly.

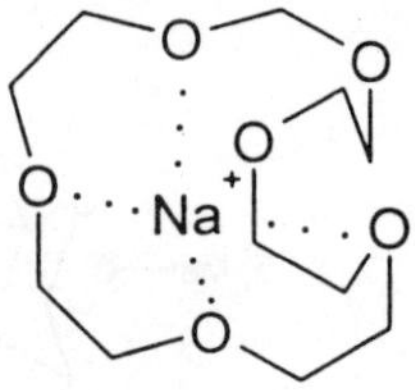

76

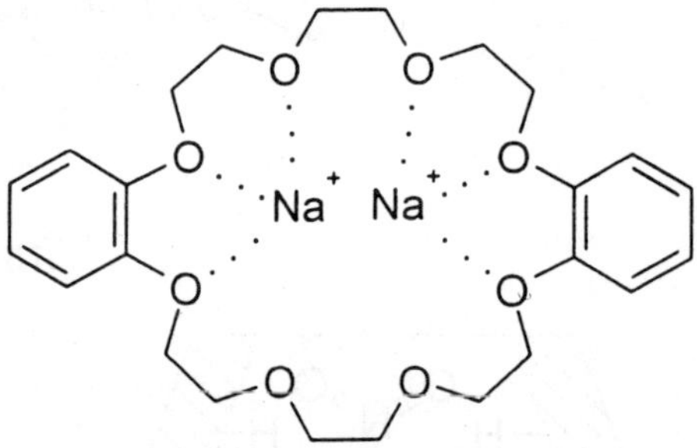

77

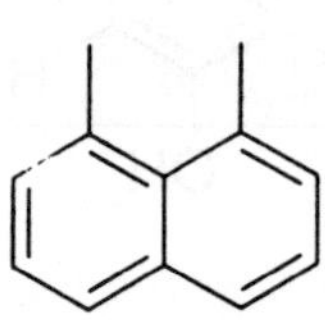

78

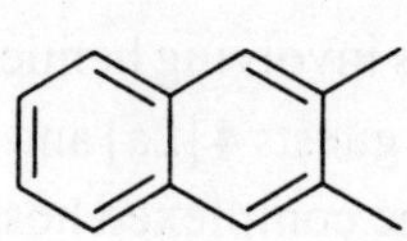

79

As discussed in some detail in Section 6.2, the crystal forces are of magnitude comparable to those inducing complex formation. Thus a complex in the solid state can have a different structure from that in solution. The complex of *p*-nitrophenol **74** with hexakis(2,6-di-O-Me)-α-cyclodextrin **75** illustrates this point, exhibiting strikingly different modes of entrance of the guest into the host cavity in these two states (Figure 3.4) [39]. Nevertheless, numerous solid state studies of the complexes structures [40] provide proofs of existence of the inclusion complexes under consideration and the details of their structure in the solid state.

It is obvious that beautiful formulae and molecular models of high symmetry do not necessarily correspond to the structure of real molecules. For instance, **48** in its uncomplexed state assumes the conformation schematically presented in formula **48b** with methylene groups filling the void, whilst in the complex with K^+ **48a** the macrocyclic ring assumes D_{3d} symmetry [41]. Similarly the complex of **76** [42] with sodium ion exhibits irregular structure different from that of the complex with K^+ (Figure 3.1) [43]. In agreement with the induced fit model of the complexation, in the above cases the guests enforce changes in the host macrocycles. On the other hand, in the dibenzo-24-crown-8 **77** complex with two Na^+ ions [44] the macrocycle assumes the 'extended' conformation. Two *ortho*-nitrophenolate counteranions each provide two additional (nitro and phenoxide) donor sites and the charge neutralization in this complex.

Sometimes the same host and guest can form complexes with varying stoichiometries that differ in stabilities. 1,8-dimethylnaphthalene **78** complex with β-CD **11** was found to be about 100 times stronger than the corresponding complex with 2,3-dimethylnaphthalene **79** and other dimethylsubstituted isomers [45]. Using a low temperature NMR technique the former complex was found to exist in solution as a mixture of two: that of 1 : 2 stoichiometry and the more usual 1 : 1 complex typical for other dimethylsubstituted isomers [46].

Several inclusion complexes with cations, anions and neutral molecules are discussed in Chapter 7 and in other parts of this book.

3.4 Dynamic Character of Inclusion Complexes

Molecules are composed of nuclei with their electron clouds executing vibrations around their respective equilibrium positions corresponding to an optimum geometry. They move as a whole and, in addition to vibrations, their fragments rotate. For instance, owing to internal rotations around Csp^3-Csp^3 bonds there are 33 isomers (some of them indistinguishable) of *n*-hexane C_6H_{14} [47]. Another example of the dynamic equilibrium is that owing to ring inversion in cyclohexane resulting in two isomers of the monosubstituted six-membered ring [48]. Sometimes the barriers to internal rotations are very small, allowing for almost unhindered movement. This is the case in cyclopentane, which executes pseudorotation [49], and in toluene $C_6H_5CH_3$ with a barrier of 0.014 kcal/mol [50] much lower than kT describing vibrational energy at room temperature. Thus, molecules are well defined, but dynamic, entities.

Supramolecular assemblies are even less rigid. As discussed in Chapter 1, they are held together by weaker forces than those maintaining molecular integrity. Thus the inclusion complex H·G in solution exists in equilibrium with its constituent parts (H and G)

$$H{\cdot}G \rightleftharpoons H + G. \qquad (3.1)$$

Two quantities characterize the equilibrium and its measurements: free energy difference between free and complex species ΔG and the rate of the complex formation and decomposition.

If the processes of complexation are assumed to be bimolecular ones, then the following equilibria hold for the simplest complexes [51]

of 1:1 $$H + G \rightleftharpoons HG, \qquad (3.2)$$

of 1:2 $$HG + G \rightleftharpoons HG_2, \qquad (3.3)$$

and of 2:1 host-to-guest stoichiometry

$$H + HG \rightleftharpoons H_2G. \qquad (3.4)$$

The corresponding stepwise binding constants (also called formation constant K_f, association constant K_a, or stability constant K_s) are given by the equations

$$K_{11} = [HG]/([H][G]), \qquad (3.5)$$

$$K_{12} = [HG_2]/([HG][G]), \qquad (3.6)$$

$$K_{21} = [H_2G]/([H][HG]), \qquad (3.7)$$

where the brackets denote molar concentrations in Mol^{-1} units. The K constants are not the real thermodynamic quantities but the concentration quotients. However, the difference between these values can be neglected in most cases taking into account the experimental accuracy and precision of most reported measurements.

In a general form the complex H_mG_n formed directly from the free, unassociated species obeys the equilibrium

$$mH + nG \rightleftharpoons H_mG_n \qquad (3.8)$$

with overall binding constant

$$\beta_{mn} = [H_m G_n]/([H]^m[G]^n) \qquad (3.9)$$

that can be expressed as a product of stepwise constants K_{ij}. In most determinations only the overall binding constant is measured, and its separation into the stepwise constants is not possible. The determination of stepwise K_{11} and K_{12} stability constants for the 1:2 complexes of camphor enantiomers **28** with α-cyclodextrin **13** allowed Dodziuk and coworkers to demonstrate the cooperativity of their formation, since the former constants were more than 1000 times smaller than the latter ones [18b].

As stated above, the stepwise binding constants are not thermodynamic constants. Therefore using the ΔG° symbol for the free energy change calculated from these constants although not precise is commonly done. By measuring the temperature dependence of the equilibrium constant or carrying out calorimetric measurements enthalpy and entropy changes upon complexation can be obtained (the choice of units is discussed by Connors [51]).

The stability constant of the complex in solution K depends on the free energy difference ΔG and temperature T [52]

$$K = e^{(-\Delta G/kT)} \qquad (3.10)$$

where k is the Boltzmann constant. According to the Gibbs-Helmholtz equation

$$\Delta G = \Delta H - T\Delta S, \qquad (3.11)$$

where ΔH and $T\Delta S$ denote the enthalpy and entropy change upon complexation, respectively. Therefore,

$$RT\ln K = -\Delta H + T\Delta S. \qquad (3.12)$$

The rate constant in conjunction with the time scale of an experimental techniques determines whether the free and complexed species can be observed separately or only the average value of a quantity characterizing the system under investigation can be observed. For instance, the exchange amongst free and complexed species for inclusion complexes with cyclodextrins **11**, **13** and **68** is usually too fast to be observed by the NMR technique. Therefore, in this case the complex formation manifests itself by, usually small, shifts of NMR signals of H_3 and/or H_5 cyclodextrin protons pointing inside the molecular cavity accompanied by some shifts of the guest signals. Separate signals of free and complexed species have been observed in case of CD complexes of rotaxane type like **12** in which the exchange is practically excluded owing to the bulkiness of the terminal groups [53]. An interesting example of a simultaneous observation of the signals pertaining to complexes of different stoichiometries in proton and carbon NMR spectra of **11** with **78** was reported by Dodziuk and coworkers. The rate constant for the formation of the 1:2 complex in this case is such that in addition to the more usual average signals of the 1:1 complex and its free constituent parts, the separate signals of the 1:2 complex could be observed at 225K [46]. The flip of the aromatic ring of benzaldehyde **73** in the complex with **11** in the solid state was mentioned earlier in this Section. Fascinating examples of the guest mobilities in cage-like hosts and in clathrate hydrates are discussed in Sections 3.5 and 8.3.3.

3.5 The Complexes Involving Induced Fit and Without It: Endohedral Fullerene, Hemicarcerand and Soft Rebek's Tennis Ball-Like Hosts

As discussed in Chapters 1 and 2, during the complexation process there is usually a mutual adaptation of host and guest bearing the name 'induced fit'. The fit involves sometimes subtle changes in molecular geometry allowing maximum attraction of the host and guest. The changes of the geometry of the structure of **48** on complexing K^+ discussed in Section 3.3 illustrate this point. The geometry adjustment can be quite intricate as is the case with the enzyme hexokinase and foldamers wraping themselves around the respective guest molecules creating the host cage only in the presence of the guest [2]. One cannot speak about the simple fitting in such cases. On the other hand, rigid cage-like molecules may undergo only minor adjustment of their geometry on guest inclusion. The guests enter inside their cages either during their formation, as is the case of endohedral fullerene complexes **41** presented in Sections 2.3 and 7.5, or they are 'produced' inside a 'molecular flask' like cyclobutadiene **4** in hemicarcerand **5**, discussed in Chapter 1 and Section 7.3. Restriction of the guests' movements by the corresponding host cage in these complexes is much smaller than those discussed earlier. Pertinent examples are discussed, for instance, by Cram [54]. An exciting example of the guests mobility was recently demonstrated by Akasaka, Nagase and coworkers [55]. By using ^{13}C and ^{139}La NMR techniques they have shown

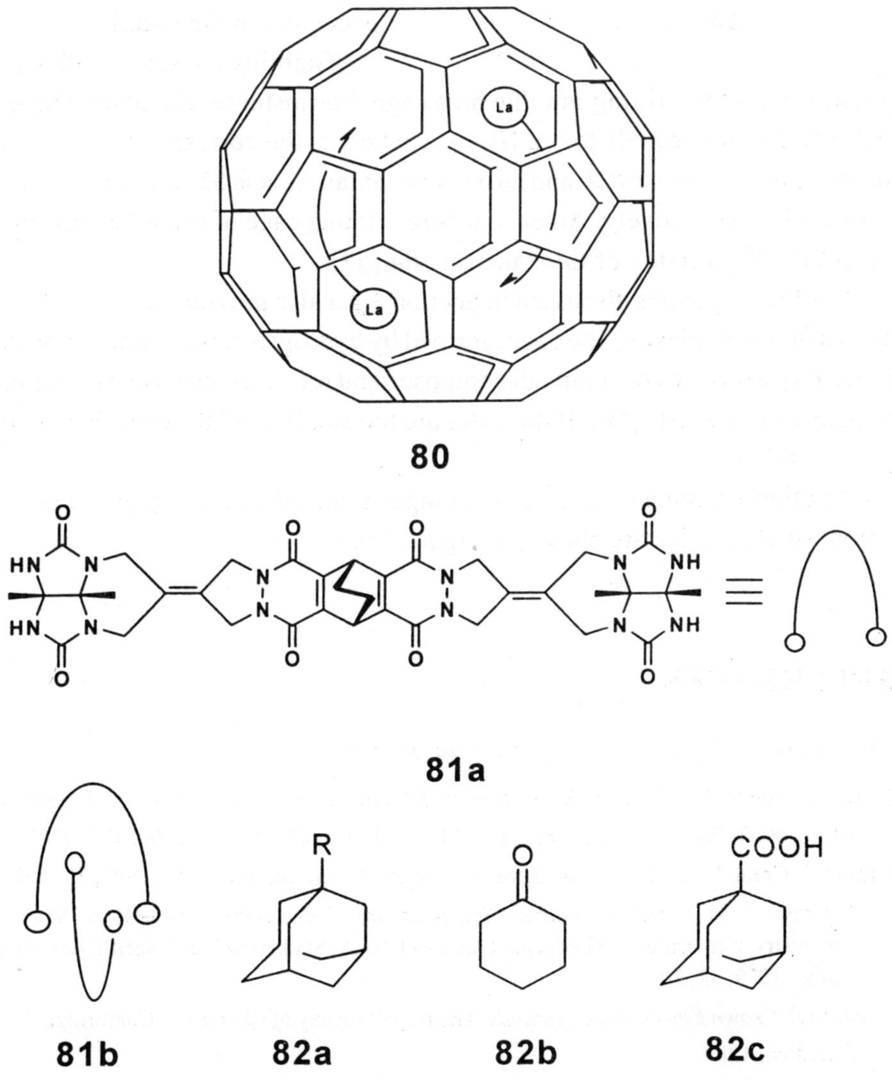

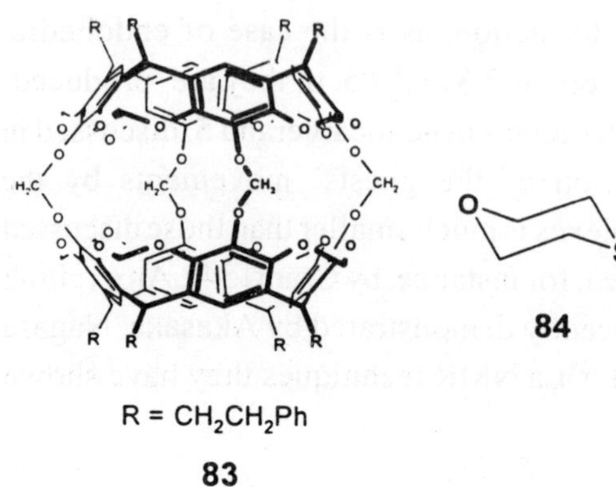

that two lantanide ions rapidly move around inside C_{80} **80** cage.

An interesting example of less rigid cages is also provided by molecular self-assembling capsules obtained in the Rebek group [56]. **81** was found to form dimers held together by hydrogen bonds between the donors at its ends and the acceptors in the middle.

Stability constants with various guests reflect their fitting into the host cage. For instance, for adamantane **82a** stability constant equals to $5.2 \cdot 10^5$ M^{-1}, whereas the corresponding values for smaller cyclohexanone **82b** and larger adamantanoic acid **82c** drops to $1.7\ 10^2$ and $1.3\ 10^2$ M^{-1}, respectively. An even less restricting cage of carceplex **83** enables execution of inversion of thioxane ring **84** [57].

Clathrate hydrates discussed in Section 8.3.3 also provide exciting examples of dynamic complexes. The cages formed by hydrogen bonded water molecules in these systems are constantly decomposed and reformed, but they are stabilized by appropriate guests [58]. If the latter are too small to fill the cage they, in turn, move inside it.

Another example of structural changes induced in the complex involving porphyrin derivative are shown in Figure 1.6.

REFERENCES

1. C. J. Pedersen, J. Am. Chem. Soc, 1967, 89, 7017.

2. (a) T. A. Steitz, M. Shoham, W. S. Burnett, Jr., Phil. Trans. Roy. Soc. Series B, 1981, 293, 43; (b) R. B. Prince, S. A. Barnes, J. S. Moore, J. Am. Chem. Soc., 2000, 122, 2758.

3. (a) D. J. Cram, M. E. Tanner, R. Thomas, Angew. Chem. Int. Ed. Engl., 1991, 30, 1024, D. J. Cram, J. M. Cram, *Container Molecules and Their Guests*, The Royal Society of Chemistry, Cambridge, United Kingdom, 1994; (b) A. Stadler-Szöke, J. Szejtli, Acta Pharm. Hung., 1979, 49, 30.

4. G.Gokel, *Crown Ethers and Cryptands*, The Royal Society of Chemistry, Cambridge, United Kingdom, 1989.

5. C. D. Gutsche. *Calixarenes*. The Royal Society of Chemistry. Cambridge. United Kingdom. 1989.

6. F. Cramer. Angew. Chem., 1952, 64, 437.

7. Cyclodextrins for which the complexation of neutral molecule was proved in innumerable studies are not even mentioned in the review 'Enantioselective and diastereoselective molecular recognition of neutral molecules' by T. H. Webb, C. S. Wilcox. Chem. Soc. Rev., 1993. 22, 383.

8. J. Szejtli. *Cyclodextrin Technology*, Kluwer Academic Publishers. Dordrecht, 1988.

9. (a) B. Perly, F. Djeddaïni, P. Berthault, in *New Trends in Cyclodextrins and Derivatives*. D. Duchene. Ed., Edition de Sante, Paris, France. 1991. p. 201; (b) Ref. 9a p. 179; (c) F. Djeddaïni, B. Perly. in Ref. 9a p. 217; (d) E. B. Brouver. J. A. Ripmeester, G. D. Enright, J. Incl. Phenom. Mol. Rec. Chem., 1996, 24, 1; Y. Inoue, F.-H. Kuan, Y. Takayashi, R. Chujo. Carbohydrate Res., 1985, 135, c12.

10. (a) J. D. Lamb. R. M. Izatt. J. J. Christensen. D. J. Eatough. in *Coordination Chemistry of Macrocyclic Compounds*, G. A. Melson, Ed., Plenum Press. New York. 1979. p. 145; (b) R. M. Izatt, Chem. Rev. 1985, 85, 271.

11. (a) J.-M. Lehn, *Supramolecular Chemistry. Concepts and Perspectives*. VCH, Weinheim, 1995; (b) p. 16.

12. M. R. Truter. Struct. Bonding, 1973, 16, 71; M. Dobler, *Ionophores and Their Structures*, J. Wiley, New York, 1981.

13. C. O. Roth, J.-M. Lehn, unpublished work, cited in Ref. 11a.

14. E. Graf, J.-P. Kintzinger, J.-M. Lehn, J. LeMoigne, J. Am. Chem. Soc, 1982, 104, 1672.

15. B. Dietrich, J.-P. Kintzinger, J.-M. Lehn, B. Metz. A. Zahidi, J. Phys. Chem., 1987, 91, 6600.

16. B. Dietrich, J. Guilhem. J.-M. Lehn, C. Pascard, E. Sonveaux. Helv. Chim. Acta, 1984, 67, 91.

17. J. Rebek. Jr., Angew. Chem. Int. Ed. Engl., 1990. 29. 245.

18. (a) C. Moeder, T. O'Brian, R. Thompson, G. Bicker, J. Chromatography A, 1996, 736, 1; (b) H. Dodziuk, A. Ejchart, O. Lukin, M. O. Vysotsky, J. Org. Chem., 1999. 64. 1503.

19. (a) see Section 7.8; (b) D. M. Rudkevitch, Z. Brzozka, M. Palys. H. C. Visser, W. Verboom, D. N. Reinhoudt, Angew. Chem. Int. Ed. Engl., 1994, 33. 467.

20. A. D. Hamilton, J.-M. Lehn, J. L. Sessler, J. Am. Chem. Soc. 1986. 108. 5158.

21. B. Dietrich, J.-M. Lehn. J.-P. Sauvage, Tetrahedron. Lett.. 1969. 2885.

22. Ref. 4, p. 19.

23. G. Zupancic, M. Sopcic, Synthesis, 1979, 123.

24. D. M. Dishong, C. J. Diamond, M. I. Cinoman. G. W. Gokel, J. Am. Chem. Soc, 1983, 105, 586.

25. K. Koenig, G. M. Lein. P. Stuckler. T. Kaneda. D. J. Cram. J. Am. Chem. Soc. 1979. 101. 3553; Ref. 2b, p. 28.

26. Ref. 2. Chapter 5.

27. R. Warmuth, Angew. Chem. Int. Ed. Engl., 1997. 36. 1347.

28. P. R. Sundararajan, V. S. R. Rao, Carbohydr. Res., 1970, 13. 351.

29. T. Nakagawa. K. Ueno. M. Kashiwa. J. Watanabe. Tetrahedron Lett.. 1994. 35. 1921.

30. L. F. Lindoy. *The Chemistry of Macrocyclic Ligand Complexes*. Cambridge University Press. Cambridge, 1989.

31. D. J. Cram, Angew. Chem. Int. Ed. Engl., 1986, 25, 1039.

32. (a) G. W. Gokel, D. M. Dishong, R. A. Schultz, V. J. Gatto, Synthesis, 1982, 997; (b) E. Kimura, in Top. Curr. Chem.. 1985, 128, 113.

33. T. Fujita, J.-M. Lehn, Tetrahedron Lett.. 1988, 29. 1709; F. Diederich. *Cyclophanes*. The Royal Society of Chemistry. Cambridge, United Kingdom. 1989.

34. Ref. 11a, p. 14.

35. Ref. 9b; E. B. Brouver. G. D. Enright. C. I. Ratcliff. J. A. Ripmeester, Supramol. Chem.. 1996, 7, 79.

36. M. G. Usha, R. J. Wittebort. J. Am. Chem. Soc., 1992, 114. 1541.

37. F. Vögtle. *Supramolecular Chemistry*. Wiley. 1991, p. 43.

38. Ref. 11a, p. 18, 20.

39. Y. Inoue, Y. Takahashi, R. Chujo. Carbohydr. Res., 144, 1985. C9.

40. See data collected in crystallographic Cambridge Data Base that is a source of comprehensive information on the (mainly) X-ray and neutron diffraction studies of the structure of free ligands and inclusion complexes in the solid state. F. H. Allen, S. Bellard, M. D. Brice, B. A. Cartwright, A. Doubleday, H. Higgs, T. Hummelink, B. G. Hummelink-Peters, O. Kennard, W. D. S. Motherwell, J. R. Rodgers, D. G. Watson, Acta Crystallogr. Sect. B: Struct. Crystallogr. Cryst. Chem., 1979, B35, 2331.

41. (a) J. D. Dunitz, M. Dobler. P. Seiler. R. P. Phizackerly, Acta Crystallogr. Sect. B, 1974. 30. 2733; (b) J. D. Dunitz. P. Seiler, Acta Crystallogr. Sect. B. 1974, 30. 2739.

42. M. Dobler, J. D. Dunitz. P. Seiler, Acta Crystallogr. Sect. B, 1974, 30, 2741.

43. P. Seiler, J. D. Dunitz, M. Dobler. Acta Crystallogr. Sect. B. 1974, 30, 2744.

44. D. L. Hughes. J. Chem. Soc. Dalton Trans., 1975, 2374.

45. D. Sybilska, M. Asztemborska, A. Bielejewska, J. Kowalczyk. H. Dodziuk. K. Duszczyk, H. Lamparczyk, P. Zarzycki, Chromatogr., 1993, 35. 637.

46. H. Dodziuk, J. Sitkowski. L. Stefaniak. D. Sybilska. J. Jurczak. K. Chmurski, Pol. J. Chem.. 1997, 71, 757.

47. S. Tsuzuki. L. Schäfer, H. Goto, E. D. Jemmis, H. Hosoya. K. Siam. K. Tanabe. E. Osawa. J. Am. Chem. Soc., 1991, 113, 4665.

48. H. Dodziuk, *Modern Conformational Analysis. Elucidating Novel Exciting Molecular Structures*, VCH Publishers, New York, 1995, p. 138.

49. T. B. Malloy, Jr., L. E. Bauman, L. A. Carreira, Top. Stereochem., 1979, 11, 97.

50. H. D. Rudolf, A. Jaeschke, P. Wendling, Ber. Bunsenges. Phys. Chem., 1968, 70, 1172A as cited by A. P. Zens, P. D. Ellis, J. Am. Chem. Soc., 1975, 97, 5685.

51. K. A. Connors, Chem. Rev., 1997, 97, 1325.

52. W. Burgemeister, R. Winkler-Oswatitsch, Top. Curr. Chem., 1977, 69, 91; R. M. Izatt, J. S. Bradshow, S. A. Nielsen, J. D. Lamb, J. J. Christensen, D. Sen, Chem. Rev., 1985, 85, 271.

53. A. Harada, J. Kamachi, Nature, 1992, 356, 325; G. Wenz, F. Wolf, M. Wagner, S. Kubik, New J. Chem., 1993, 17, 729.

54. Ref. 2b p. 139, 203.

55. T. Akasaka, S. Nagase, K. Kobayashi, M. Wälchli, K. Yamamoto, H. Funasaka, M. Kako, T. Hoshino, T. Erata, Angew. Chem. Int. Ed. Engl., 1997, 36, 1643.

56. J. M. Rivera, T. Martin, J. Rebek, Jr., J. Am. Chem. Soc. 1998, 120, 819.

57. R. G. Chapman, J. C. Sherman, J. Org. Chem., 2000, 65, 513.

58. K. A. Udachin, G. D. Enright, C. I. Ratcliffe, J. A. Ripmeester, J. Am. Chem. Soc., 1997, 119, 11486.

Chapter 4

MESOSCOPIC STRUCTURES AS AN INTERMEDIATE STAGE BETWEEN MOLECULES (MICRO SCALE) ON THE ONE HAND AND BIOLOGICAL CELLS (MACRO SCALE) ON THE OTHER

4.1 Introduction

Mesoscopic structures [1,2] occupy an intermediate position between the micro scale represented by small molecules and the macro scale to which polymers and complex biological assemblies (e.g., biological cells and their constituent parts) belong. As discussed earlier, the properties of an assembly differ from those of its parts, justifying studies of these objects in an emerging separate science supramolecular chemistry. The latter has become a kind of hot topic today and research dubbed by this name covers significantly different objects using techniques which sometimes have very little in common. This allows one to expect that the research area called supramolecular chemistry at present will eventually split into three domains on the basis of the size of the assemblies under investigation and specific methods used for their studies. The first one will deal with host-guest (or inclusion complexes) and other small assemblies consisting of few molecules. The second which could be called aggregate science encompasses research of larger molecular assemblies such as thin molecular films (Langmuir and Langmuir-Blødgett layers), micelles, vesicles (also called liposomes), fibers and liquid crystals. The third domain, crystal engineering, today just in its infancy, should allow one to obtain crystals with the properties desired to be used in optoelectronics, information storage systems, sensors and in many other fields. In this chapter molecular aggregates of the order of magnitude intermediate between simple assemblies consisting of few molecules and macroscopic objects will be presented. As mentioned earlier, their study requires the application, and even development, of novel experimental techniques mostly different from standard physical methods used for the determination of structure in classical organic chemistry that are also applied in the studies of small inclusion complexes.

4.2 Medium Sized Molecular Aggregates

Most molecular assemblies are usually formed by amphiphilic molecules (also called surfactant or detergent molecules) consisting of a polar 'head' and of one or more nonpolar 'tails'. Head groups can be cationic, anionic, non-ionic

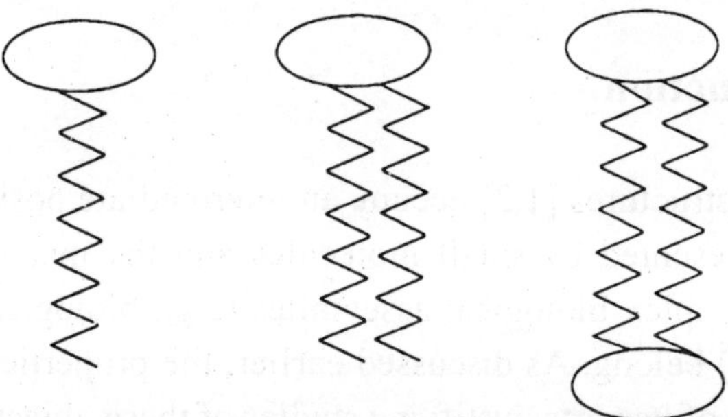

Figure 4.1. Schematic view of bilayer forming one- and two-tail one-headed amphiphiles and two-headed bolaamphiphile that can form a monolayer.

(neutral) or zwitterionic. Types of amphiphilic molecules are presented in Figure 4.1. Among them there are also bolaamphiphiles such as **85** having two head groups connected by two (usually hydrocarbon) chains [3]. As schematically shown in Figure 1.4, in a polar solvent amphiphilic molecules tend to associate exposing their polar ends to the solvent since polar-nonpolar interactions are unfavourable (see however, the discussion on the micelles structure in Section 4.2.3). In such a way molecular mono- and polylayers, as well as micelles and vesicles, are formed. In nonpolar solvents, analogous structures, i. e., reverse micelles with the opposite arrangement of polar groups pointing inwards, form. One example of a reverse micellar system was briefly presented in Chapter 1. If, in addition to water and surfactant, halogenated compounds or long chain alcohols are present in solution, then structurally ill-defined but thermodynamically stable microemulsions involving surfactant and the alcohol are formed. A cosurfactant can, eventually, also take part in the formation of the so called oil-in-water (o/w) microemulsions. On the other hand, water-in-oil (w/o) microemulsions are created in hydrocarbon solutions which, in addition to surfactant, contain water and long-chain alcohol.

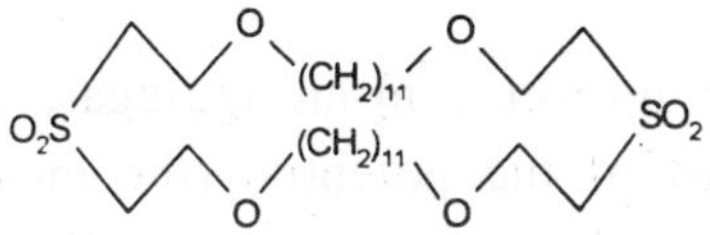

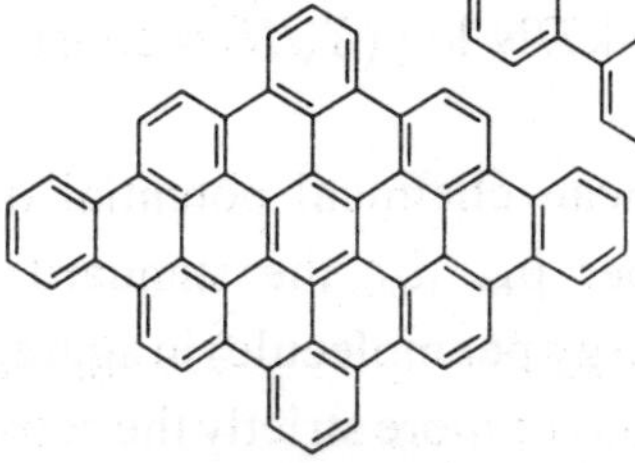

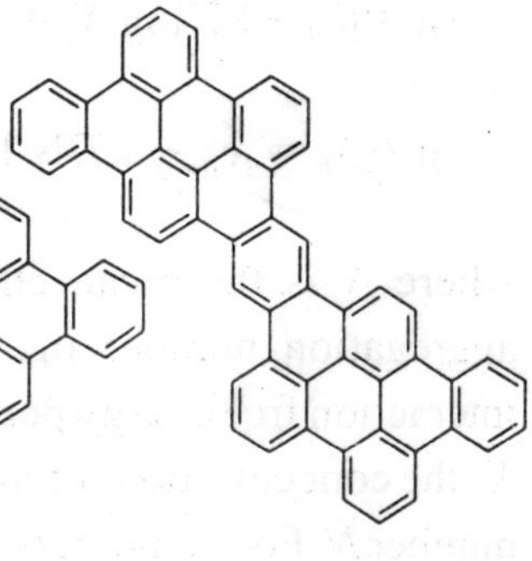

Micelles and vesicles can be formed above a certain concentration. For instance, small micelles are formed above critical micellar concentration, cmc. (The latter abbreviation is often used for critical vesicle concentration, too. However, sometimes a more general term critical aggregate concentration, cac is also applied.) Bilayers of specific amphiphiles with two 'tails' are typical of the central part of cell membranes discussed in some detail in the next chapter. Studying artificial mono- and bilayers (uniform or with built in pores) is indispensable for gaining information about the structure and functioning of cell membranes involving the transport through them.

Another type of ordered monolayer has been obtained by vapour deposition of polycyclic aromatic hydrocarbons **86** [4a] onto a freshly cleaved surface of a single crystal of the layer chalcogenide MoS_2 or by deposing dendrimers on a gold surface [4b]. Close packing in a plane enabling monolayer formation was achieved by directed syntheses of two isomeric **86a** and **86b** C_{54} aromatic hydrocarbons. According to equilibrium thermodynamics [5] a system of molecules in solution forming aggregated structure should have the same chemical potential of all identical molecules. Thus

$$\mu = \mu_{01} + kT \log X_1 = \mu_{02} + 1/2\, kT \log 1/2\, X_2 = \mu_{03} + 1/3\, kT \log 1/3\, X_3 = ...$$

or

$$\mu = \mu_N = \mu_{0N} + kT/N \log (X_N/N) = \text{const}, \quad N = 1, 2, 3, ...$$

where N is the mean chemical potential of a molecule in an aggregate of aggregation number μ_N, μ_{0N} the standard part of the potential (the mean interaction free energy per molecule) in aggregates of aggregation number N, and X_N the concentration (or more strictly the activity) of molecules in aggregates of number N. For monomers $N = 1$ and μ_{01} and X_1 correspond to isolated molecules.

The latter equation can be rearraged to the form

$$X_N = N\{X_1 \exp[(\mu_{01} - \mu_{0N})/kT]\}^N,$$

which together with the equation for the total solute concentration C rearranges to

$$C = X_1 + X_2 + X_3 + ... = \sum_{N=1}^{\infty} X_N \, .$$

If the rate of formation of N-th aggregate, i.e., its rate of association is equal to $k_1 X_{N1}$ whilst the corresponding rate of dissociation is given by $k_N(X_N/N)$ then, on the basis of the law of mass action the equilibrium constant

$$K = k_1/k_N = \exp[-N(\mu_{0N} - \mu_{01})/kT].$$

The necessary condition for the existence of stable aggregates is the presence of cohesive energy enabling the aggregate formation. The major factors forcing self-assembly of amphiphiles into micelles, bilayers, or other well-defined aggregates arise (in a polar sovent) from the hydrophobic attraction at the hydrocarbon/water interface leading to the association and the hydrophilic, ionic (or steric) repulsion of the head groups which oppose it. The delicate balance between these two competing interactions determines the interfacial area a per molecule (that is, the effective headgroup area) exposed to water.

4.2.1 Langmuir and Langmuir-Blødgett Films and Other Self-assembling Layers

By pouring a spoon of oil on a surface of a pond in the middle of the 19th century Benjamin Franklin probably obtained the first intentionally created molecular monolayer. There is no uniformity in the notation of mono- and polylayers. According to Fuhrhop and Köning [2a] monolayers on water bear the name Langmuir films whilst those on the solid surfaces are called Langmuir-Blødgettt films. On the other hand, Lednev and Petty [4c] speak about Langmuir monolayers and Langmuir-Blødgett multilayers. Fuhrhop and Köning [2a] claim that such films are not molecular assemblies at all but represent separate liquid or solid phases exhibiting domain structures typical for bulk phases. The films spread on a (water) surface can be compressed, forcing into existence a regular arrangement of amphiphilic molecules (Figure 4.2).

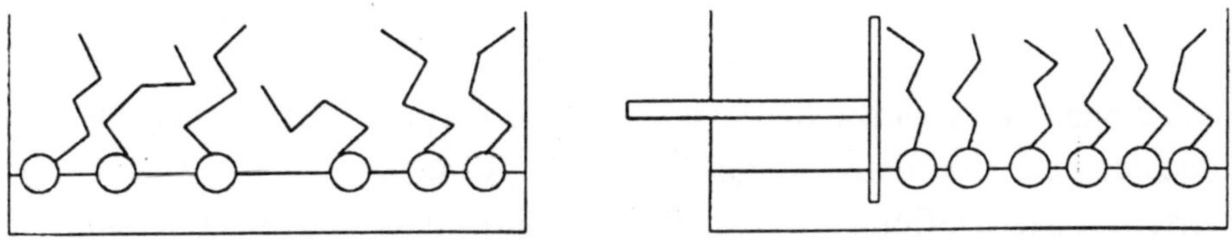

Figure 4.2. Uncompressed and compressed film of amphiphilic molecules on a surface.

Importantly, even linear molecules are not necessarily oriented perpendicularly to the surface in the compressed monolayer [6]. Such an aggregate can be stabilized by covalent bonding to a solid, often gold, surface and polymerization. One example of this kind, a film of 1-octadecanethiol $C_{18}H_{37}SH$ on gold (Figure 4.3), was probed by measuring time-resolved emission spectra of a tethered chromophore 1-pyrenebutanethiol **87**. The results obtained indicated that: (a) the long axis of alkanethiols forms an average tilt angle of 26^0; and (b) that within about the first 10 CH_2 groups the local environment is very rigid whilst the rest of the molecules are much more mobile. This example shows how the structure and dynamics of a monolayer can be studied by analyzing the response of a chromophore embedded in the monolayer. The tilted arrangement of amphiphiles (in contrast to that with an orientation of molecules perpendicular to the film surface) analogous to that of 1-octanethiol is typical for many films. The first are

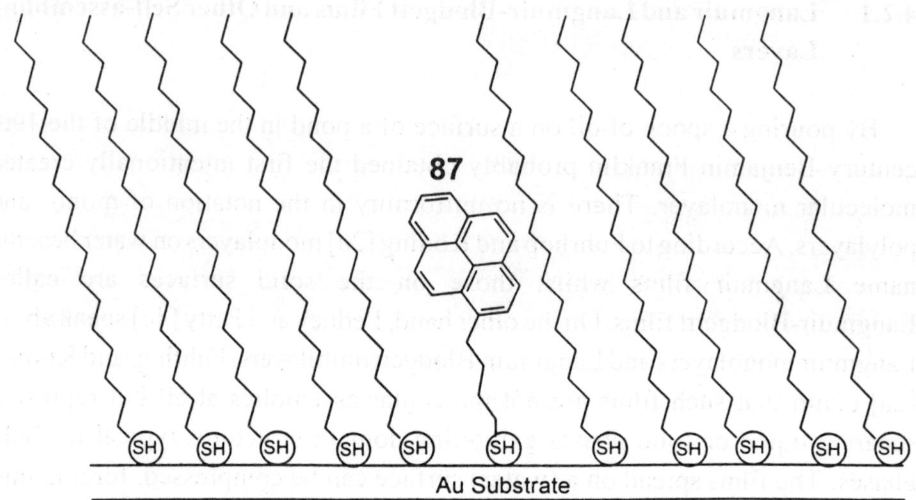

Figure 4.3. Tethered pyrene inserted in a monolayer on gold.

called J-aggregates whilst the layers with the perpendicular arrangement of molecules bear the name H-aggregates [7]. Kunitake and coworkers, who reported the first artificial bilayer membrane [8a], undertook a detailed study on the molecular design of molecules forming bilayers [8b]. The ability to build aggregates was analysed for more complicated molecules capable of the layer formation. The resulting aggregates were observed mainly by electron micrographs.

Amphiphilic monolayers are two-dimensional model systems which are extensively analyzed not only to gain a basic understanding of self-organization and structure-property relationships at an interphase. In addition, they are frequently studied, since one expects that they will find applications as lubricants and anti-corrosive coatings [9]. As mentioned earlier, the study of molecular aggregates requires the application of specific, sometimes specially developed experimental methods. In the case of mono- and multilayers they include, amongst others, Brewster Angle Microscopy [10], the Langmuir Trough (also called Langmuir Balance) [11], light scattering [12], Differential Scanning Calorimetry [13], techniques making use of second harmonics (second harmonics generation SHG, second harmonics microscopy SHM, Fourier Transform Infrared FTIR) [14] and special X-ray techniques (Small-Angle X-ray Scattering

SAXS [15a], Near-Edge X-ray Absorption Fine Structure spectroscopy, NEXAFS [15b], and X-ray Photoelectron Spectroscopy, XPS [15c]). In the case of larger assemblies, such as giant vesicles [16], modifications of microscopic technique provide interesting information. A discussion of these methods falls outside the scope of the present book. However, it should be mentioned that the Langmuir trough allows one to measure isotherms relating surface area per molecule A and interfacial pressure π while Brewster Angle Microscopy provides information on the morphological features of amphiphilic monolayers. Using the latter method, one can establish that monolayer structures are not uniform. They are, rather, characterized by domains of various sizes and shapes [17]. Today we are far from understanding the origin of the domains' formation.

4.2.2 Mono- and Bilayer Lipid Membranes

Monolayer lipid membranes, MLM, are composed of bolaamphiphiles schematically shown in Figure 4.1 which are more water soluble since they have two polar head groups, compared with single headed amphiphiles forming bilayer lipid membranes, BLM, (Figure 4.4). BLM constitute the central part of cell membranes (briefly discussed in the next Chapter) which play an essential role in all living processes. MLM and BML are huge aggregates which should exhibit very strong inter-membrane van der Waals attraction, leading to their precipitation. The latter

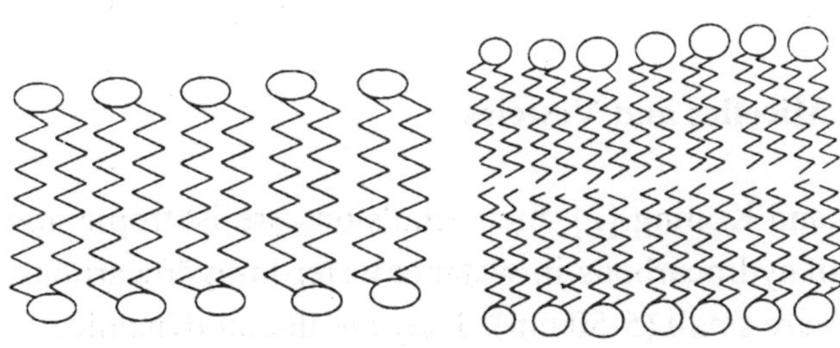

Figure 4.4. Mono- and bilayer membranes.

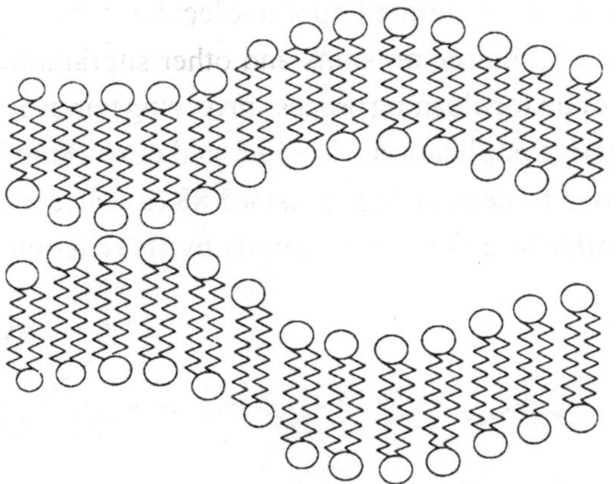

Figure 4.5. Thermal undulation of bilayer membrane.

usually does not occur since the membranes undergo thermal fluctuations called undulations (Figure 4.5) [18] preventing their further aggregation, thus counteracting the precipitation. One can stabilize a vesicle membrane by inserting hydrophobic side chains of a comb shape polymer into it or by polymerizing other suitable inserted groups (Figure 4.6) [19]. Such polymers mimic stabilizing role of cytoskeleton in cell membranes (see next Chapter).

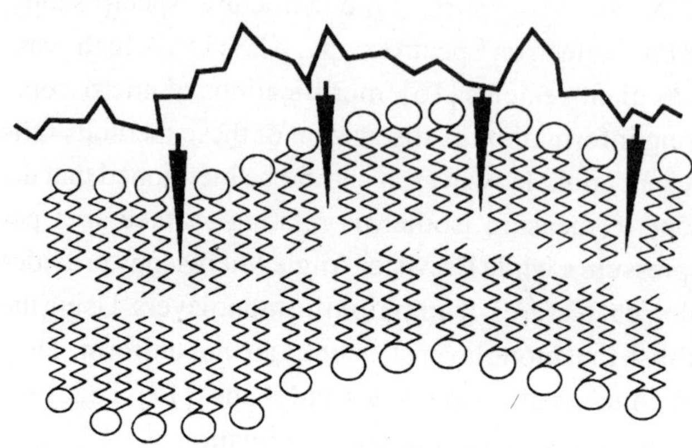

Figure 4.6. Stabilization of membrane by a comb shaped polymer,

4.2.3 Microemulsions, Micelles and Vesicles

According to Fuhrhop and Köning [2] microemulsions are isotropic and optically clear dispersions of hydrocarbons-in-water or water-in-hydrocarbons, where oil or water droplets are small (5-50 nm). They are thermodynamically stable and, interestingly, they remain clear indefinitely. Microemulsions are formed spontaneously by mixing water, hydrocarbon, surfactant and cosurfactant in specific proportions. They have no defined supramolecular structure. As briefly discussed in Section 6.3.5, microemulsions and other supramolecular assemblies provide an ideal medium for cleaning process involving removal of fat as well for some other prospective applications. Menger and coworkers [20a] proposed to use microemulsions to destroy half-mustard **88** (a warfare agent much less dangerous, but similar in action to, mustard) by its reaction with

88 **89**

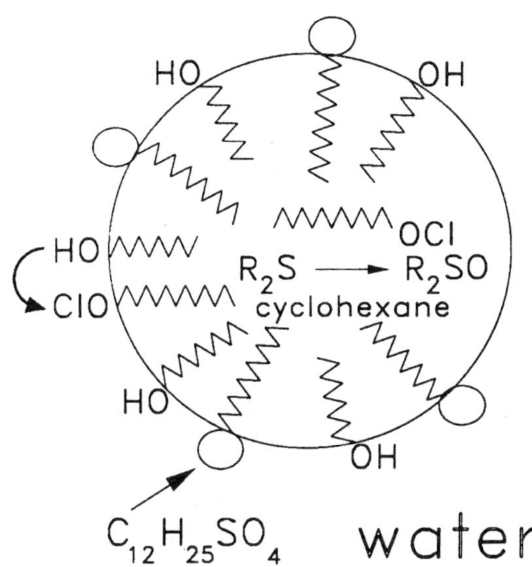

Figure 4.7. Scheme of deactivation of half-mustard.

hypochlorite in microemulsion formed by water, cyclohexane, sodium dodecyl sulfate $C_{12}H_{25}SO_4Na$ (surfactant, often denoted SDS) **89** and 1-butanol (cosurfactant). As schematically shown in Figure 4.7, the half-mustard is oxidized exclusively to nontoxic sulfoxide on a reasonable scale by this simple, inexpensive, mild and rapid method. 5% aqueous solution of hypochlorite (which is used in domestic bleach) was added to half mustard first. Then the mixture was dissolved in a microemulsion formed by water, cyclohexane, a common amphiphile SDS **89** (surfactant), and 1-butanol (cosurfactant) in a specific proportion. Substrate with the oxidant, water, hydrocarbon, surfactant and cosurfactant represent a complicated system in which all components act cooperatively. Water dissolves the inorganic oxidant, while cyclohexane acts as a solvent for half mustard. Both these immiscible components are combined with SDS. The role of butanol-1 that acts as cosurfactant lies in filling the space between the charged SDS molecules.

The effectiveness of the method is most probably based on the fact that alkyl hypochlorite is formed at the oil/water interface where the cosurfactant alcohol resides. The oxidation that follows takes place either inside or on the surface of oil droplet. The rate of the reaction can result from a large hydrocarbon/water contact area permitting interaction between oil-soluble sulfide with interfacial cosurfactant that served as an intermediary. An extension of this procedure to mustard deactivation has also been proposed [20b]. Such systems could be also applied to the degradation of several environmentally contaminating materials.

The formation of microemulsions, micelles and vesicles is promoted by unfavourable interactions at the end sections of simple bilayer membranes. There is no simple theory of solute-solvent interactions. However, the formation of aggregates of nonpolar molecules in water is known to be owed to the entropic effect on the basis of the estimation of the free energy change ΔG of the transfer of bulk alkane into water. The low solubility of alkanes and the entropic character of ΔG of solvation result in the so-called hydrophobic effect. Micelles and vesicles shown in Figure 1.4 differ in their size and stability. Micelles are small short living species containing 50-100 amphiphillic molecules while considerably larger vesicles are so stable that they can be separated by chromatography. As briefly discussed in Section 6.3.5, small nonpolar molecules can reside within micelles and vesicles in water, thus these aggregates can act as surfactants or drug carriers. In this way the inclusion of nonpolar molecules inside micelles leads to their increased solubility in water. For instance, SDS solubilizes benzene in water, resulting in about 40 benzene molecules per micelle [21]. Interestingly, NMR shifts seem to situate most of the benzene molecules not inside the micelles in a nonpolar environment but at the micelle/water interface [22]. This brings us to the fascinating problem of the micelles structure. Certainly their schematic representation in Figure 1.4 is an oversimplification and representing the shapes of surfactant molecules in form of cones is misleading [23]. Several models of micelle structure have been proposed: sphere-like model with statistical distribution of conformations of the dynamic hydrocarbon chains [24], the 'block' model [25] and 'reef' or 'rugged' model [26]. Still another, more complicated model of a helical micelle structure was proposed for rigid facial amphiphiles like deoxycholic acid **90** with the lipophilic sides of the acid orientated inside the ultrathin cylinder formed by the helically arranged molecules of the acid. Surprisingly, such a model exhibits the nonpolar face of the amphiphile orientated toward the bulk aqueous medium [27]. Experimental results involving, amongst others, NMR and laser Raman spectra, diffusion and light scattering data, and high resolution neutron scattering do not allow for unequivocal choice of the model. This is certainly the result of the dynamic character of the micelles' structure which have millisecond

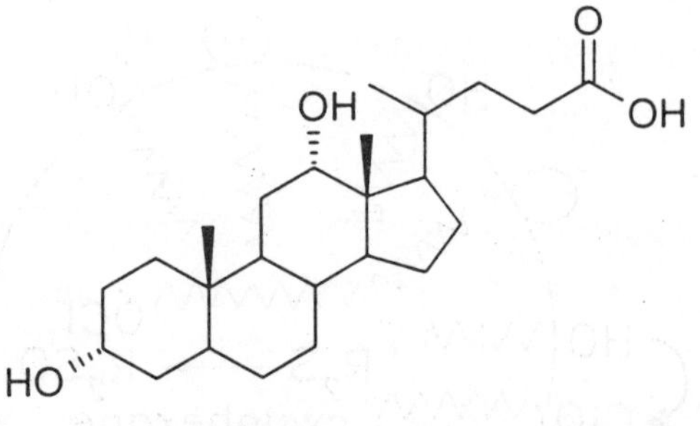

OH OH HO O

90

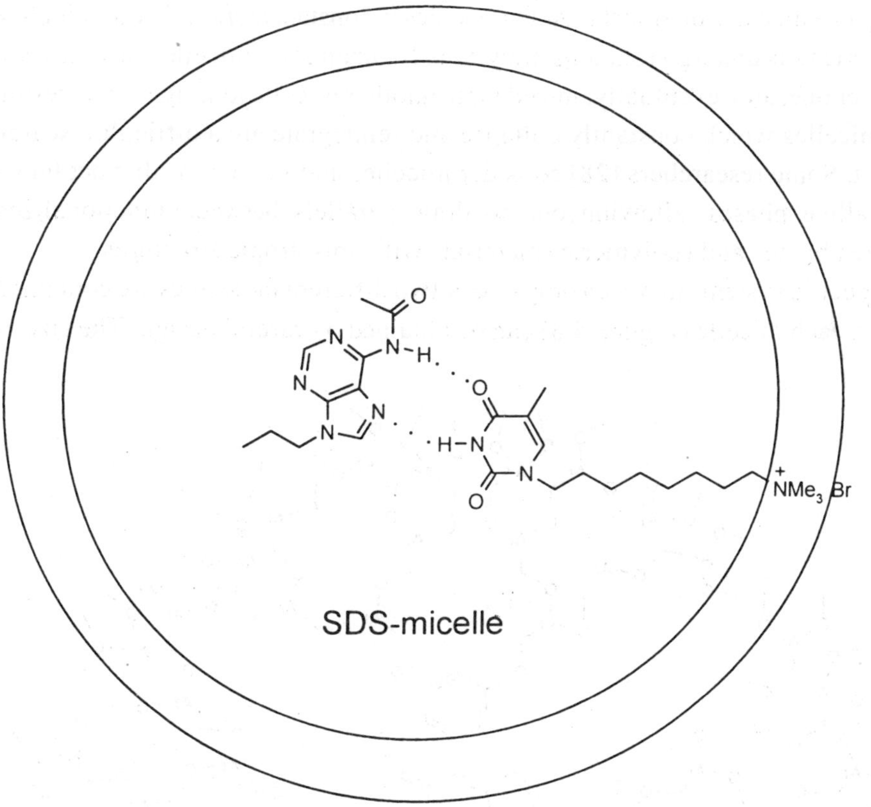

SDS-micelle

Figure 4.8. Hydrogen-bonded system of two molecules inside a micelle.

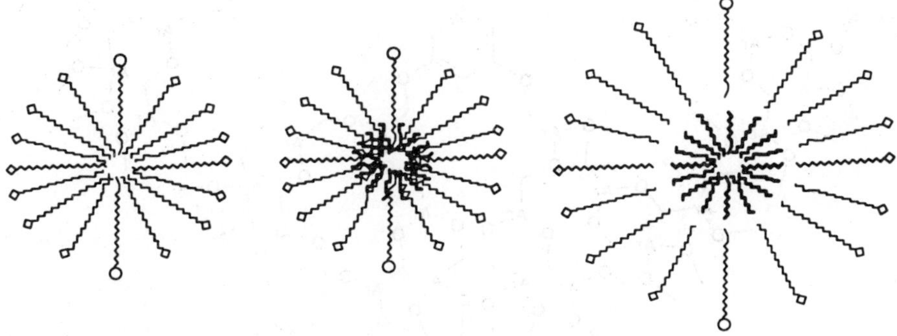

Figure 4.9. Expanded micelles leading to pore enlargement.

lifespan with individual surfactant molecules in constant motion. The amphiphile aliphatic tails undergo rotations, they can also abandon a micelle, move toward another one, and, eventually, intrude into another micelle joining it. As a result, the micelles which constantly collapse and reintegrate are a difficult research object. Some researchers [28] consider micelles and vesicles as distinct liquid crystalline phases, allowing one to draw parallels between functionalized biomembranes and (polymeric) materials with anisotropic structures.

Specific systems in which only one or two different molecules are contained within each micelle (Figure 4.8) can be obtained by careful design. The size of

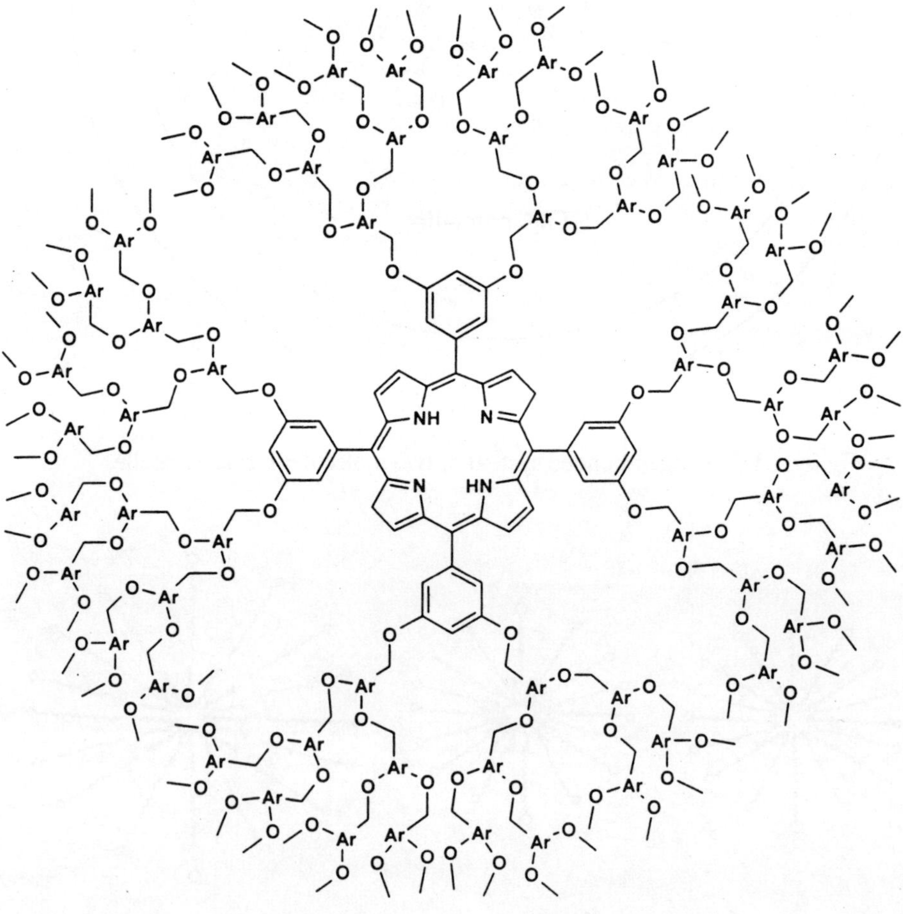

Figure 4.10. Fourth-generation dendrimer with porphyrin core.

the molecules and the micelles lifetime have to be fine-tuned for this purpose.

An interesting example of extension of the size of micelles is provided by the system composed of hexadecyltrimethylammonium bromide acting as a surfactant, to which *n*-alkane acting as solubilizing agent was added [29].

The process utilizing supramolecular organization involves pore expansion in silicas. A schematic view of such micelles built from the pure surfactant and those involving in addition *n*-alkane is shown in Figure 4.9. Another example of pore creation provides a cross-linking polymerization of monomers within the surfactant bilayer [30]. As a result vesicle-templated hollow spheres are created. Dendrimers like that shown in Figure 4.10 exhibit some similarity to micellar structures and can host smaller molecules inside themselves [2c]. Divers functionalized dendrimers that are thought to present numerous prospective applications will be presented in Section 7.6.

Larger and more stable vesicles are sealed, extremely thin (< 10nm), mostly spherical, membranes characterized by the aggregation number of 10^4-10^5. They can be divided into six groups according to their considerably varying sizes from 15 to 1000 nm. Starting from small unilamellar vesicles (SUV) typical for phospholipids and other amphiphiles (e.g., 15 nm egg lecithin); through intermediate sized unilamellar vesicles (IUV) with diameters of ca. 100 nm; multilamellar vesicles (MLV) of considerable sizes range (100-1000 nm) consisting of at least 5 layers; to large unilamellar vesicles (LUV) with diameters of ca. 1000 nm. Even larger are giant vesicles that were studied by Menger as primitive models of biological cells [16].

Supporting the bilayer surfaces by synthetic polymer scaffolds (Figure 4.6) enables one to enhance the vesicle stability and to control the permeability of their membranes. The polymers can be embedded within the vesicles among

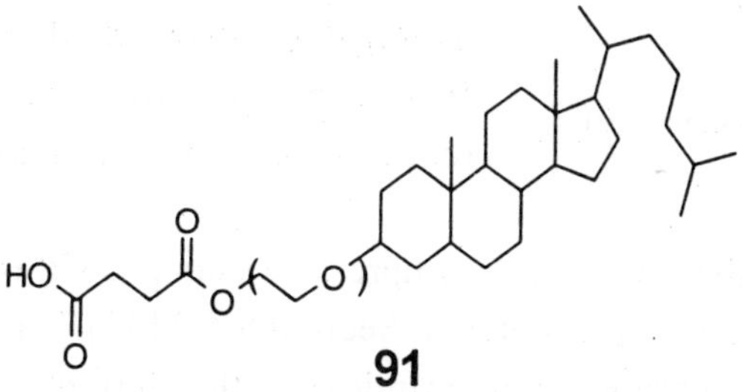

HO O O O O

91

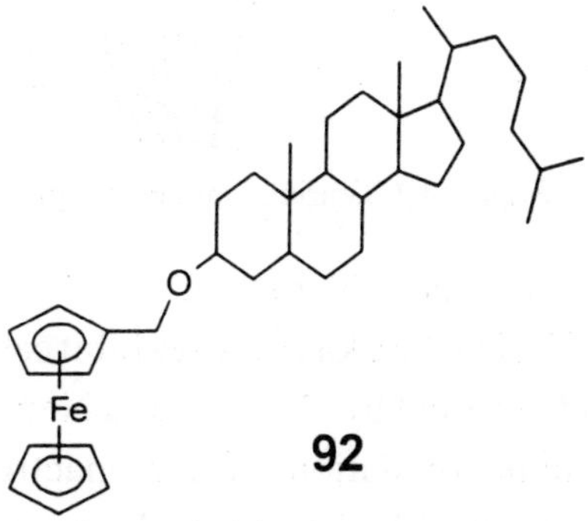

O Fe

92

others by incorporating them during the vesicle formation or by polymerization of hydrophobic monomers solubilized in the interior of the surfactant bilayer [19].

The association of amphiphiles is strongly solvent-, temperature-, pH- and concentration-dependent. The studies of various aggregates formed by a cholesterol derivative **91** [31] illustrate this point (Figure 4.11). The dissociation and reassembling of tobacco mosaic virus into its constituent parts discussed in Section 5.2.1 provides an elegant example of a more complicated assembly process that also depends on these factors.

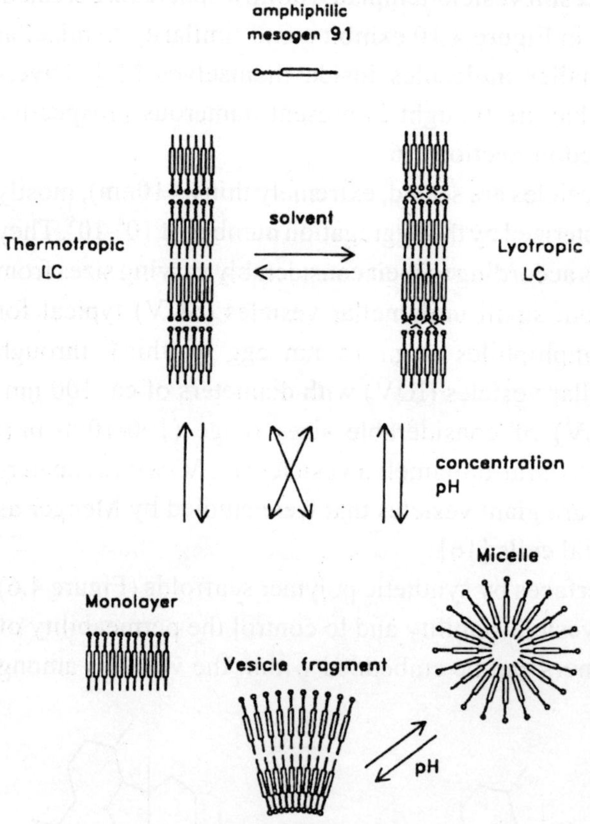

Figure 4.11. Various modes of aggregation of **91**.

Vesicles can be formed not only by simply adding a surfactant (eventually with a cosurfactant) but also upon sonification. As shown by Gokel and coworkers [32], their reversible collapse can be controled chemically by influencing the polarity of head groups. For instance, cholestanyl ferrocenylmethyl ether **92** does not form any aggregates but its corresponding ferrocenium salt obtained by oxidation afforded vesicles upon sonification. However, the aggregates collapsed when ferrocenium ion was reduced to its normal relatively nonpolar state.

Vesicles are known to exhibit catalytic activity. An impressive example of this kind reported by Grooves and Neumann is presented in Section 6.4.2 [33]. The catalytic oxidation of the inactive carbon atom mimicking the action of cytochrome P-450 enzyme by a porphyrin derivative in the presence of vesicles was rationalized by assuming embedding of both the porphyrin catalyst and substrate into the vesicle membrane.

4.2.4 Nanotubes

Self-assembled objects of various nature bear the name nanotubes. Some of them are covalently bound structures while other are self-assembled aggregates.The best known among them are probably elongated fullerenoid structures called carbon nanotubes obtained, in addition to spheroidal fullerenes, like C_{60} **40**, and its higher analogues like C_{70} and C_{80}, **93** and **80**, by evaporating carbon in an arc discharge [34]. In view of prospective applications mostly associated with their ability to include other molecules, the latter finding attracted huge interest. At first, non-uniform multi-wall nanotubes in the form of a number of tubes inserted one into another like Russian dolls and capped with several half-spheres have been produced [35]. Then, by optimizing the procedure not only simple single walled nanotubes, SWNT, have been obtained [36] but the cap could be removed [37] allowing one to insert small crystals [38a] or metal cluster [38b] into the nanotubes.

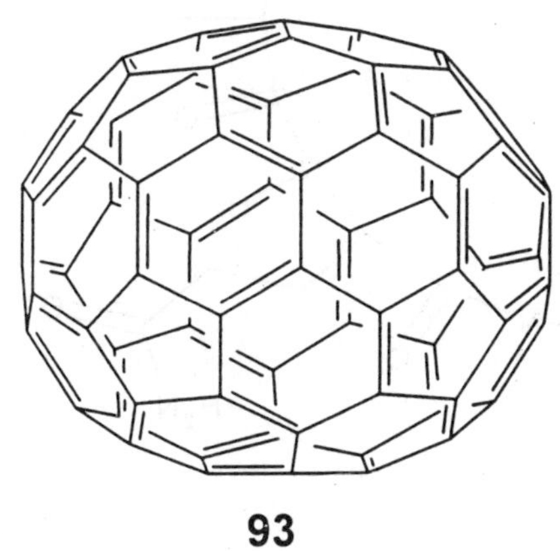

93

C_{60} is known to play a unique role within the fullerene family. Its analog, “a perfect nanotube”, has 13.8 Å in diameter, is capped by half of a giant C_{240} fullerene and has enough space inside to host C_{60} [39]. Seven to 14 single walled nanotubes, SWNTs, can form fibers which can be used to store H_2 inside or in the interstitial spaces between bundled nanotubes [40]. At present the process is not very efficient but its optimization could lead to fuel cells needed for vehicles driven by electricity. Another possibility of SWNTs’ application is to use them as quantum wires. Tans and coworkers [41] have shown that one can handle a single nanotube and measure its conductance. Moreover, an electronic device consisting of a single nanotube has recently been reported by the same group [42].

94

96

Technical problems related to carbon nanotubes and their applications are discussed in some detail in Section 7.5.

A different kind of self-assembled nanotubes (Figure 4.12) are spontaneously built from cyclic peptides made up of alternating *D*- and *L*-amino acids. As predicted by DeSantis group in the mid 1970s [43], the octapeptide *cyclo*-[-(*D*-Ala-Glu-*D*-Ala-Gln)$_2$] **94** [44a] is approximately planar with NH and C=O bonds orientated perpendicularly to the mean plane of the macro-ring thus

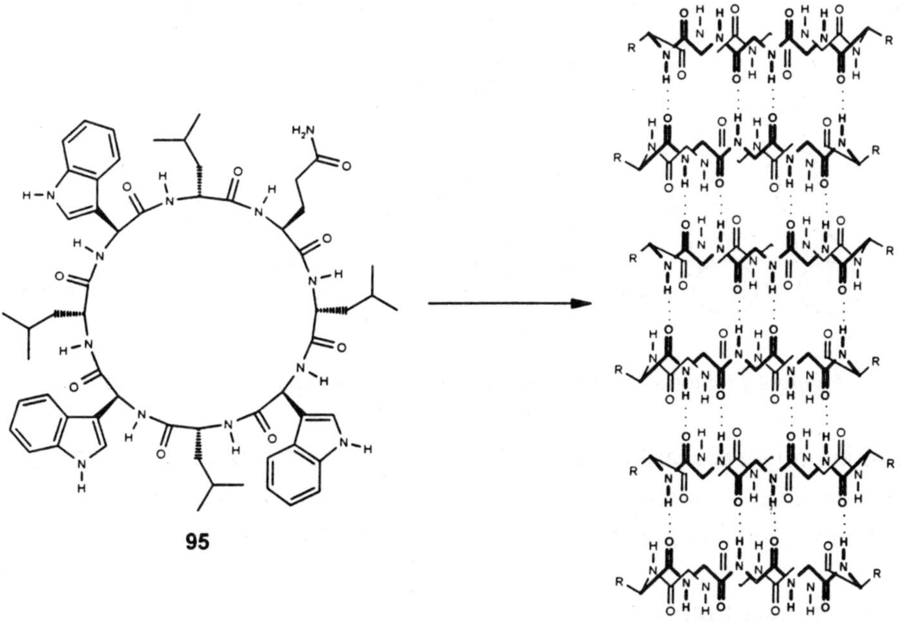

95

Figure 4.12. Self-assembled nanotube built from cyclic peptides.

favouring formation of hydrogen bonds. Hydrophobic side chains in *cyclo*-[-(Trp-*D*-Leu)$_3$-Gln-*D*-Leu] **95** obtained by Ghadiri and coworkers enabled the incorporation of such nanotube into a membrane where **95** can play a role of trans-membrane channel capable of proton transport [45] (Figure 4.13). The channel pore diameter leading to a selective transport of ions depends on the macrocycle size. A modified peptide as a transmembrane channel was also reported by Meillon and Voyer [46].

A similar type of macrocycle stacking leading to the nanotube formation was recently reported by the Stoddart group [44b]. The latter authors synthesized cyclodextrins analogues **96** built of alternating *D*- and *L*- sugar units. The

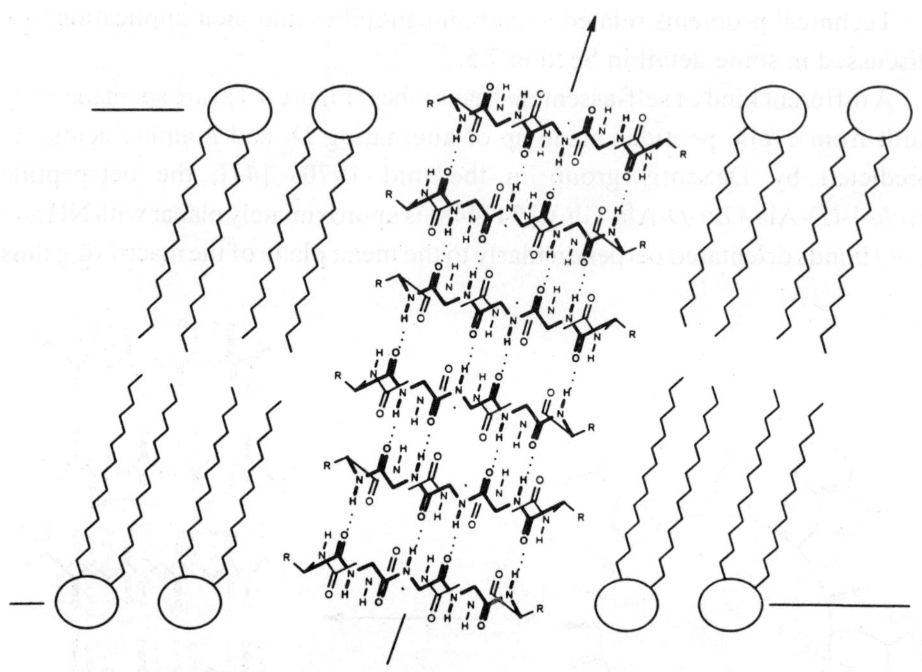

Figure 4.13. Self-assembled peptide nanotube incorporated into a membrane.

resulting oligosaccharides also self-assemble into nanotubes. Another type of self-assembled nanotube formed by a cyclodextrin derivative and 1,6-diphenyl-1,3,5-hexatriene was reported by Pistolis and Malliaris [47].

4.2.5 Fibers [48]

The solvophobic effect and head group repulsion promote micelles and vesicle formation. These effects and the dynamic character of self-assembled nanostructures protect them from further aggregation and crystallization. Hydrogen bonding counteracts the repulsion and by including secondary amides as head groups, one obtains vesicular tubules for amphiphiles with low cmc ($<10^{-5}$ M) or thinner micellar rods for amphiphiles exhibiting higher cmc of 10^{-4}-10^{-2} M. The process is slow, involving the formation of simpler helical structures first. It may take even a month to get fibers as was the case for glutamic acid diester **97** [48].

Hydrogen bonds can be reversibly broken by raising the temperature above the phase transition of the bilayers (Tc = 34^{0}C) and fibers slowly grow up upon

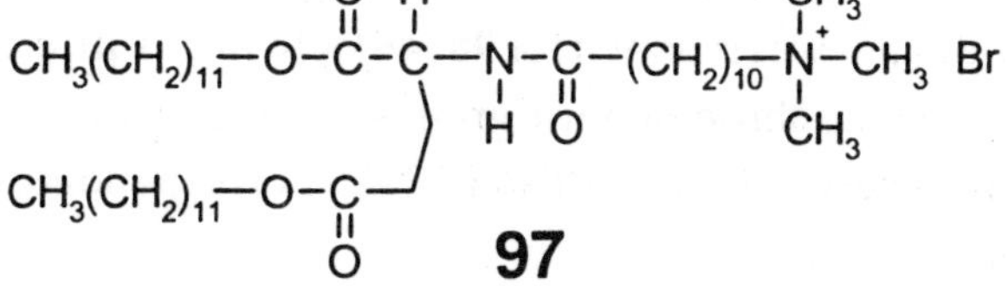

97

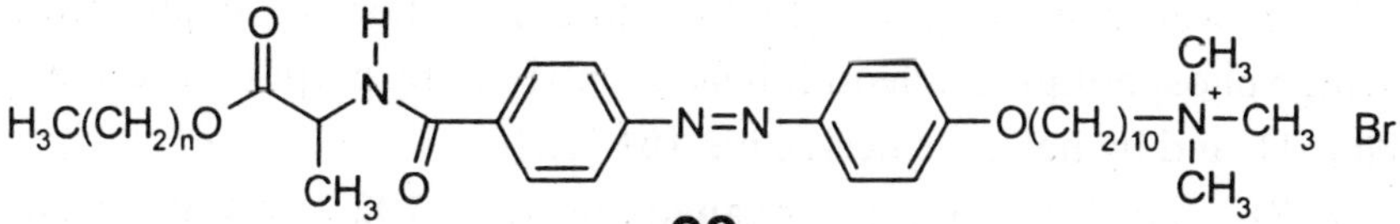

98

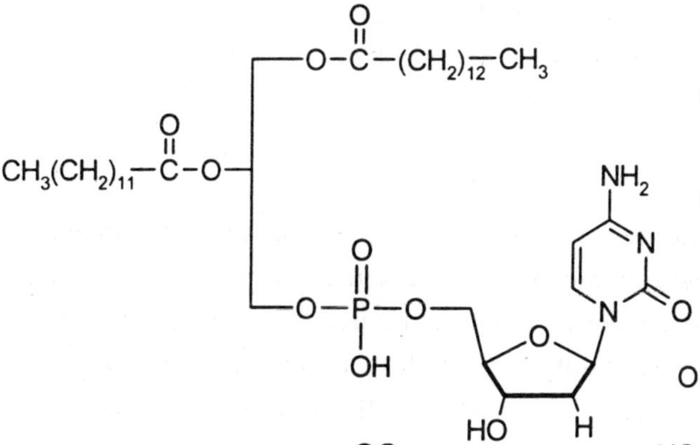

99

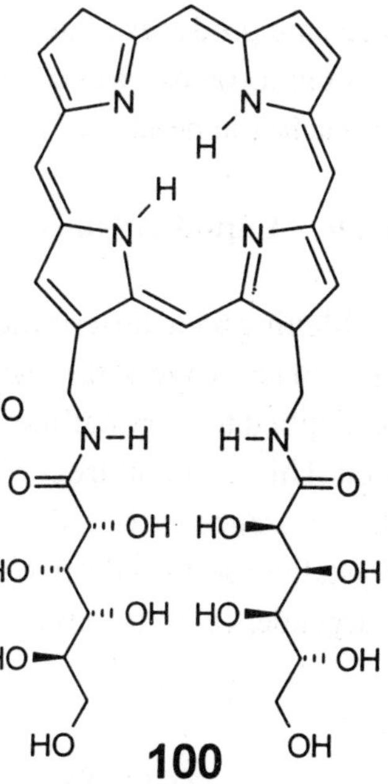

100

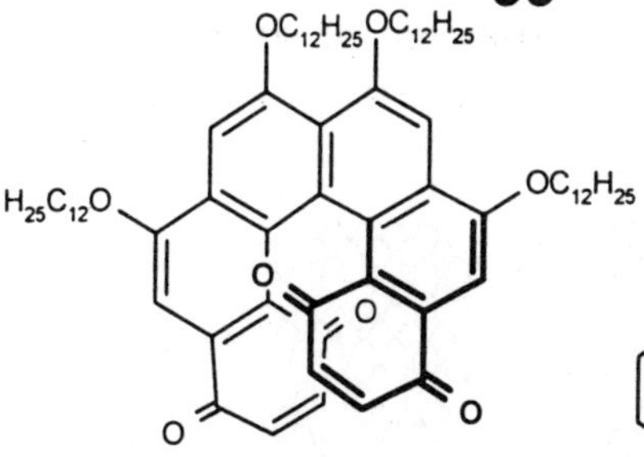

101

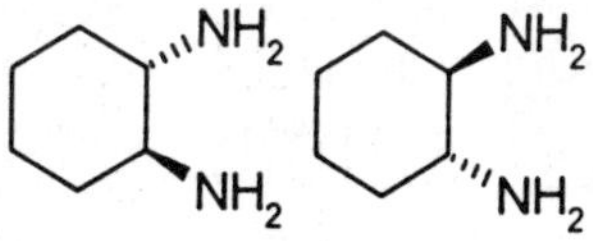

102

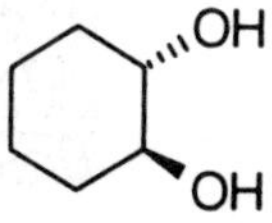

103

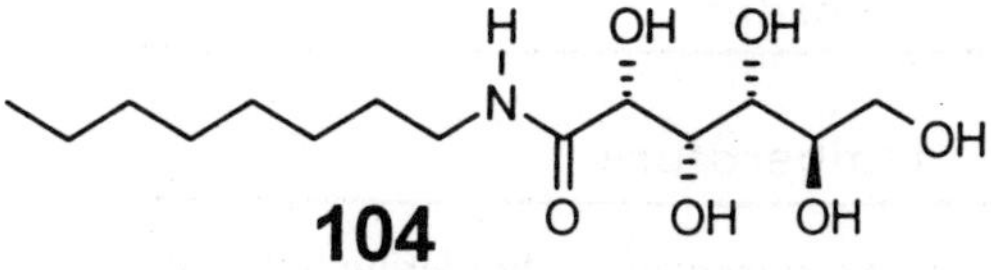

104

cooling. An interesting dependence of UV absorption wavelengths on the length of the amphiphiles tail was reported for **98** [49]. Owing to differences in the close packing of azobenzene chromophoric groups, the aggregate aborbs at 320 nm for $n = 9$, 11 and 13 whereas for $n = 10$ and 12 the corresponding absorption band appears at 350 nm.

Fibers were also found to be formed, amongst others, by amphiphiles like **99** having a phospholipid nucleotide head group [50], by porphyrin derivatives like **100** [51], and by helicene derivative **101** [52].

(*R,R*)- and (*S,S*)-cyclohexanediamines **102** self-assemble with cyclohexane-1,2-diol **103** forming helical fibers of 40-70 nm length whilst the racemate produces platelets [53].

Fuhrhop and coworkers [54] discovered higher aggregates - ropes - consisting of quadruple helices of N-octyl-*D*-gluconamide **104**.

4.2.6 Liquid crystals [55]

Melting point describing the temperature at which the solid to liquid transition takes place is one of the main characteristics of chemical substances. When heat is supplied to a crystalline species its temperature usually rises until it starts to melt. This temperature corresponding to the melting point is maintained until all the substance is liquified. During this process the long-range order of the crystalline solid is destroyed. Simultaneously anisotropy of the crystal, that is, a dependence of its optical and some other properties on the direction of, for

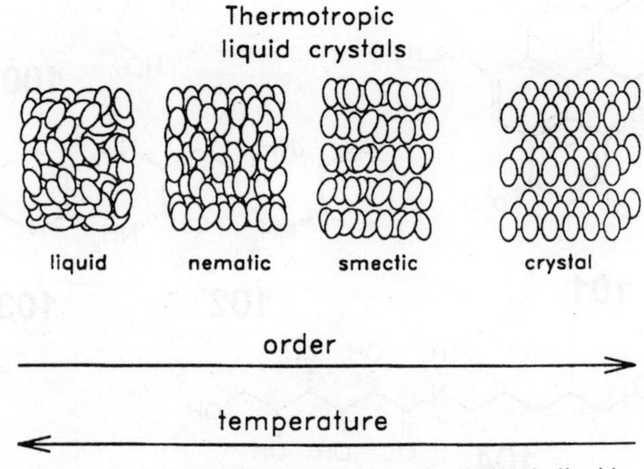

Figure 4.14. Schematic presentation of ordering in liquid. liquid crystal, and crystal.

105a R = CH_3

105b R = [phenyl]

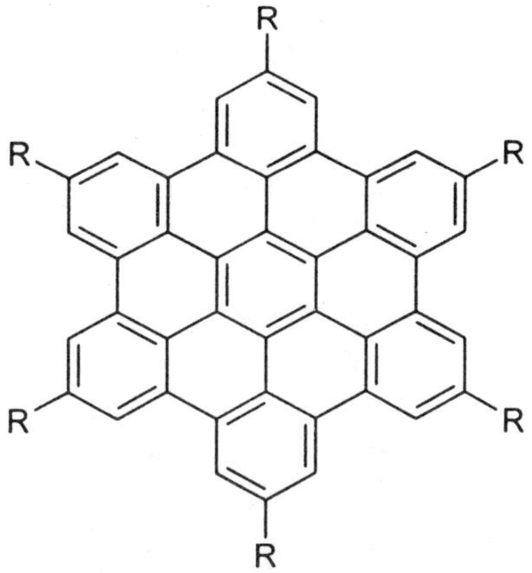

106 (R = H)
106a (R = n-$C_{12}H_{25}$)
106b (R = n-$C_{14}H_{29}$)
106c (R = n-$C_{16}H_{33}$)

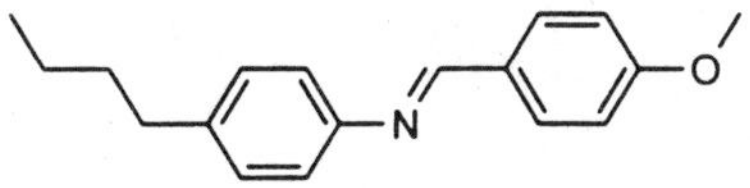

107

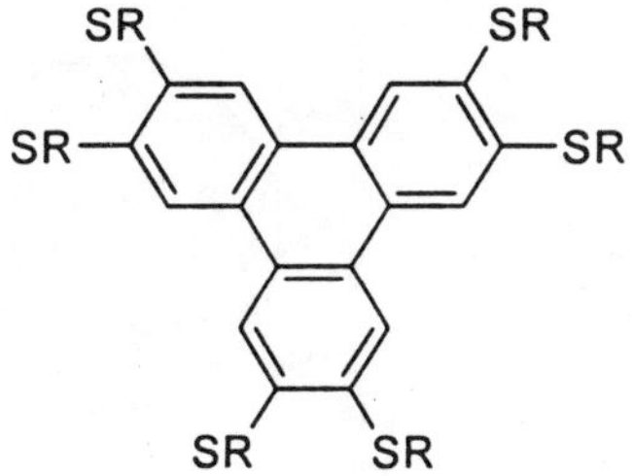

R = C_6H_{13}

108

instance, incident light, subsides during the melting process, which leads to an isotropic liquid. However, this is not always the case. To the great surprise of the Austrian botanist Reinitzer [56], two cholesteryl esters **105a** and **105b** [57] did not give a clear, transparent liquid on melting. The 'liquid', later called a thermotropic liquid crystal, was anisotropic until the clearance temperature was reached. Thermotropic liquid crystals are built from form-anisotropic molecules bearing the name *mesogens*. By varying the temperature between those corresponding to the crystalline and liquid phases one can generate a nematic phase with a preferred orientation of the longitudal molecular axis and one of the smectic phases characterized by a layered arrangement of mesogens orientated perpendicularly or at an angle to the layers. Thermotropic liquid crystals can be

109

110

also built from disc shaped molecules like **106** [58] forming molecular stacks. Another kind of liquid crystal requiring a solvent for formation bears the name lyotropic liquid crystals (shown in Figure 4.14). For almost 80 years studies of these and other substances exhibiting similar behaviour were thought to be a mere curiosity of no practical significance. The title of an article in the German journal News from Chemistry, Techniques, and Laboratory in 1966 "Liquid Crystals - an Area of Research with Little Use?" reflects these opinions [59]. The situation has changed rapidly with the development of the first compound exhibiting liquid crystal behaviour at room temperatures 4-methoxybenzylidene-4'-n-butylaniline **107** [60]. At present there are huge industrial application of liquid crystals in wrist watches and high contrast computer displays. Many other applications of ferroelectric liquid crystals as sophisticated optical computing systems [61], flat panels for high definition computer and video displays, and materials exhibiting nonlinear optical properties are foreseen in the near future. Notably, some high strength polymeric materials, such as the famous Kevlar, are spun out of lyotropic liquid crystal.

There is a significant degree of π-orbital overlap between the adjacent aromatic ring of discotic columns built of **108**, thus they were thought to form quasi one-dimensional conductors. Instead, they proved to be insulators. However, an electron is extracted from the triphenylene core by dissolving a small amount of an electron acceptor, such as $AlCl_3$, in the liquid-like hydrocarbon matrix in which the stacks are immersed. This charge transfer results in the creation of a positive hole which is highly mobile along the stack, leading to quasi one-dimensional conductivity [62]. Colour-responsive liquid crystals built of cholesterol and its crown ether derivative in a helical arrangement were obtained by Shinkai group [63] whilst a quasi-polymeric hydrogen-bonded nematic mesophase **109** was reported by Lehn and coworkers [64]. An interesting combination of fullerene, ferrocene and cholestenyl units in one molecule also results in thermotropic liquid crystalline properties **110** [65].

Interestingly, tobacco mosaic virus, discussed in Section 5.2.1, owing to its rodlike shape forms lyotropic liquid crystalline phases which can be analysed theoretically [66].

REFERENCES

1. An excellent introduction into micro-, meso- and macroscale in chemistry and biology is given in D. A. Tomalia, A. M. Naylor, W. A. Goddart III, Angew. Chem. Int. Ed. Engl., 1990, 29, 138.

2. J.-H. Fuhrhop, J. Köning, *Membranes and Molecular Assemblies. The Synkinetic Approach*, The Royal Chemical Society, Cambridge, 1994, (a) p. 149; (b) p. 30; (c) p. 39.

3. J.-H. Fuhrhop, U. Liman, V. Koesling, J. Am. Chem. Soc., 1988, 110, 6840.

4. (a) M. Müller, J. Petersen, R. Strohmaier, C. Günther, N. Karl, K. Müllen, Angew. Chem. Int. Ed. Engl., 1996, 35, 886; (b) H. Tokuhisa, M. Zhao, L. A. Baker, V. T. Phan, D. L. Dermody, M. E. Garcia, R. F. Peez, R. M. Crooks, T. M. Mayer, J. Am. Chem. Soc., 1998, 120, 4492; (c) I. K. Lednev, M. C. Petty, Adv. Mater., 1996, 8, 615.

5. J. N. Israelachvili, *Intermolecular and Surface Forces*, Academic Press, London, 1992.

6. D. S. Karpovich, G. J. Blanchard, Langmuir, 1996, 12, 5522.

7. K. Okuyama, Y. Soboi, N. Iijima, K. Harabayashi, T. Kunitake, T. Kajiyama, Bull. Chem. Soc. Jpn., 1988, 61, 1485; G. Xu, K. Okuyama, M. Shimomura, Mol. Cryst. Liq. Cryst., 1992, 105, 213; K. Okuyama, H. Watanabe, K. Harabayashi, T. Kunitake, T. Kajiyama, N. Yasuoka, Bull. Chem. Soc. Jpn., 1986, 59, 3551.

8. (a) T. Kunitake, Y. Okahata, J. Am. Chem. Soc., 1977, 99, 3860; (b). T. Kunitake, Angew. Chem. Int. Ed. Engl., 1992, 31, 709.

9. A. Ulman, Chem. Rev., 1996, 96, 1533.

10. D. Hoenig, D. Moebius, J. Phys. Chem., 1991, 95, 4590.

11. B. S. Murray, P. V. Nelson, Langmuir, 1996, 12, 5973.

12. B. J. Berne, R. Pecora, *Dynamic Light Scattering*, Wiley, New York, 1975; H. Versmold, Light Scattering, Chapter 7 in *Studies in Physical and Theoretical Chemistry*, vol. 74, Elsevier, 1991.

13. M. J. Rucco, G.G. Shipley, Biochem. Biophys. Acta, 1983, 735, 305.

14. M. S. Johal, A. N. Parikh, Y. Lee, J. L. Casson, L. Foster, B. I. Swanson, D. W. McBranch, D. Q. Li, J. M. Robinson, Langmuir, 1999, 15, 1275.

15. (a) T. P. Rieker, S. Misono, F. Ehrburger-Dolle, Langmuir, 1999, 15, 914; (b) J. Stöhr, NEXAFS Spectroscopy, Springer Series, in *Surface Science 25*, Springer, New York, 1992; (c) D. Brigs, M. P. Seah, *Practical Surface Analysis and X-ray Photoelectron Spectroscopy*, 2nd Ed., Wiley, Chichester, 1990.

16. F. M. Menger, K. D. Gabrielson, Angew. Chem. Int. Ed. Engl., 1995, 34, 2091.

17. D. Vollhardt, Adv. Colloid and Interface Sci., 1996, 64, 143.

18. M. D. Mitor. J. F. Faucon. P. Meteard, P. Bothorel. Adv. Supramol. Chem., 1992. 2. 93; W. Helfrich. Z. Naturforsch., 1973, 28c, 693.
19. H. Ringsdorf, B. Schlarb, J. Venzmer. Angew. Chem. Int. Ed. Engl., 1988, 27, 113.
20. (a) F. M. Menger. A. R. Erlington, J. Am. Chem. Soc., 1990, 112, 8201; (b) F. M. Menger. M. Rourk. Langmuir. 1999, 15, 309.
21. F. M. Menger. Acc. Chem. Res., 1979, 12, 111.
22. J. C. Eriksson, G. Gilberg, Acta Chem. Scand., 1966. 20. 2019.
23. Ref. 2, p. 28.
24. (a) K. A. Dill, P. J. Flory, Proc. Natl. Acad. Sci. USA, 1981, 78, 676; (b) D. W. R. Gruen, J. Colloid Interface Sci., 1981, 84, 281.
25. P. Fromherz, Ber. Bunsenges. Phys. Chem., 1981, 85, 891.
26. G. Conte, R. Di Blasi, E. Giglio, A. Parretta, N. V. Pavel, J. Phys. Chem., 1984, 88, 5720.
27. G. Esposito. E. Giglio. N. V. Pavel, A. Zanobi. J. Phys. Chem., 1987. 91. 356.
28. J.-H. Fuhrhop, M. Baccouche, Liebigs Ann. Chem., 1976. 2058; J. S. Nowick, J. S. Chen, G. Noronha, J. Am. Chem. Soc., 1993, 115, 7636.
29. N. Ulagappan, C. N. R. Rao, Chem. Commun., 1996. 2759.
30. J. Hotz. W. Meier, Langmuir, 1998, 14, 1031.
31. G. Decher, *Dissertation*, University of Mainz, 1986; cited in H. Ringsdorf, B. Schlarb, J. Venzmer. Angew. Chem. Int. Ed. Engl., 1988. 27. 113.
32. A. Nakano, Q. Xie, J. V. Mallen, L. Echegoyen. G. W. Gokel, J. Am. Chem. Soc., 1990, 112, 1287.
33. J. T. Grooves, R. Neumann, J. Org. Chem., 1988, 53, 3891
34. Buckyball is an americanized version of the somewhat cumbersome name buckminsterfullerene, coined by Kroto for C_{60} to honour the architect Buckminster Fuller who had built domes consisting of hexa- and pentagons. H. W. Kroto, Angew. Chem. Int. Ed. Engl., 1992, 31, 111.
35. S. Iijima, Nature, 1991, 354, 56.
36. S. Iijima, T. Ishihashi, Nature, 1993, 363, 603.
37. H. Hiura. T. W. Ebbesen, K. Tanigaki, Adv. Mater., 1995. 7, 275.
38. (a) M. Ata, K. Yamaura, A. J. Hudson, Adv. Mater., 1995, 7. 286; (b) G. Che, B. B. Lakshmi, C. R. Martin, E. R. Fischer, Langmuir, 1999, 15, 750.
39. P. Ball. Nature. 1996. 382, 207.
40. A. C. Dillon, K. M. Jones, T. A. Bekkedahl, C. H. Kiang. D. S. Bethune, M. J. Heben. Nature. 1997. 386, 377.

41. S. J. Tans, M. H. Devoret, H. Dai, A. Thess, R. E. Smalley, R. J. Geerligs, C. Dekker, Nature, 1997, 386, 474.

42. S. J. Tans, A. R. M. Verschueren, C. Dekker, Nature, 1998, 393, 49.

43. P. DeSantis, S. Morozetti, S. Rizzo, Macromolecules, 1974, 7, 52.

44. (a) M. R. Ghadiri, J. R. Granja, R. A. Milligan, D. E. McRee, N. Khazanovich, Nature, 1993, 366, 324; (b) P. R. Ashton, S. J. Catrill, G. Gattuso, S. Menzer, S. A. Nepogodiev, A. N. Shipway, J. F. Stoddart, D. J. Williams, Chem. Eur. J., 1997, 3, 1299; G. Gattuso, S. Menzer, S. A. Nepogodiev, J. F. Stoddart, D. J. Williams, Angew. Chem. Int. Ed. Engl., 1997, 36, 1451.

45. M. R. Ghadiri, J. R. Granja, L. K. Buehler, Nature, 1994, 369, 301.

46. J.-C. Meillon, N. Voyer, Angew. Chem. Int. Ed. Engl., 1997, 36, 967.

47. G. Pistolis, A. Malliaris, J. Phys. Chem., 1996, 100, 15562.

48. N. Nakashima, S. Asakuna, T. Kunitake, J. Am. Chem. Soc., 1985, 107, 509.

49. N. Yamada, N. Kawasaki, J. Chem. Soc. Chem Commun., 1990, 568.

50. G. M. T. Wijik, F. W. J. Gadella, K. W. A. Wirtz, K. Y. Hestetker, M. von den Bosch, Biochem., 1992, 31, 5912.

51. J.-H. Fuhrhop, U. Bindig, U. Siggel, J. Chem. Soc. Chem Commun., 1994, 1583.

52. A. J. Lovinger, C. Nuckolls, T. J. Katz, J. Am. Chem. Soc., 1998, 120, 264, 1944.

53. S. Hanessian, A. Gomtsyan, M. Simard, S. Roelens, J. Am. Chem. Soc., 1994, 116, 4495.

54. S. Svenson, J. Köning, J.-H. Fuhrhop, J. Phys. Chem., 1994, 98, 1022.

55. (a) R. Bissell, N. Boden, Chem. Brit., 1995, January, 38; (b) P. G. De Gennes, *The Physics of Liquid Crystals*, Clarendon Press, Oxford, 1974; (c) G. Gomper, M. Schick, *Self-Assembling Amphiphilic Systems in Phase Transitions and Critical Phenomena*, C. Domb, J. Lebowitz, Eds., Academic Press, NY, 1994.

56. F. Reinitzer, Monatsh. Chem., 1888, 9, 421.

57. F. Vögtle, *Supramolecular Chemistry*, J. Wiley, New York, 1991, p. 232.

58. A. M. van de Craats, J. M. Warman, K. Müllen, Y. Geerts, J. H. Brand, Adv. Mater., 1998, 10, 36.

59. Nachr. Chem. Techn.,1966, 14, 29.

60. H. Kelker, B. Scheurle, Angew. Chem. Int. Ed. Engl., 1969, 8, 884.

61. D. M. Walba, Ferroelectric Liquid Crystals, in *Advances in the Synthesis and Reactivity of Solids*, 1991, 1, 173.

62. D. Adam, P. Schumacher, J. Simmerer, L. Häussling, K. Siemensmeyer, K. H. Etzbach, H. Ringsdorf, D. Haarer, Nature, 1994, 371, 141.

63. T. Nishi A. Ikeda, T. Matsuda, S. Shinkai, J. Chem. Soc. Chem. Commun., 1991, 339.

64. M. Kotera. J.-M. Lehn. J.-P. Vigneron, J. Chem. Soc. Chem. Commun.. 1994. 197.

65. R. Deschenaux. M. Even. D. Guillon, J. Chem. Soc. Chem. Commun.. 1998. 537.

66. (a) L. Onsager. Ann. NY Acad. Sci.. 1949. 51. 627; (b) R. Oldenbourg. X. Wen; R. B. Meyer. D. L. D. Gaspar. Phys. Rev. Lett., 1988. 61. 1851.

Chapter 5

BETWEEN CLASSICAL ORGANIC CHEMISTRY AND BIOLOGY. UNDERSTANDING AND MIMICKING NATURE

5.1 Introduction

In the last hundred years a considerable amount of information has been collected by chemists on small and middle sized molecules owing to the development of very precise analytical methods enabling the determination of their structure and reactivity. On the other hand, the structure and functioning of large natural objects such as cells and cellular networks have been studied by completely different methods typical for biology. An exciting field of biochemistry bridging the gap between these fields was started not long ago. It involves the study of the structure of biopolymers and their functioning at the molecular level. On the other hand, the structure of molecular aggregates, transport through membranes, recognition phenomena between simple molecules, as well as the study of artificial biomimetic structures, are the subject of supramolecular chemistry. It should be stressed that there is no clearcut division between biochemistry and biomimetic supramolecular chemistry; the two fields partly, but strongly, overlap.

One of the most exciting phenomenon in biochemistry is molecular recognition enabling outstanding selectivity and efficiency of chemical reactions in living organisms. As briefly discussed in Chapters 2 and 3, the revolutionary 'key and lock' model of the recognition was introduced by Emil Fischer in 1894 [1]. It consists in the highly specific interaction of a larger molecule, the substrate, with a smaller one, the receptor, which fits into the cavity of the former. This model was later replaced by the more subtle "induced fit" model [2] in which both molecules have to undergo conformational or other changes [3] to lower the barrier for the complex formation. A remarkable example of induced fit (presented in Chapter 1) provides a porphyrin dendrimer (shown in Figure 1.6) changing its conformation to accept the guest molecule. Several other examples of induced-fit complexation are presented throughout this book. However, this fruitful concept can hardly be applied to the hexakinase enzyme creating the host site by wrapping itself around the guest (Section 2.1). On the other hand, the difficulties encountered sometimes by choosing host and guest in the solid state are discussed in Section 6.2.

Only a few examples from the vast area of biochemistry and biomimetic chemistry will be presented in this chapter to substantiate the claim that supramolecular chemistry is situated between chemistry and biology.

5.2 The Role of Self-Organization and Self-Association in the Living Nature

Living organisms are built of highly organized assemblies of organic molecules performing very efficiently specific tasks of enormous complexity. These assemblies are characterized by high level of organization and considerable mobility. At present we are only beginning to understand the structure and functioning of living organisms at molecular level [4]. Here the role of supramolecular assemblies in Nature will be illustrated only by few examples exhibiting their basic importance.

5.2.1 Tobacco Mosaic Virus

Let us consider the structure of the tobacco mosaic virus first. As shown schematically in Figure 5.1, it is composed of a single strand of ribonucleic acid, RNA, covered by a sheath formed from 2130 identical protein units. Thus the whole virus constitutes a rather simple supramolecular assembly. By changing

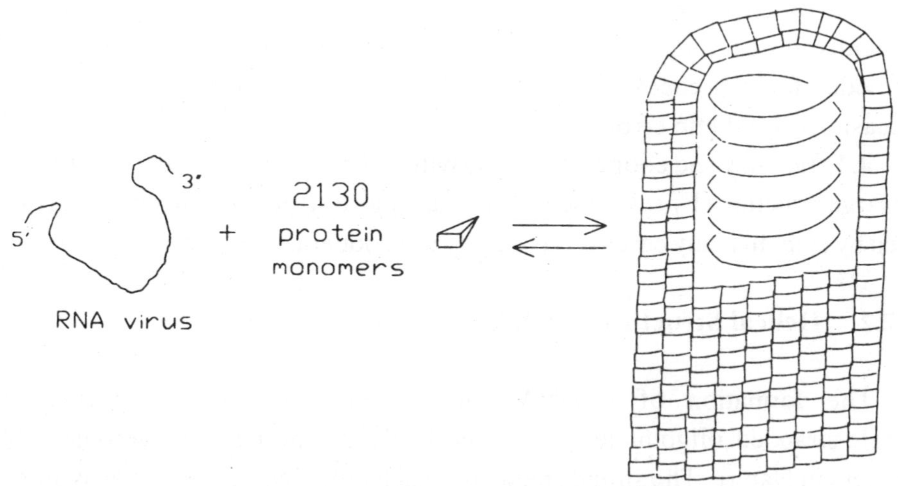

Figure 5.1. The scheme of formation of tobacco mosaic virus.

physical or chemical stimuli this assembly can be split into its component parts. Moreover, Fraenkel-Conrat and Williams showed that such a living structure can not only dissociate into the corresponding RNA and proteins [5] but, amazingly, the virus can be recomposed *in vitro*, that is in the test tube. In addition, the concentration, time, and pH dependences of the reassembly process are typical for chemical reactions implying that a living organism [6a], a virus, can be formed artificially from molecules which in principle could be synthesized in a test tube. The importance of this fascinating discovery can be compared only with that of the Wöhler synthesis of the organic molecule, urea, from the inorganic reagents in 1825 [6b]. The latter observation provided the decisive argument against the living force 'élan vital' which had earlier been considered indispensable for the existence of a living creatures.

The process of assembling of tobacco mosaic virus exhibits several characteristic features ensuring its effectiveness [7]:

1. Information: A large structure is built from only one or a few types of subunits. In this way the amount of genetic information is greatly reduced.

2. Control: The highly selective fit of constituent parts ensures control of the assembly process. Simultaneously, making use of numerous weak interactions in the association of subunits not only allows for the control of the process and but also enables a response to the changes in environment.

3. Error checking: The complementarity of shape and electrostatic field of subunits constituting the assembly forms the basis for the recognition process. It provides for self-checking and excludes defective subunits thus correcting occasional synthetic errors.

4. Efficiency and cooperativity: As will be discussed below, the construction of large structures by self-assembly is much more effective than obtaining them step by step. In many cases the assembly is a cooperative all or nothing process.

5.2.2 Helical Structure of DNA

The formation of a DNA double helix from the complementary single-stranded oligonucleotides is probably the best known self-assembly process in Nature. The strands recognize each other by very selective hydrogen bonds between adenine **111** and thymine **112** (or uracil **113** for RNA) or guanine **114** and cytosine **115** base pairs (Figure 5.2) first found in 1953 by Watson and Crick [8] and later confirmed in numerous crystal structure analyses [9]. A detailed study of an assembly of the model nucleic acids containing either 17 adenine residues or the same number of uracil residues has shown that the formation of the first three to four base pairs is necessary to trigger the process [10]. This first stage, nucleation, is unfavourable in terms of free energy but the subsequent base-pairing provides large and negative contribution to the overall

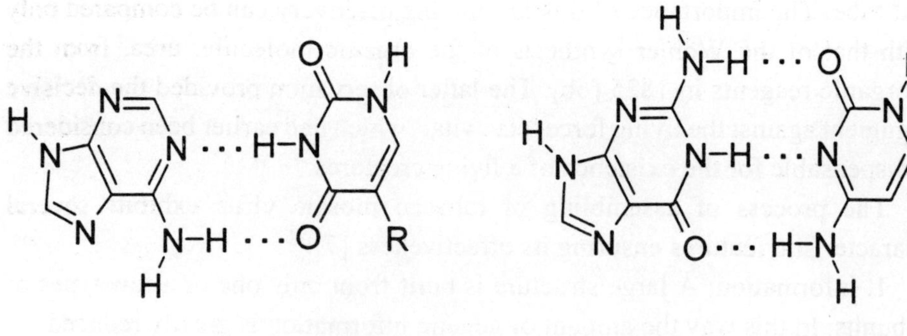

111 **112** R = CH_3 **113** R = H **114** **115**

Figure 5.2. Adenine-thymine(uracil) (left) and guanine-cytosine (right) dimers.

free energy. Thus once the initial nucleus is formed the further self-assembly proceeds cooperatively until all possible base pairs are formed. The double helix formation is all or nothing process since it cannot stop before completion of the pairing. Several molecules intertwining upon complexation with a metal ion, thus spontaneously forming double-, triple-, or quadruple-strand helicates, will be discussed in Section 8.4.2.

5.2.3 Cell membranes

Cell membranes play an essential role in all basic biological phenomena [11]. A biomembrane consists of three interconnected layers (Figure 5.3): the glycocalix in which recognition phenomena on the cell surface take place, the lipid bilayer with built in protein channels enabling, amongst others, transport through membrane, and the cytoskeleton stabilizing the membrane. Carbohydrates forming glycocalix take part in intercellular communication, signal transduction, cell adhesion, infection, cell differentiation, development and metastasis. In addition to its function as a filter, the middle bilayer participates in signal transduction and cell motion.

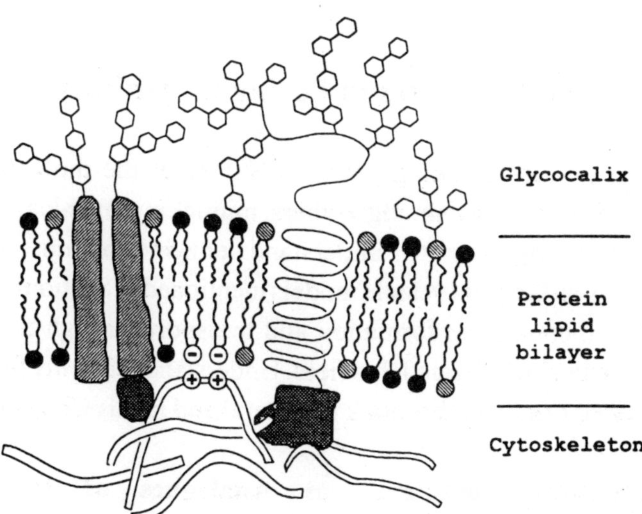

Figure 5.3. Schematic view of cell membrane.

Biological membranes are highly complicated entities performing various functions. Thus also in this case one has to study simpler supramolecular systems (like those shown in Figure 4.1) to better understand their operation. A spontaneous formation, i e., self-assembly, of lipid bilayers was presented in the previous Chapter. As discussed in Section 4.2.2, one can mimic the bilayer support realized by the cytoskeleton by inserting hydrophobic side chains of an appropriate polymer into the membrane (Figure 4.6). Multiple recognition phenomena executed by sugars on a cell membrane surface have been recently reviewed by Whitesides and coworkers [12a], whilst the use of smaller molecules mimicking carbohydrates was discussed by Sears and Wong [12b]. Recent development in the domain of recognition of carbohydrates through noncovalent interactions was summarized by Davis and Wareham [13]. These studies are not only of basic importance enhancing and deepening our understanding of the operation of cell membranes but they also have great practical significance since bacterial and viral infections, cancer metastasis, inflammatory reactions, and other similar processes are based on biological recognition by carbohydrates at the cell surface.

Interestingly, complicated systems of parallel bilayers form organelles involved in photosynthesis which bear the name chloroplasts [14].

The above examples of self-assembling supramolecular systems in Nature point to the basic importance of the understanding of their structure and functioning.

5.3 Modeling Processes in Living Organisms

Modeling processes taking place in Nature at the molecular level is a formidable task. It involves, amongst others, mimicking chemical reactions in the organisms, allowing one to understand enzyme action. Studying cell transformations on the giant vesicle models mentioned in Section 4.2.3, modeling self-replication, membrane formation and transport through them, and many other phenomena also belong to this field. Understandably, only few such studies can be discussed briefly in this book in general and in this Chapter in particular.

5.3.1 Host-guest Complexes as Analogues of the Interacting Substrate-Receptor Unit in Biochemistry

An amazing selectivity and efficiency characterizes chemical reactions carried out in Nature that are catalyzed by specific enzymes. The fact that living organisms are built almost exclusively (see, however, [15]) from *L*-aminoacids points to their enormous enantio-specificity. Topoisomerases enabling the synthesis of knotted or catenated circular DNAs (Section 2.3) represent another example of the thought inspiring enzymes action. The spatial folding of one or more peptide chains forming the tertiary structure of enzymes allows for the formation of reaction sites in which the reactants perfectly fit not only spatially but also stereo-electronically. Macrocyclic hosts catalyzing chemical reactions mimic the receptor site of an enzyme whilst the guest plays the role of the substrate. Few examples of enzyme mimics will be discussed in Section 5.3.5.

As discussed in Section 6.3.3, rational drug design [16] plays an ever increasing role in the development of more effective new drugs. Simplifying the problem, one can say that the development consists in finding molecules that enter the active site, blocking the malfunctioning of the organism. This means that the shape and electrostatic field of the active site and those of the drug molecule under development should complement each other. In such situation one speaks about their stereo-electronic fit. The task is less complicated when the active site is known. Otherwise it can be modelled on the basis of the stereo-electronic structure and pharmacological activity of known drugs [16a].

5.3.2 Principles of Molecular Modeling of the Origin of Life

It should be stressed that the most widely accepted models of the origin and evolution of life are based on the concept of self-assembling molecular systems. The models rooted in information theory involve [17]:

(1) The association of small molecules into more complicated "smart" biomolecules and assemblies that possess retrievable molecular information for specialized functions.

(2) Successful molecular evolution that puts strict demands on accurate transfer of molecular information executed by highly precise control of size, shape, topology, flexibility, and surface chemistry at the molecular level. The latter five features have been dubbed critical molecular design parameters.

(3) The control of these parameters allowing for the development of highly complicated structures of living organisms which are able to reproduce, to pass on genetic information to the next generations, and to carry out various specialized functions.

These simple principles together with natural selection have led to the enormous complexity of life forms. As discussed in the review of Tomalia et al.

[18] in which microscopic, mesoscopic, and macroscopic dimensions of some chemical and biological objects are given, there is a remarkable correspondence between the size of a system and the complexity of its structure on one hand and its complicated functioning on the other. As mentioned earlier, modeling of natural systems is one of the most important tasks of supramolecular chemistry.

5.3.3 Modeling of Self-replication

Self-replication is probably the most essential property of living organisms [19]. Can it be modeled by artificial systems? The closest to the prebiotic life modeling are probably condensation reactions of trideoxynucleotides like **116** reported by Kiedrowski and coworkers [20], which in the presence of 1-(3-dimethylaminopropyl)-3-ethylcarbodiimide lead to a hexadeoxynucleotide template promoting the reaction. Self-replicating micelles and vesicles have been described by Luisi and coworkers [21]. The latter authors reported the growing number and decreasing size of "newly born" aggregates using, amongst others, light scattering and luminescence techniques. Another approach to the study of self-replication processes was adopted by Menger and coworkers [22]. The latter authors observed birthing, i.e., partitioning, and growth of giant vesicles by a modified microscope technique. Interestingly, Maoz, Sagiv et al. [23] have shown that even amphiphilic monolayers can self-replicate.

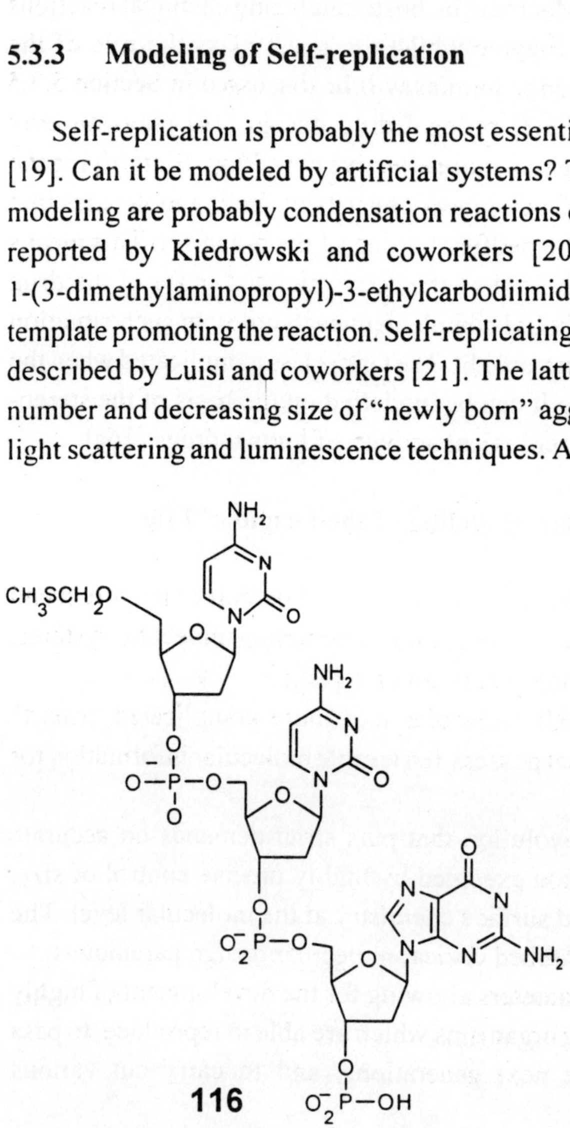

116

The Rebek group reported an increase of the yield of a reaction of the Kemp acid derivative **117** owed to template action of the product **118a** [24] (Figure 5.4). The effect observed was

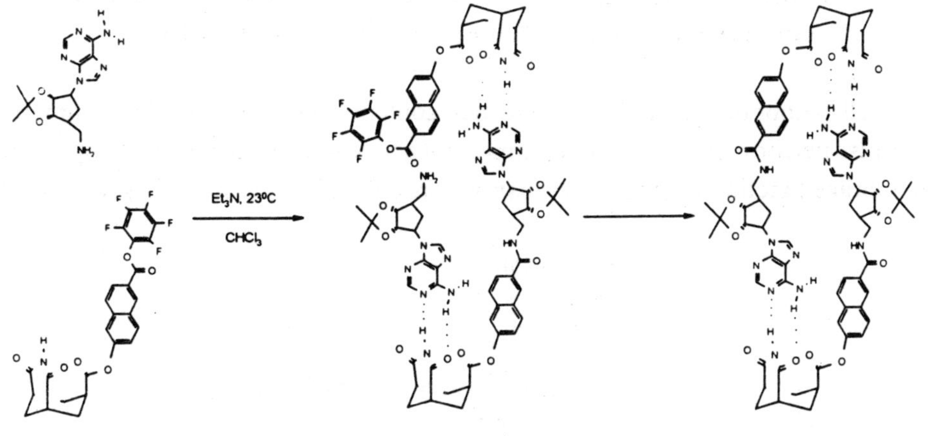

117 **118a** **118b**

Figure 5.4. Rebek's self-replicating system.

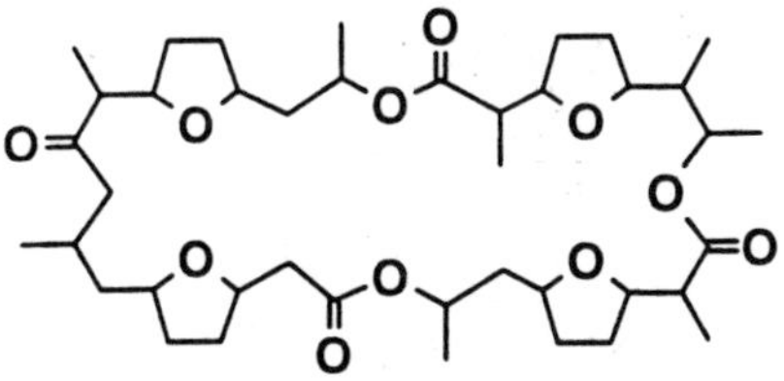

119 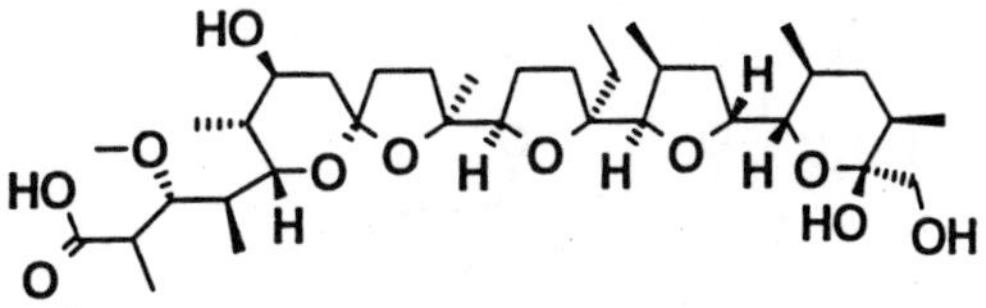 **120**

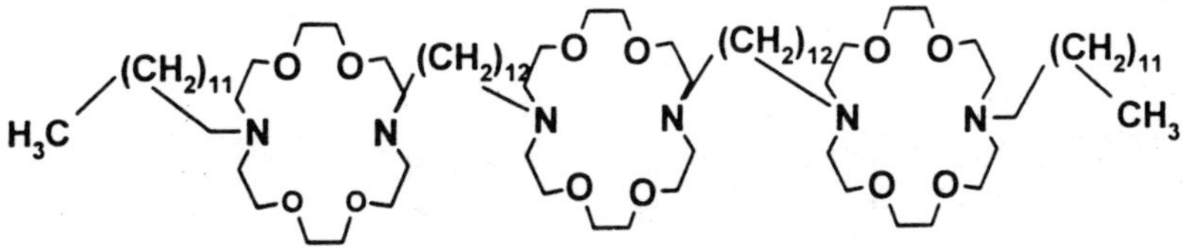

121

interpreted in terms of the preorganization of substrates enforced by the reaction product consisting in the favourable formation of the intermediate hydrogen-bonded complex **118b**. This interpretation was questioned by the Menger group [25] who claimed that simple catalytic action of amides was responsible for the yield increase. The controversy seems still open to discussion in spite of further arguments by the Rebek and Reinhoudt groups [26].

5.3.4 Transport through Membranes. "Transport antibiotics": Valinomycin, Nonactin, Monensin and Their Mimics

Cell membranes not only separate their content from the outside environment, they also perform numerous other important activities. One of the most important membrane functions consists in transport through it. The passive transport based

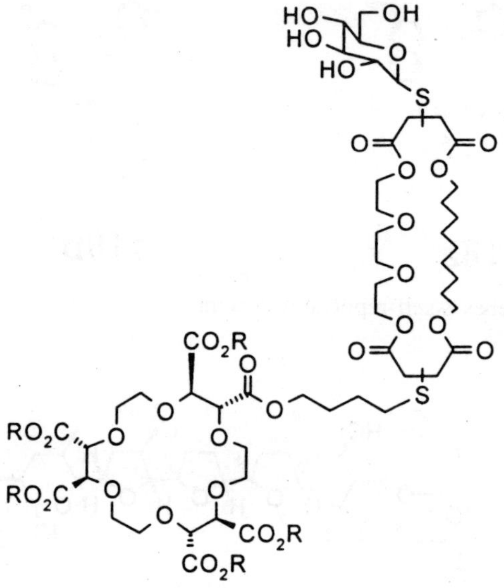

122 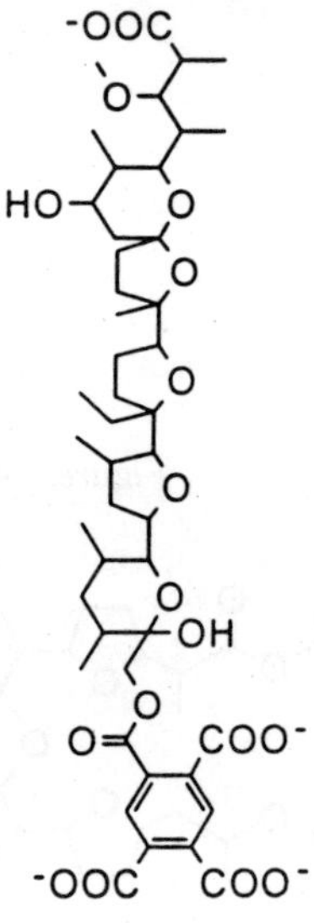 **123**

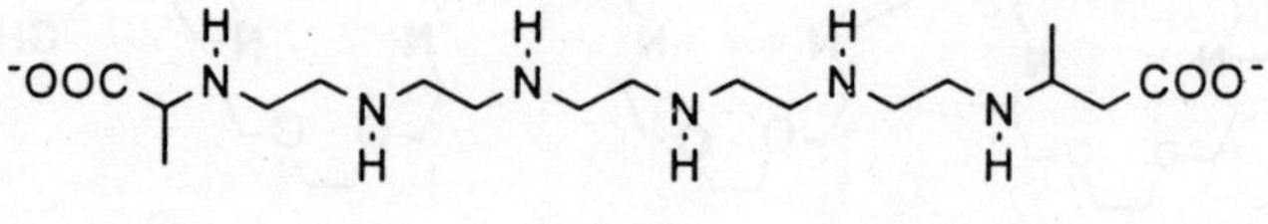

124

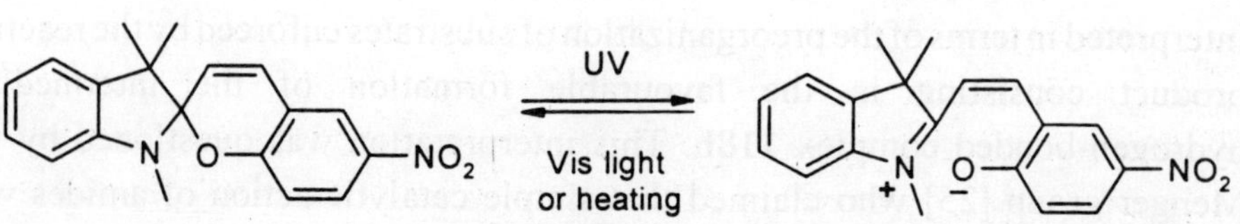

UV

Vis light or heating

125

on concentration gradient is seldom effective and the active one coupled to another thermodynamically favoured process takes place in living cells. Membrane pores play an active role in such a transport. Valinomycin **51**, nonactin **119** and monensin **120** [27] are so called "transport antibiotics" which bind and transport cations through membranes. Crown ethers derivatives like **121** [28] and bolaamphiphile **122** [29], monensin based bolaamphiphile **123** [30], Fe II transporting oligoamine with anionic head groups (also bolaamphiphiles) **124** [31] and some other molecules have been incorporated into vesicle membranes mimicking cell border. **122** not only executed proton flow out of the vesicle but it also exhibited the transport inhibition in the absence of the simultaneous transport of K^+ ions through the membrane. A transport of a

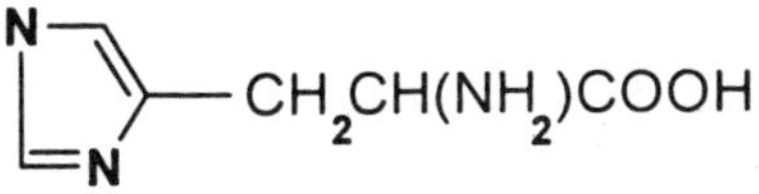

$CH_2CH(NH_2)COOH$

126

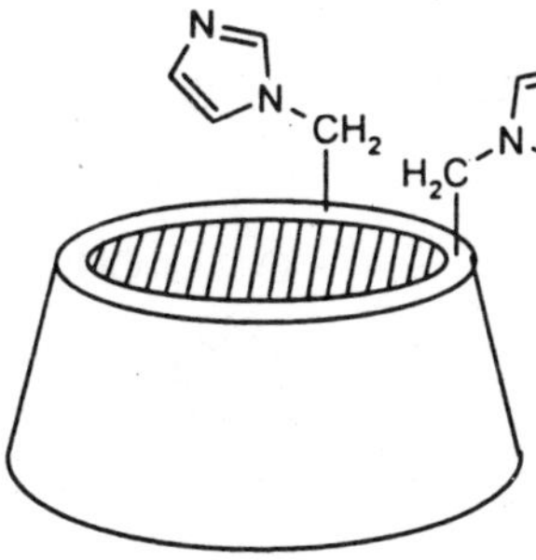

127a

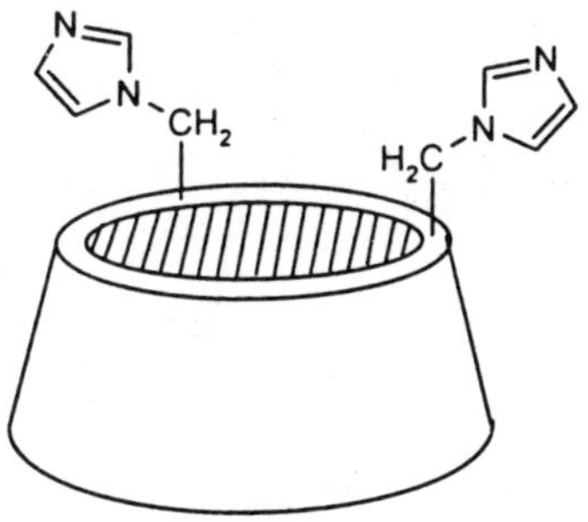

127b

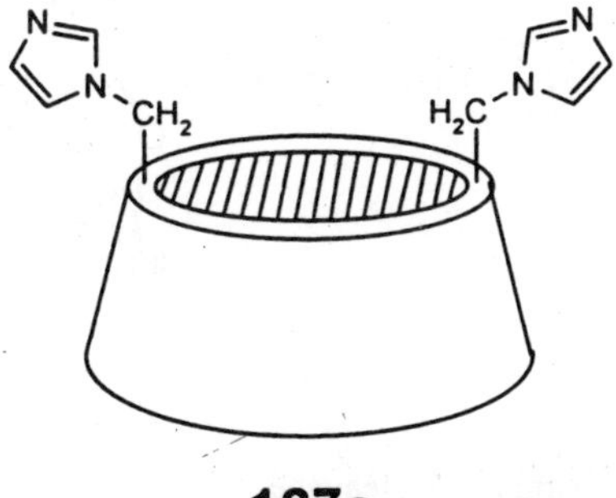

127c

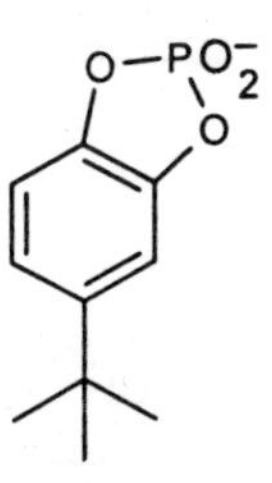

128a

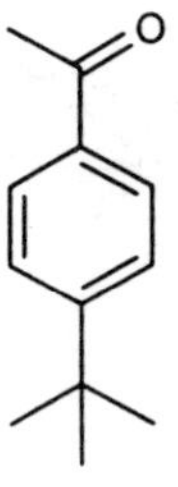

128b

tryptophane derivative in the presence of zwitterionic dye **125** resulted in the transport through the membranes upon the UV irradiation [32].

Another type of self-assembling channel transporting sodium and potassium through lipid membrane represent nanotubes built of eight *D* and *L* aminoacids residues discussed in Section 4.2.4 (Figure 4.13) [33].

5.3.5 Cyclodextrins as Enzyme Mimics [34]

Catalytical aspects of supramolecular chemistry will be discussed in some detail in the next chapter. In this section only cyclodextrins as enzyme models will be briefly presented. Several examples of cyclodextrins' catalytic activity have been reported. The acceleration of the reaction rate upon adding of CDs is

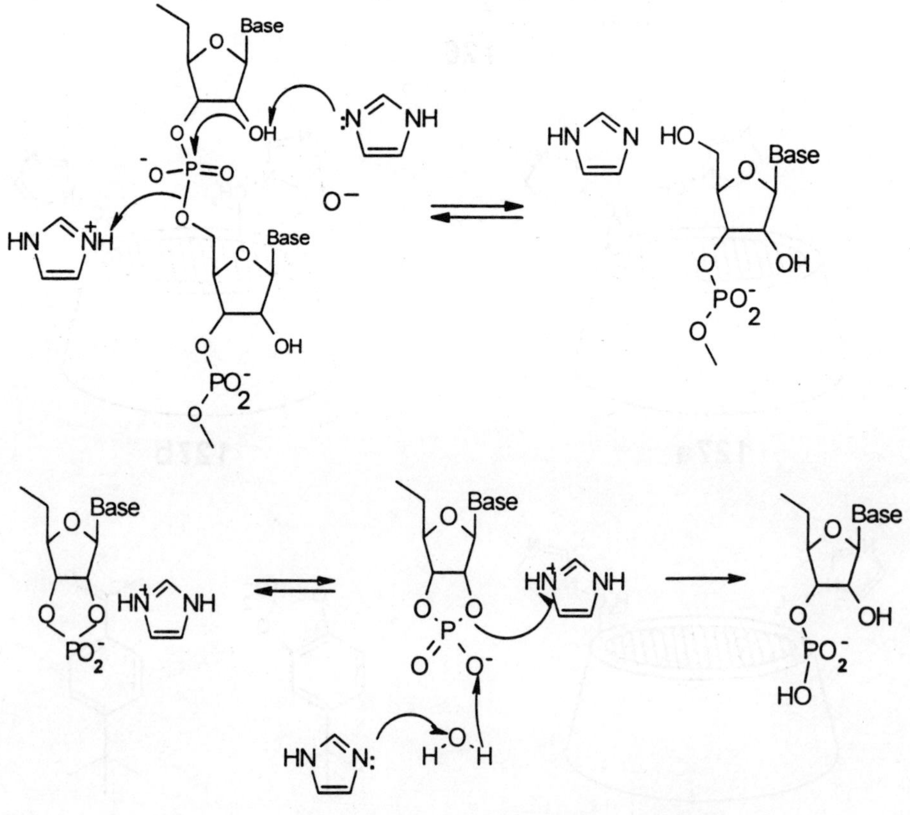

Figure 5.5. The classic mechanism of enzymatic RNA cleavage.

usually moderate [34]. However, in a few cases figures as many as several hundreds thousands or even millions have been found [35a]. Thus the catalytic activity of these oligosaccharides and their derivatives was studied with the aim of elucidating the mechanism of enzyme action. For instance, ribonuclease A is known to catalyze the cleavage of RNA with subsequent cyclization of cyclic phosphate ester followed by its hydrolysis using imidazole groups of Histidine-12 and Histidine-119 (histidine **126**) residues. To model the action of such bifunctional enzymes the Breslow group synthesized three regioisomers of disubstituted β-cyclodextrins **127a-127c** and studied their influence on the hydrolysis of **128a**. Both the inclusion of the aromatic ring of **128a** into the cyclodextrin cavities and the mutual orientation of imidazole rings in the latter molecules with respect to the substituents allowed one to test the classic mechanism of the enzyme action presented in Figure 5.5. If the latter mechanism were to operate, then the isomer **127c** with an almost linear arrangement of the imidazole groups should be the most efficient catalyst. However, the **127a** isomer proved to be the best leading to the abandonment of the classic mechanism found in textbooks. The same model cyclodextrins were also used to study enolization of ketone **128b** in order to mimic simultaneous bifunctional mechanism of catalysis of many enzymes. In this case, the reaction did not proceed without cyclodextrin and the rates were comparable to that of monosubstituted cyclodextrin when the isomers **127a** and **127b** were added. Only with the **127c** catalyst was the reaction significantly faster, thus it best fitted the transition state for bifunctional catalysis of enolization. More complicated cyclodextrin derivatives were used as models of nuclease, ligase, phosphatase and phosphorylase enzymes [35b].

5.3.6 Porphyrins Involving Systems Modeling Photosynthesis

The synthesis of carbohydrates (usually starch or sugar) from H_2O and CO_2 using light energy is one of the most common photosynthetic processes in Nature [36]. It is a complicated process involving light energy for transferring electrons from water to $NADP^+$ **129** and the generation of adenosine triphosphate ATP **15**. Chlorophyll **130** having a porphyrin core is the light receptor driving the process. Absorbtion of light by an isolated chlorophyll molecule produces its excitation, i.e., one of its electrons is raised to a higher energy state. In a photosystem built of several chlorophyll molecules the electron is passed on to a neighbouring

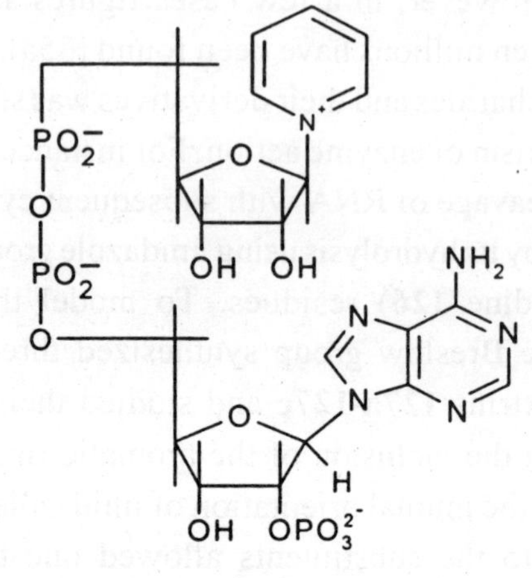

129

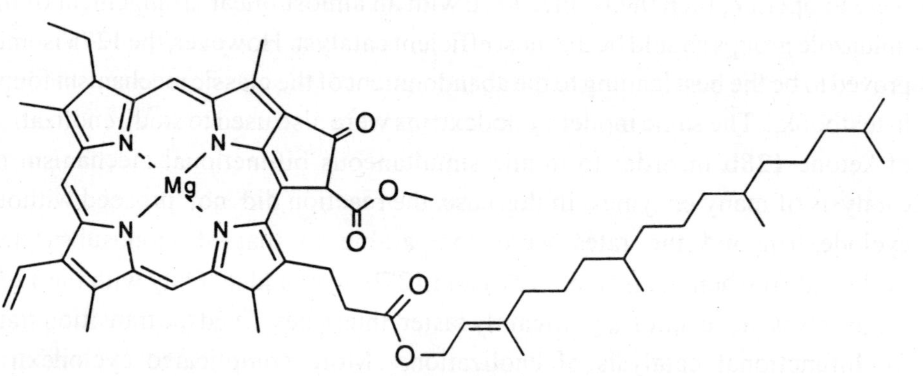

130

molecule. Such a supramolecular photosystem forms a crucial part of pigment protein complexes, bound to membranes, that bear the name reaction center. Recent X-ray analysis of the structure of the reaction center protein of some bacteria [37] allowed one to model electron transfer taking part in photosynthesis. One of the numerous examples of such modeling is provided by a study of porphyrins containing rotaxanes by Sauvage group [38]. The latter authors synthesized and analyzed photo-induced electron transfer in bis-porphyrin **131a**, rotaxane **131b** presented in Figure 5.6. They found a very fast transfer from Zn to Au in **131c** (1.7 ps) which involved the central Cu atom, a slower one in **131b**

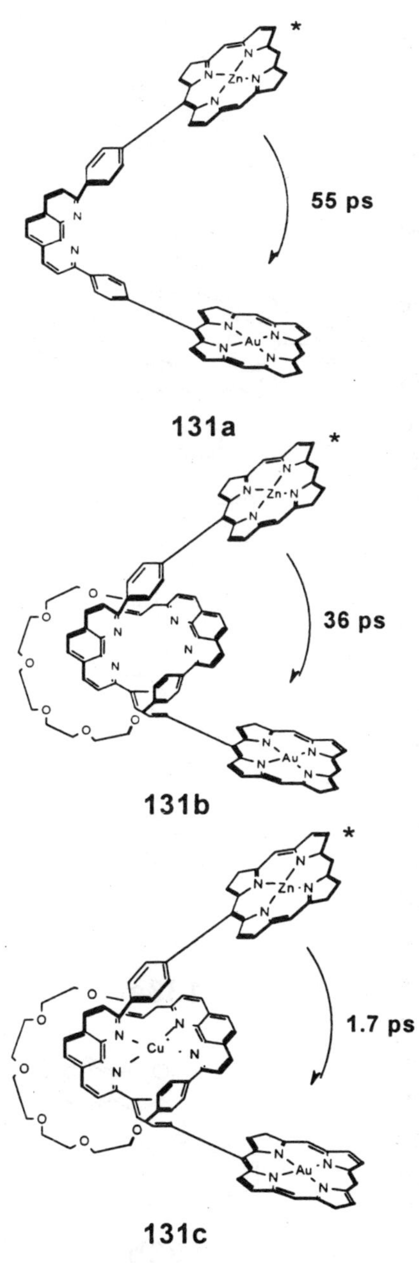

Figure 5.6. Electron transfer in linear diporphyrine **131a** and two rotaxanes **131b** and **131c**. involving it.

lacking the mediating atom and a much slower process in **130** (55 ps). Such studies provide not only a better understanding of the electron transfer process in photosynthesis but they are also a prerequisite for the design of systems enabling conversion of the solar energy for future useage. Some other supramolecular systems for artificial photosynthesis are presented in the Wasielewski review [39a] whilst dioxygen carriers like cyclidene **132a** pertaining to O_2 transport and storage have been reviewed by Bush and Alcock [39b].

132a

Several porphyrin derivatives mimicking O_2 carriers myoglobin and hemoglobin have been reported [40].

5.3.7 Light Driven Proton Pump

In the purple bacterium *Halobactium halobium*, light driven proton transfer through membranes is executed by bacteriorhodopsin [41a]. This protein consisting of seven helical fragments criss-crosses the membrane seven times,

132b

forming a cluster of seven helices spanning the membrane with chromophore molecule retinal **132b** embedded within it. Photoisomerization of all-*trans* retinal to its 13-*cis* isomer causes large movement of the helices to the position favouring proton transfer. Mimicking this action by the pump incorporated into the lipid bilayer of a vesicle wall [41b] is probably one of the greatest achievement of biomimetic chemistry in recent years. The pump **133a** which represents a molecular device composed of only one molecule (the so called C-P-Q triad) consists of a quinone Q acting as the electron acceptor, free base porphyrin P and carotene C as the primary and final electron donors, respectively. The triad is embedded into the vesicle with the smaller lipophilic carotenoid unit situated in the inner layer whilst bulkier porphyrin and quinone parts occur in the less dense packed region closer to the outer surface. Photoexcitation and proton transfer it causes proceed in seven steps. In the first step, C-P-Q is converted to the charge separated diradical state $C^{\cdot+}$-P-$Q^{\cdot-}$. Then an electron is transferred from $Q^{\cdot-}$ to the lipid-soluble 2,5-diphenylbenzoquinone Qs **133b** yielding $Qs^{\cdot-}$. In the third step uncharged semiquinone $QsH^{\cdot}$ is formed when the latter accepts a proton from the external aqueous solution. The basic function of a proton shuttle is carried out when the semiquinone radical diffuses through the membrane in the fourth step. It is then oxidized to $Qs^{\cdot+}$ by the carotenoid radical cation encoutered in the interior layer of the membrane (Step 5) and releases the proton into the aqueous medium (Step 6). The cycle is closed by random diffusion of the regenerated Qs (Step 7).

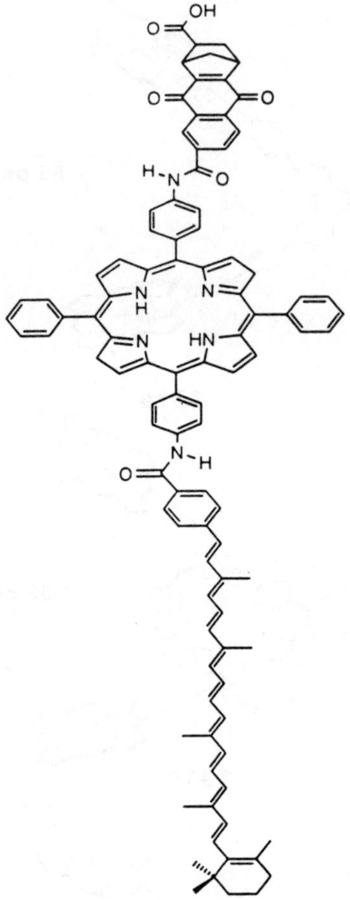

133a

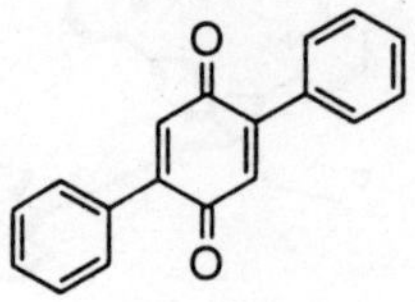

133b

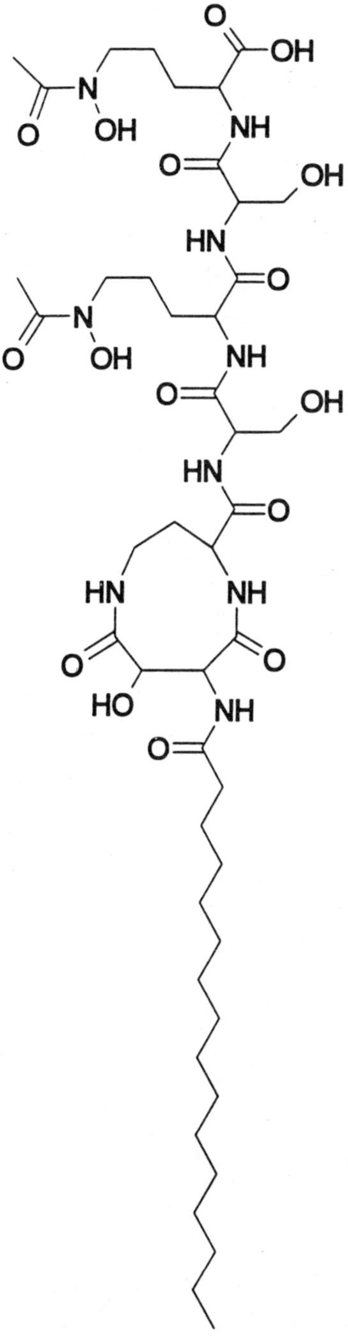

5.3.8 Iron Sequestering Agents Promoting Microbial Growth Siderophores

134

Siderophores are highly specific agents performing a very important role in sequestering low-soluble iron ions and accumulating them in living organisms [42]. Together with their synthetic analogues they provide a good example of the systems of biological significance bearing also supramolecular aspects. Modeling of their functioning allows one to understand better the operation of living organisms. Almost all of them require iron for metabolic processes. However, salts of ferric ion are insoluble at physiological pH and low molecular weight (500-1000 Daltons) strong ferric ion chelating agents sideophores sequester iron and accumulate it in living organisms. They are produced by micro-organisms in cases of iron deficiency and promote microbial growth by solubilizing and assimilating this conspicuous ingredient. They also execute some other biological functions. Pathogenic micro-organisms exposed to blood produce siderophores that compete for iron with the human protein transferrin [43a]. On the other hand, the secretion of strongly iron-binding siderophores protects humans from potentially harmful micro-organisms by making iron unavailable for them [43b]. The biosynthesis of chlorophyll by plants requires iron. Thus siderophores, more precisely called phytosiderophores in this case, are also found in plant roots.

An interesting example of siderophores was recently found in marine bacteria. The other amphiphilic molecules like **134** consist of a polypeptide polar end that binds to Fe(III) and a hydrophobic fatty acid tail [44]. They form spontaneously micelles in iron-free solution. However, in the presence of Fe(III), large spherical vesicles 200 nm in diameter are formed. Siderophores have fiound some practical applications. For instance, a siderophore is employed clinically in the treatment

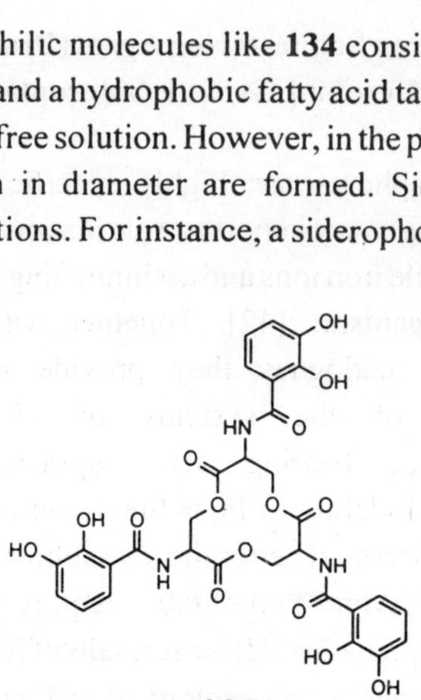

135 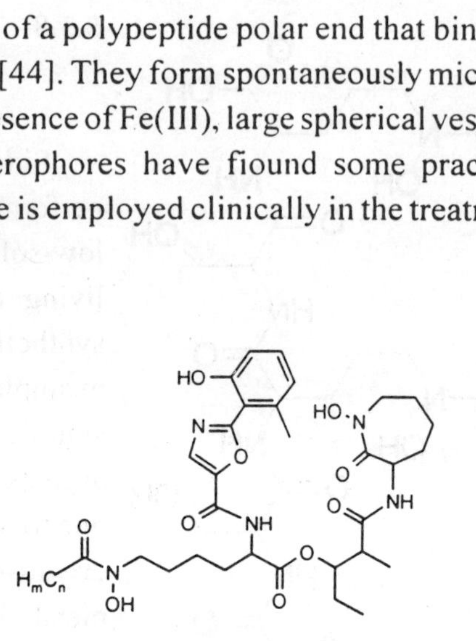 **136**

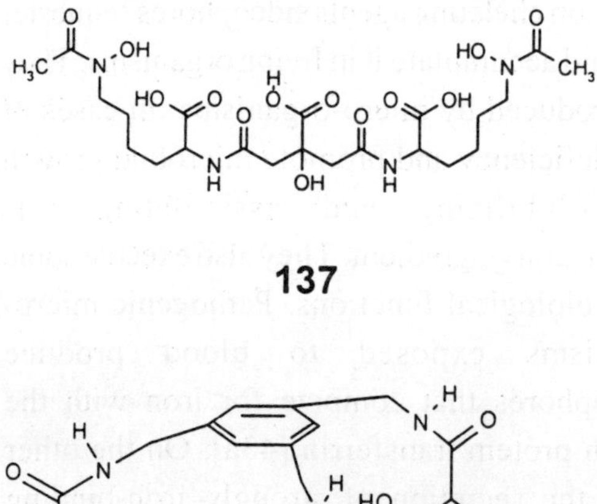

137 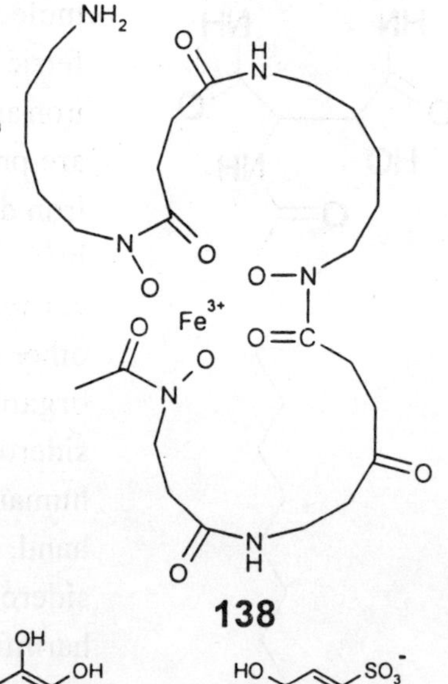 **138**

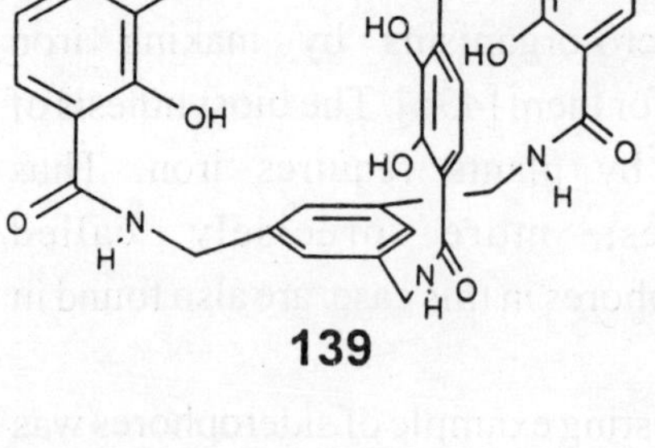

139 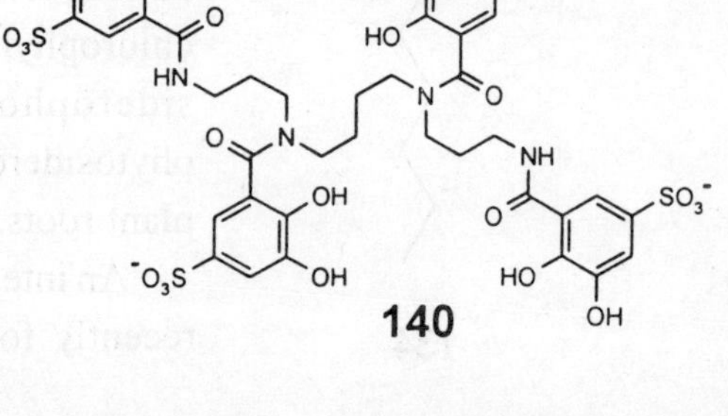 **140**

of acute iron poisoning by removing it from the body [45]. Studying siderophore complexes with ferric ion and their synthetic analogues improves the understanding of the stereochemical recognition of ferric siderophore complexes by membrane receptors as well as their mode of operation by iron removal from mammalian iron transport. Enterobactin **135**, mycobactin **136**, aerobactin **137**, ferrioxamine **138** and numerous other naturally occuring siderophores have been isolated (mostly.from bacteria) and characterized. Their most important and common feature is formation of six-coordinated octahedral complexes with ferric ion of great thermodynamic stability although mixed forms of coordination have sometimes been found. The complexes with iron (III) are mostly of 1:1 stoichiometry.

The latter is many orders of magnitude too low for microorganisms that need at least a total concentration of ca. 5.10^{-7} mol/l. Siderophores are potent chelating agents capable of collecting iron from the environment and promote its transport through the membranes into the microbial cells. Stability constants of the naturally occurring siderophore complexes with iron were very big reaching the value of 10^{52} for enterobactin **135**, mycobactin **136**, aerobactin **137**, ferrioxamine B **138** [46]. Only recently synthetic siderophores (like **139**) characterized by a significantly larger Fe (III) sequestering capability have been obtained [47]. For the most effective of them, the stability constant of ca. 10^{59}, that is, seven orders of magnitude bigger than that of enterobactin, has been found.

Recent studies have shown that the mechanism of iron sequestering in bacteria is highly complicated, involving complexing of ferric ion Fe^{3+} by potent siderophores and the transport of such a complex through the cell membrane [48]. Synthetic siderophores like **139** are highly capable complexing agents. However, to our knowledge their transport through cell membrane has not been analysed yet.

Octadentate ligands with four catechol units like **140** exhibiting eight coordination sites are capable of effective complexation of not only iron ions, but also much larger Pu^{4+} [49].

REFERENCES

1. E. Fischer, Ber., 1894, 27, 2985.
2. D. E. Koshland, Jr., Angew. Chem. Int. Ed. Engl., 1994, 33, 2475.
3. W. H. Elliott, D. C. Elliott, *Biochemistry and Molecular Biology*, Oxford University Press, Oxford, 1997, p. 6; B. E. Bernstein, P. A. M. Michels, W. G. J. Hol, Nature, 1997, 385, 275.
4. H. Kuhn, J. Waser, Angew. Chem. Int. Ed. Engl., 1981, 20, 500; M. Eigen, Naturwiss., 1971, 10, 465.
5. H. Fraenkel-Conrat, R. C. Williams, Proc. Natl. Acad. Sci. USA, 1955, 41, 690.
6. (a) Strictly speaking viruses are not considered as living organisms by some scientists since they can reproduce only in another organism; (b) Until Wöhler's synthesis of urea living creatures were thought to be different from the rest of Nature owing to a vital force they contained.
7. J. S. Lindsey, New J. Chem., 1991, 15, 153.
8. J. D. Watson, F. H. C. Crick, Proc. Roy. Soc. Lond., 1954, A223, 80.
9. See, for instance, G. A. Jeffrey, W. Saenger, *Hydrogen Bonding in Biological Structures*, Springer, Berlin, 1991.
10. D. Pörschke, M. Eigen, J. Mol. Biol., 1971, 62, 361; M. E. Craig, D. M. Crothers, P. Doty, J. Mol. Biol., 1971, 62, 383; W. Saenger, *Principles of Nucleic Acid Structure*, Springer, New York, 1984, p. 116.
11. (a) H. Ringsdorf, B. Schlarb, J. Venzmer, Angew. Chem. Int. Ed. Engl., 1988, 27, 113; (b) J.-H. Fuhrhop, J. Köning, *Membranes and Molecular Assemblies: The Synkinetic Approach*, The Royal Society of Chemistry, Cambridge, 1994.
12. (a) M. Mammen, S. K. Choi, G. M. Whitesides, Angew. Chem. Int. Ed. Engl., 1998, 37, 2755; (b) P. Sears, C. H. Wong, Angew. Chem. Int. Ed. Engl., 1999, 38, 2301.
13. A. P. Davis, R. S. Wareham, Angew. Chem. Int. Ed. Engl., 1999, 38, 2978.
14. W. H. Elliott, D. C. Elliott, *Biochemistry and Molecular Biology*, Oxford University Press, Oxford, 1997, p. 188.
15. H. Dodziuk, *Modern Conformational Analysis. Elucidating Novel Exciting Molecular Structures*, VCH Publishers, New York, 1995, p. 239, 240.
16. S. P. Gupta, Chem. Rev., 1987, 87, 1183; *ibid.*, 1989, 89, 1765; Ref. 15, Section 12.1.1.
17. D. Philp, J. F. Stoddart, Angew. Chem. Int. Ed. Engl., 1996, 35, 1154.

18. D. A. Tomalia, A. M. Naylor, W. A. Goddard, III, Angew. Chem. Int. Ed. Engl., 1990, 29, 138.
19. L. E. Orgel, Nature, 1992, 358, 203; A. Eschenmoser, E. Loewenthal, Chem. Soc. Rev., 1992, 1; (b) S. Hoffmann, Angew. Chem. Int. Ed. Engl., 1992, 25, 1032.
20. (a) G. von Kiedrowski, Angew. Chem. Int. Ed. Engl., 1986, 25, 932; (b) T. Achilles, G. von Kiedrowski, Angew. Chem. Int. Ed. Engl., 1993, 32, 1198.
21. K. Morigaki, S. Dallavalle, P. Walde, S. Colonna, P. L. Luisi, J. Am. Chem. Soc., 1997, 119, 292 and references cited therein.
22. F. M. Menger, K. D. Gabrielson, Angew. Chem. Int. Ed. Engl., 1995, 34, 2091.
23. R. Maoz, S. Matlis, E. DiMasi, B. M. Ocko, J. Sagiv, Nature, 1996, 384, 150.
24. (a) T. Tjivikua, P. Ballester, J. Rebek, Jr., J. Am. Chem. Soc., 1990, 112, 1249; (b) J. Rebek, Jr., Angew. Chem. Int. Ed. Engl., 1990, 29, 245.
25. F. M. Menger, A. V. Eliseev, N. A. Khanjin, J. Am. Chem. Soc., 1994, 116, 3613.
26. (a) E. A. Wintner, J. Rebek, Acta Chem. Scand., 1996, 50, 469; (b) D. N. Reinhoudt, D. M. Rudkevich, F. De Jong, J. Am. Chem. Soc., 1996, 118, 6880.
27. G. Decher, *Dissertation*, University of Mainz, 1986; as cited by H. Ringsdorf, B. Schlarb, J. Venzmer, Angew. Chem. Int. Ed. Engl., 1988, 27, 113.
28. S. Muñoz, E. Abel, K. Wang, G. W. Gokel, Tetrahedron, 1995, 51, 423.
29. T. M. Fyles, T. D. James, K. C. Kaye, J. Am. Chem. Soc., 1993, 115, 12315.
30. J.-H. Fuhrhop, U. Liman, H. H. David, Angew. Chem. Int. Ed. Engl., 1985, 24, 339.
31. J.-H. Fuhrhop, U. Liman, V. Koesling, J. Am. Chem. Soc., 1988, 110, 6840.
32. J. Sunamoto, K. Iwamoto, Y. Mohri, T. Kominato, J. Am. Chem. Soc., 1982, 104, 5502.
33. M. R. Ghadiri, J. R. Granja, L. K. Buehler, Nature, 1994, 369, 301.
34. R. Breslow, S. D. Dong, Chem. Rev., 1998, 98, 1997; R. Breslow, Supramol. Chem., 1995, 6, 41; R. Breslow, Science, 1982, 218, 532.
35. (a) G. Trainor, R. Breslow, J. Am. Chem. Soc., 1981, 103, 154; R. Breslow, G. Trainor, A. Ueno, J. Am. Chem. Soc., 1983, 105, 2739; (b) M. J. Han, K. S. Yoo, J. Y. Chang, T.-K. Ha, Angew. Chem. Int. Ed. Engl., 2000, 39, 347.
36. W. H. Elliott, D. C. Elliott, *Biochemistry and Molecular Biology*, Oxford University Press, Oxford, 1997.
37. (a) J. Deisenhofer, O. Epp, K. Miki, R. Huber, H. Michel, J. Mol. Biol., 1984, 180, 385; Nature, 1985, 318, 618; (b) C.-H. Chang, D. M. Tiede, J. Tang, J. R. Norris, M. Schiffer, Febs Lett., 1986, 205, 82; (c) J. P. Allen, G. Feher, T. O. Yeats, H. Komiya, D. C. Rees, Proc. Natl. Acad. Sci. USA, 1987, 84, 6162.
38. J.-C. Chambron, S. Charbon-Noblat, A. Harriman, V. Heitz, J.-P. Sauvage, Pure Appl. Chem., 1993, 65, 435; J.-P. Sauvage, Acc. Chem. Res., 1993, 31, 611.

39. (a) M. R. Wasielewski, Chem. Rev., 1992, 92, 435; (b) D. H. Bush, N. W. Alcock, Chem. Rev., 1994, 94, 585.

40. J. P. Collman, L. Fu, Acc. Chem. Res., 1999, 32, 455.

41. (a) F. Hucho, Angew. Chem. Int. Ed. Engl., 1998, 37, 1518; (b) G. Steinberg-Yfrach, P. A. Liddel, S. C. Hung, A. L. Moore, D. Gust, T. A. Moore, Nature, 1997, 385, 239.

42. K. N. Raymond, G. Müller, B. F. Matzanke, Top. Curr. Chem., 1984, 123, 49.

43. (a) H. J. Rogers, Infect. Immun., 1973, 7, 445; (b) M. N. Schroth, J. G. Hancock, Science, 1982, 216, 1376.

44. J. S. Martinez, G. P. Zhang, P. D. Holt, H. T. Jung, C. J. Carrano, M. G. Haygood, A. Butler, Science, 2000, 287, 1245.

45. *Development of Iron Chelators for Clinical Use*, W. H. Anderson, M. C. Hiller, Eds., Bethesda, Maryland, 1975.

46. W. R. Harris, C. J. Carrano, K. N. Raymond, J. Am. Chem. Soc., 1979, 101, 2213.

47. W. Kiggen, F. Vögtle, Angew. Chem. Int. Ed. Engl., 1984, 23, 714.

48. A. D. Ferguson, E. Hofmann, J. W. Coulton, K. Diederichs, W. Welte, Science, 1998, 282, 2215.

49. (a) F. R. Weitl, K. N. Raymons, P. W. Durbin, J. Med. Chem., 1981, 24, 203; (b) D. Philp, J. F. Stoddart, Angew. Chem. Int. Ed. Engl., 1996, 35, 1154.

Chapter 6

ON THE BORDER BETWEEN CHEMISTRY AND TECHNOLOGY - NANOTECHNOLOGY AND OTHER INDUSTRIAL APPLICATIONS OF SUPRAMOLECULAR SYSTEMS

6.1 Introduction

As discussed in Chapter 1, the ultimate aim of supramolecular chemistry is not only to gain an understanding of Nature but also to build devices of practical importance on the basis of supramolecular aggregates. Developing crystals with predetermined desired properties has become a vivid branch of supramolecular chemistry crystal engineering. Another promising venue consists in developing devices built of one molecule or one molecular aggregate. One of numerous exciting ventures in this area is the miniaturization of computers aiming at a "cray of the size of paperback" [1] being an ultimate challenging goal. It requires development of electronic elements (logical gates and switches) as well as connecting 'wires' of a single molecule or a supermolecule [2] and nano-sized tools capable of manipulating them. Systems making use of the generation of second (or higher) harmonics form the basis of devices that, amongst others, should find application in much more effective computer memories [3]. Supramolecular sensors [4] will undoubtedly find diverse applications, in particular in environmental protection. Systems superconducting at room temperatures allowing one to minimize energy losses by electricity transfer are at present another aim of intensive research [5]. Harvesting light energy and mimicking the process of photosynthesis [6] should also lead to more effective and environmentally friendly energy generation and use. Rational design of new drugs or new ways of drug administration is yet another target that has experienced an enormous boost thanks to the development of supramolecular chemistry. It should be stressed that at present there are very few industrial applications of supramolecular systems. Liquid crystalline displays which may soon come out of the use [7] and cyclodextrin (like **13**) complexes in pharmaceutical and food industry, in cleaning and cosmetics are those few bringing profits today [8]. Therefore most of the contents of this chapter is devoted to the presentation of ideas. They promise a lot, however, as is illustrated by the failure of numerous proposed fullerene applications discussed in Section 7.5, not all of them will be realized. It should be stressed that the prospective applications of supramolecular aggregates are numerous and diverse. Therefore, only a few of their representative examples could be presented here.

6.2 Between Chemistry and Solid State Physics - Crystal Engineering. Obtaining Crystals With Desired Properties

Obtaining crystals with desired properties, that is crystal engineering, is a formidable task [9, 10]. The anticipated turn from inorganic to organic materials for electronics, optoelectronics, information storage and other applications is associated with the ability of fine-tuning properties of organic molecules on the one hand, and with the development of theoretical methods which allow one to reliably predict them, on the other. The possibility of building crystals from two (or more) different types of molecules expands the ability to manipulate their properties. Although a model of packing of molecules in a crystal was proposed by Kitajgorodsky long ago [11], at present we are, in general, not able to predict the crystal structure formed by one type of molecules let alone the structure of a cocrystal formed by at least two types of molecules. Intermolecular forces (charge-transfer, hydrogen bonding, electrostatic and van der Waals ones [12]) are responsible for both crystallization and complex formation. Desiraju [9] states that "the interactions that bind molecules in crystals are identical to those responsible for molecular assembly in solution". He considers the crystallization process as an ultimate example of molecular recognition. Desiraju also states that "the solid state supermolecule is not only of inherent interest but also a model for

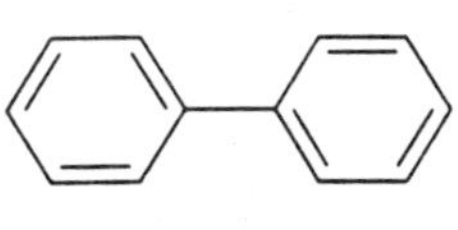

the less precise, 'looser', solution supermolecule, which can be studied only indirectly with NMR and other spectroscopic methods." We believe that such statements represent an oversimplification since: (1) the solvent plays an important role in the complexation in solution; and (2) the binding forces of many inclusion complexes are of the order of magnitude of crystal forces. In crystal, these forces 'disturb' the structure of the loosely bound solution aggregates. Thus, in a certain respect, some supramolecular complexes resemble the famous biphenyl **141** which is twisted in solution [13] and planar in the crystalline state [14]. In close similarity, different modes of entrance of the guest have been found for the complex of *para*-nitrophenol **74** with hexakis(O-methyl)-α-cyclodextrin **75** in solution by NMR technique [15] and in the solid state by X-ray analysis [16]. Moreover, owing to the time and positional averaging, X-ray structure determinations strongly underestimate internal mobility of the complexes [17- 19]. Thus the information obtained from X-ray analysis of the crystals of inclusion complexes is of great value. However, its application for the elucidation of the structure of the complexes in solutions may be limited.

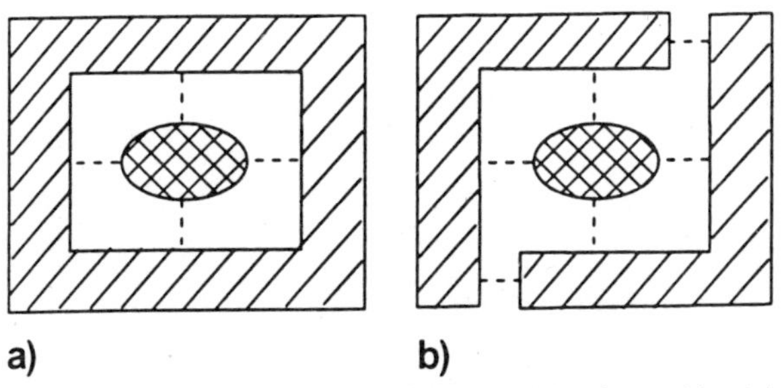

Figure 6.1. Schematic view of host cavities formed by (a) one and (b) two molecules.

Contrary to the Desiraju claim that complexes in solutions can be adequately modelled by those in the solid state, there is an essential difference between them, namely, in the former the complexes are usually in equilibrium with their free constituent parts. Their smaller rigidity reflected in the bigger mobility in solution [17-19] can facilitate, at least in some cases, better mutual adjustment of the host and guest. On the other hand, as has been noticed by Pedersen in his early study of crown ether complexes (discussed in Chapter 3), some complexes present in crystals (in which solvent molecules often participate) can be very weak or even

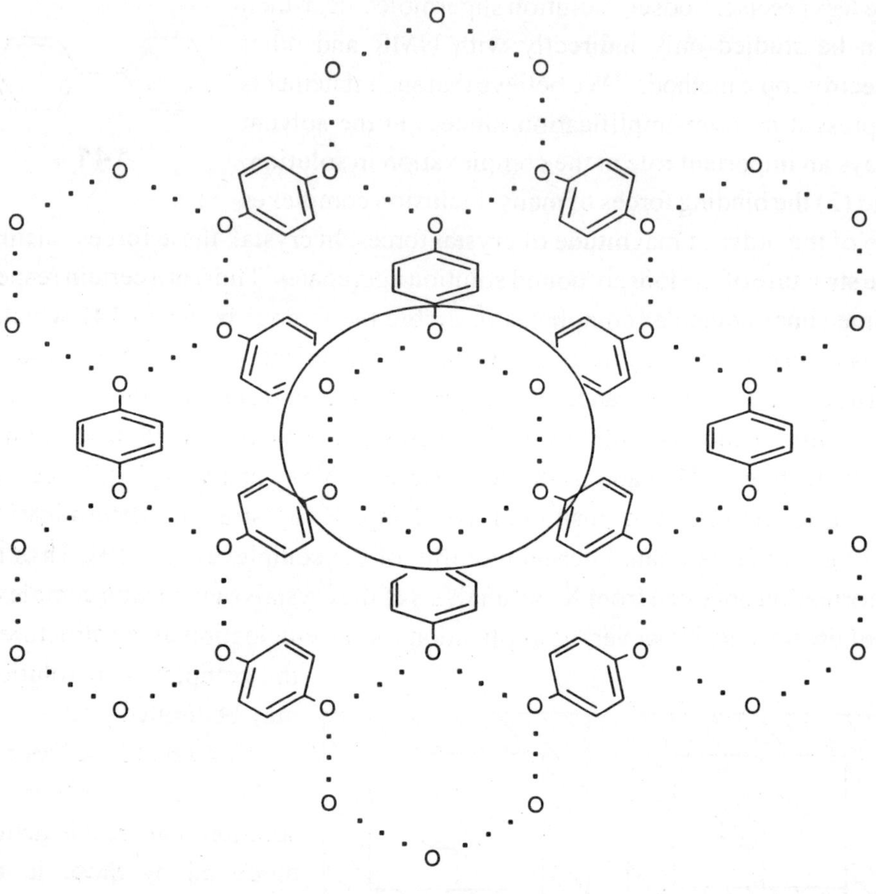

142

nonexistent in solution. Thus the Desiraju claim that crystallography is the most reliable source of information on the structure of inclusion complexes not only in the solid state but also in solutions is not always valid.

Desiraju distinguishes two types of inclusion complexes in the solid state. In the first type (Fig. 6.1) the guest molecule is embedded inside the host cavity. In the second two or more different molecules form the cavity to include another one. It should be stressed that the decision about which is the host and which is the guest is not always obvious. For instance, the complex of three hydroquir.one molecules **142** with C_{60} [20] reflects this ambiguity since the molecular mass of

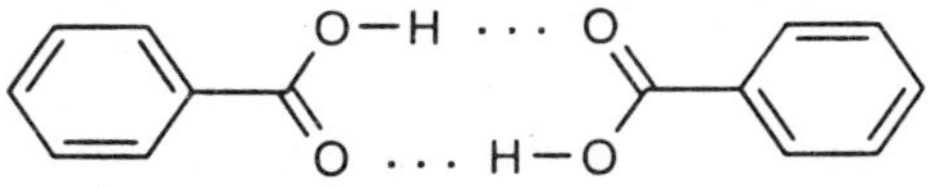

143

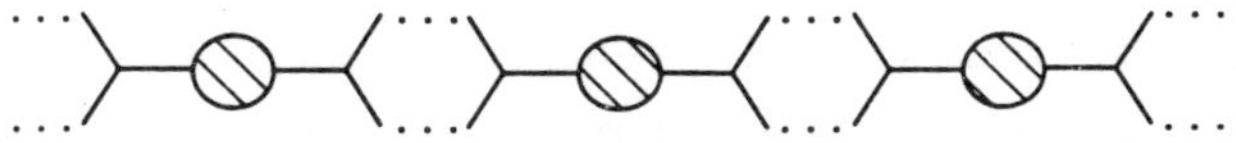

144a

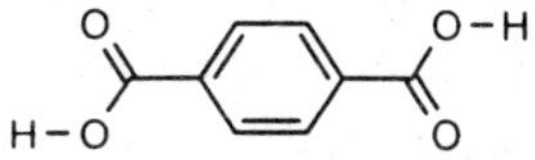

144b

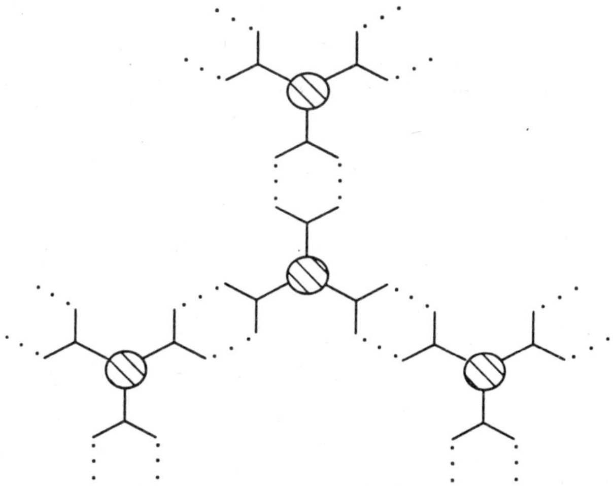

144c

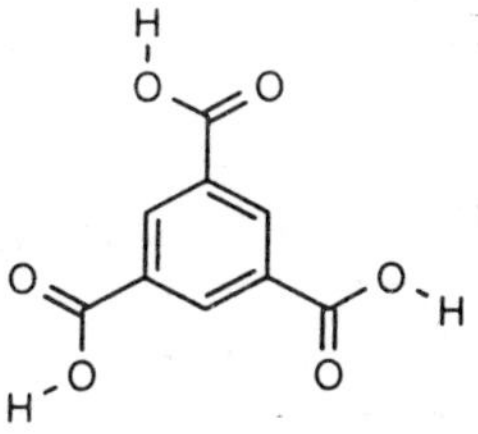

144d

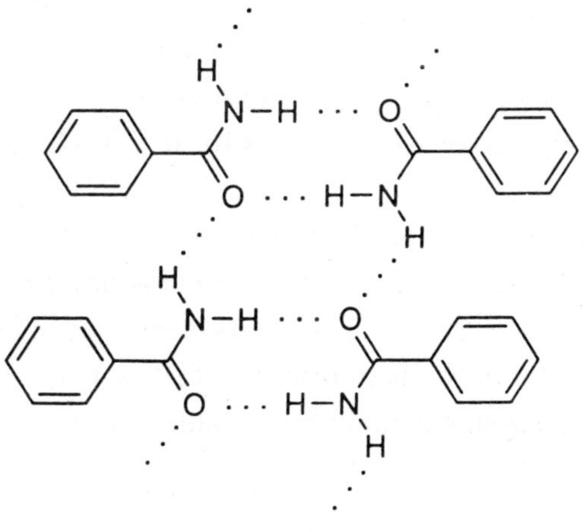

145

the guest here is more than twice as large as that of three molecules forming the host cavity.

As mentioned before, intermolecular interactions force complex formation and/or crystallization. The directionality and dependence on the distance are the most important characteristics of some of these interactions from the point of view of crystal engineering. The statistical analysis of numerous crystal data collected in the Cambridge Data Base [21, 22] (over 220 000 structures as of March 2001) enables one to study the influence of weak intermolecular interactions on molecular structure. Typical building blocks forming characteristic 'motifs' or 'patterns' are then recognized. Subsequently, by analogy with synthetic methods of organic chemistry one can (at least theoretically) obtain supramolecular aggregates and/or crystals with foreseeable spatial structures and desired properties.

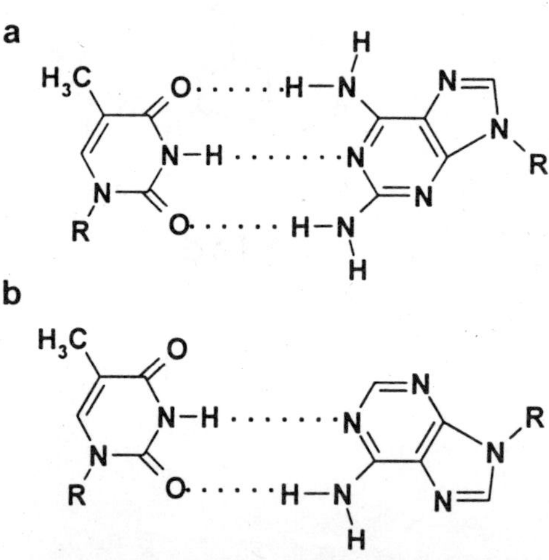

Figure 6.2. Watson-Crick (a) and reverse Watson-Crick (b) types of hydrogen bonding.

Hydrogen bonding is one of the strongest nonbonded interactions manifesting pronounced directional properties [10]. As such it is mostly used in crystal engineering. Let us look at some frequently appearing H-bond patterns in crystals. (1) Carboxylic acids are known to form very strong dimers **143** persisting even in the gas phase [23]. More complicated ribbon **144a** and planar **144c** motifs have been obtained for aromatic di- and tri- carboxylic acids **144b, d** [23]. (2) Ribbon structures are characteristic for primary amides **145** [23]. (3) Watson-Crick H-bond base pairing [24] of adenine and thymine (or uracil) **111**, **112**, **113** and that of guanine and cytosine **114**, **115** are typical of numerous crystal structures [25] (Fig. 6.2). (As mentioned in Section 5.2.2, these bonds are responsible for the helical structure of DNA.) (4) Exciting diversity of hydrogen-bonded aggregates formed by melamine **1** and cyanuric acid **2** are

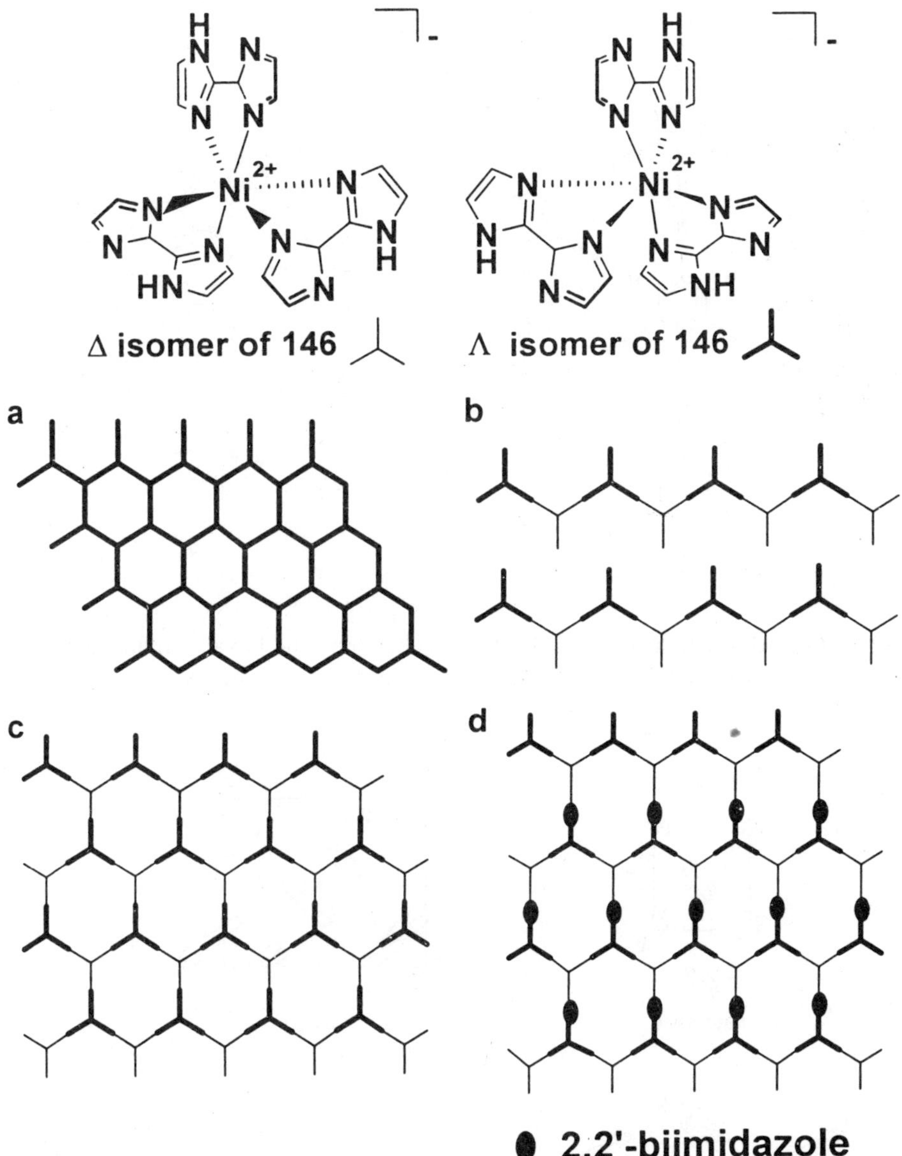

Figure 6.3. Various planar arrangements of Δ and Λ isomers of **146** that may involve (d) 2,2'-biimidazole.

discussed in Chapter 1 and Section 8.3. Some of them were found to crystallize in well-defined structural motifs [26]. One-pot cation-dependent formation of different hydrogen-bonded networks from nickel Δ and Λ complexes of 2,2'-biimidazole **146** shown in Fig. 6.3 was reported by Tadokoro and coworkers

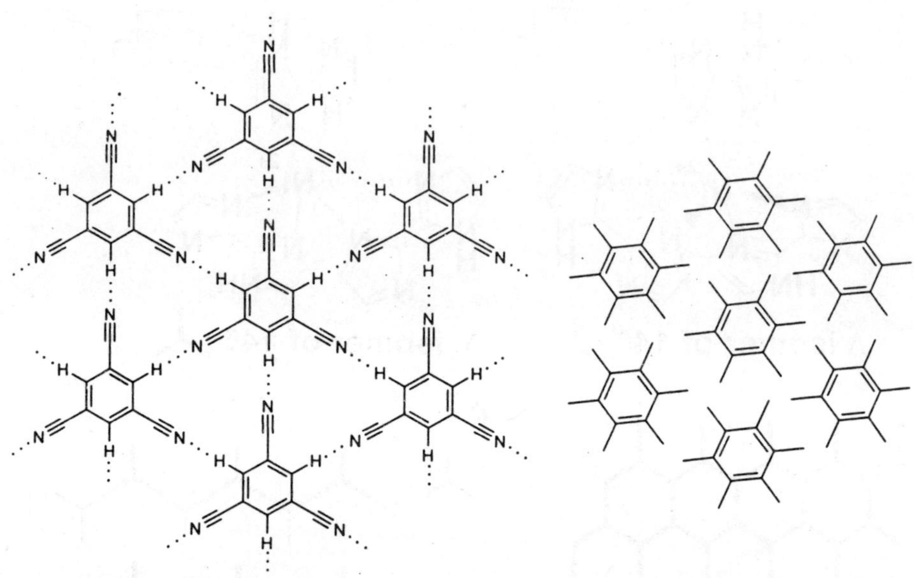

Figure 6.4. Alternate layers of 1,3,5-tricyanobenzene and hexamethylbenzene are stacked in the crystal structure of the 1:1 complex.

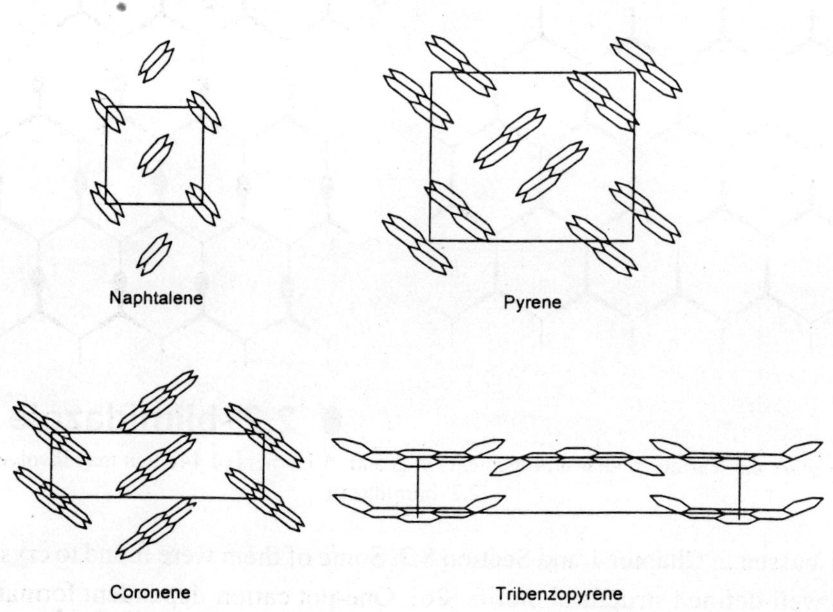

*Figure 6.5. T*he four basic aromatic crystal packings.

[27]. It should be stressed that all these H-bonds are so effective since these systems exhibit an especially favourable match of the donor and acceptor groups as concerns both their spatial arrangements and electrostatic potential fields. Multiple recognition in the latter systems allows for highly effective collective hydrogen bond formation. The above examples encompassed mostly cases of strong OH...O , NH...O, and NH...N hydrogen bonds. The existence of much weaker hydrogen bonds involving CH protons and/or π-electron systems have been proposed [28]. A hexagonal network involving relatively weak CH...N hydrogen bonds was reported by Desiraju group [29] as one example of such systems. The 1:1 crystal was built from the parallel layers of 1,3,5-tricyanobenzene and hexamethylbenzene (Fig.6.4). However, not only hydrogen bonding but also strong charge transfer interactions could be the driving force for the crystal formation in the latter case.

Wolf and coworkers [30] have shown that a combination of hydrogen bonding capability with symmetry requirements can enforce similar crystalline arrangements yielding the pattern analogous to that formed by unsubstituted melamine **1** and cyanuric acid **2** (Chapter 1) independent of the steric requirements of the substituents. The pattern can be called *molecular fabric*. Mascal and coworkers [31] have used this term for their more loose structure **147** exhibiting voids.

Analysis of the continuous donor-acceptor stacks in the solid state such as those depicted in Fig. 6.4 [32] in the solid state inspired Stoddart to use stacking interactions of aromatic rings as the factor forcing preorganization in the syntheses of topological molecules discussed in Sections 2.3 and 8.1. These stacking interactions are especially effective between π-electron acceptors and π-electron donors. However, in addition to electrostatic attractions, such interactions also involve van der Waals C...C dispersive attractions [33] between the atoms of neighbouring rings. On the basis of the statistical analysis of the data collected in Cambridge Data Base, Desiraju and Gavezzotti found that fused aromatic molecules crystallize in four basic crystal packing patterns (Fig. 6.5) and proposed empirical rules allowing one to predict the packing mode of a molecule under study [34].

Metal coordination, a powerful organizing factor allowing one to design, among others, cage-like or cylindrical nanoarchitectures will be discussed in Section 8.5 [35].

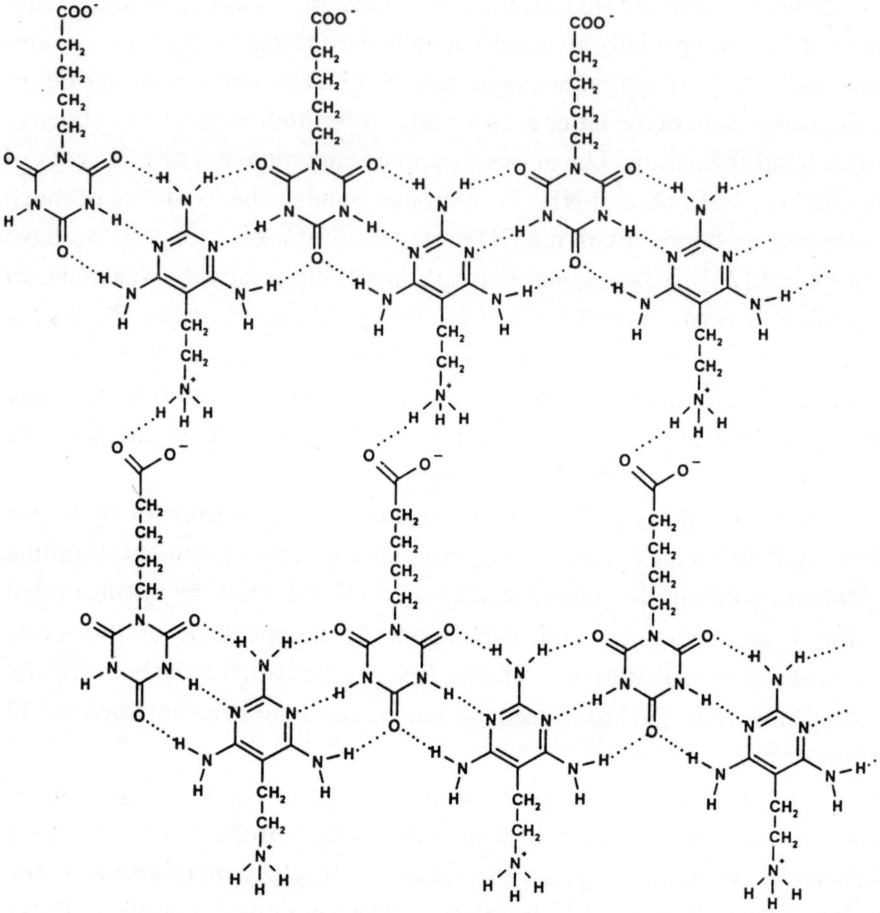

147

As discussed in detail in Ref. 36, for use in optoelectronics only systems crystallizing in non-centrosymmetric crystal lattices are of interest if the use of expensive enantiomers of chiral molecules is to be avoided. This considerably limits the available crystal lattices since most organic achiral molecules crystallize into centrosymmetric space groups. An interesting example of enantioselective inclusion complexation was reported by Gdaniec and coworkers [37].

6.3 Nanotechnology and Other Industrial Applications of Supramolecular Systems

The very trendy term nanotechnology is somewhat misleading since it is used with at least three different meanings. The first of them refers to small, metal clusters measuring no more than 100 nm which could find practical applications. This domain falls outside the scope of the present book but an amazing production of gold nanoparticles by bacteria *Pseudomonas stutzeri* AG259 [38] can be mentioned here. The second field also dubbed nanotechnology is based on a concept first proposed by Nobel laureate Richard Feynman in 1959 [39]. It deals with hypothetical devices, so-called Drexler machines [40], which can replicate themselves [41]. There are predictions stating that around the year 2030 such machines could be "biodegradable medical devices that are small enough to fit inside a cell. There is no limit to the molecular or structural defects that can be repaired nor viruses or bacteria that can be destroyed." by them. Another vision speaks about "tiny robots which may be swallowed or injected into your bloodstream, where they'd race around inside arteries and veins, cleaning out plaque, chewing up tumors, perhaps repairing faulty blood vessels." [42]. Even if the last statements promise too much too fast, as the author of this book thinks, the prospects of nano-sized molecular machines are bright. Supramolecular systems (discussed below) which, responding to photo or electric stimuli, can reversibly execute large amplitude motion represent the first step towards molecular machinery. The last meaning of nanotechnology, most closely related to supramolecular chemistry, refers to technological applications of molecular aggregates of nanometer size.

In spite of great efforts focussed on the present and prospective applications of supramolecular aggregates, no comprehensive account of this field can be found since this is a huge, rapidly developing domain. Moreover, last but not least, the publications are hampered by companies propriety interests. DuPont is mentioned in the following citation illustrating this point but such an information policy is obviously not restricted to this company: "....if DuPont is talking about it, it has absolutely no commercial interest in it whatsoever. If there were any suggestion that those compounds were interesting, you would never see a word about them in print" [43]. On the other hand, researchers abundantly propose new applications in efforts to attract funding. This could probably best be seen

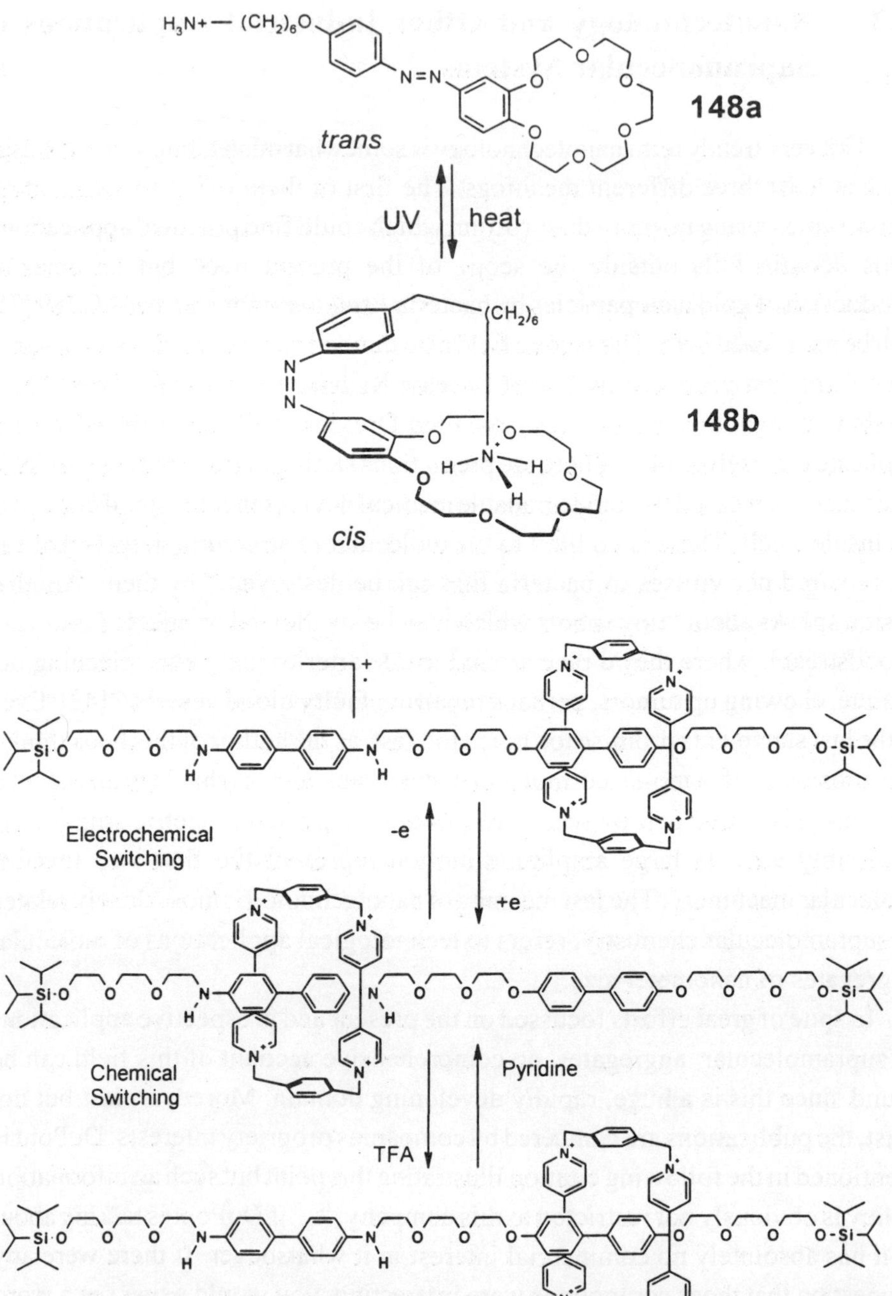

Figure 6.6. Various switching factors in **148** and **149** sensor molecules.

at early stages of fullerene studies [44]. A flood of patents was filed and proposed applications abounded after the Krätchmar group paper described C_{60} purification that allowed them to obtain gram quantities of the compound [45]. Several, sometimes fantastically sounding prospective fullerene applications have been described (see also Section 7.5) including a fullerene cage possessing an opening with a door which should enable controlled release of a drug [44]. Some other apparently more down to earth applications proved at least up to date unrealizable. Until now this has been the case of high-temperature superconductivity [46] and the application of $C_{60}F_{60}$ as a lubricant [47] that are briefly discussed in Section 7.5. Keeping in mind that very few supramolecular systems have really made their way to industry, let us look at their few present and numerous prospective applications.

6.3.1 Molecules in motion: towards machines and motors consisting of a single molecule or molecular aggregate [48]

According to Sauvage [48a] "...compounds containing interlocking rings or rings threaded onto an acyclic fragment are the ideal precursors to molecular machines, i. e. multicomponent systems for which selected parts can be set in motion while the other fragments are motionless." Several such systems possessing photo-, thermo- **148** [49a], chemo- or electroactive groups **149** [49b] (Fig. 6.6) can perform large amplitude motions. Metal complexation inducing rotaxane formation **150** and **151** also causes large amplitude molecular motion consisting in threading and dethreading of one or two rings on a molecular string [50]. Analogous

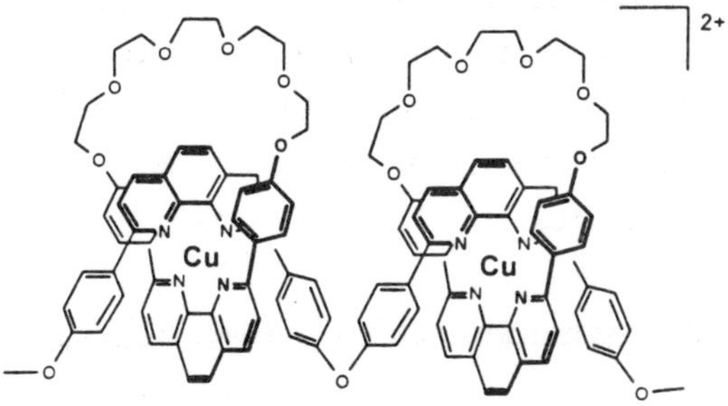

150

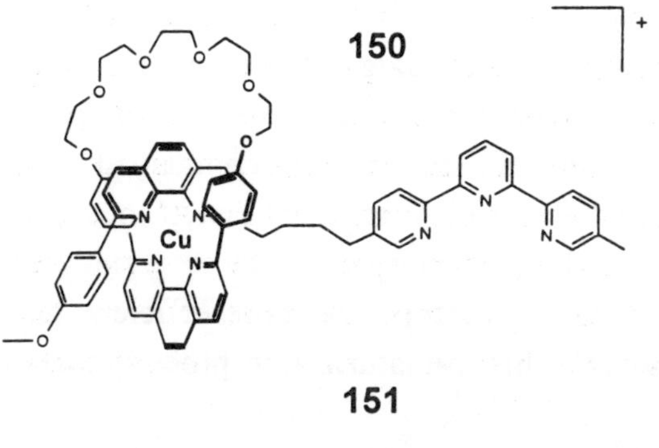

151

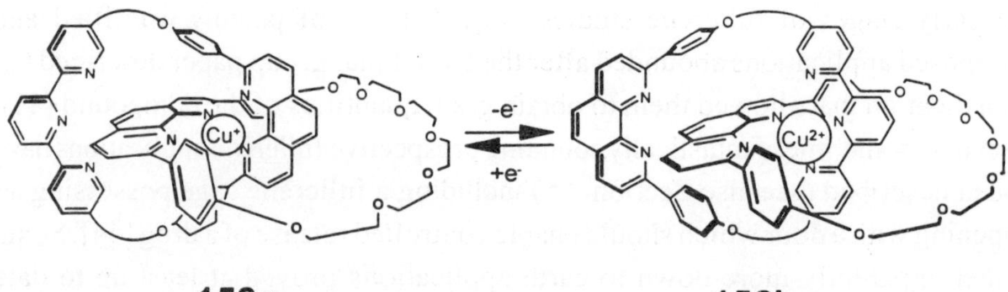

152a 152b

Figure 6.7. Redox controlled rearrangement in metallocatenanes.

electrochemically induced molecular rearrangements take place in catenane **152** (Fig. 6.7). In all examples presented in this section the shape of the molecule and its internal motion can be externally controlled by electrochemical, photochemical or other means. For instance, in **149** the threaded ring is a tetracationic electron deficient cyclophane which can interact with two electron-donating groups situated on the central rotaxane string. Of the two groups, benzidine and biphenol, the former is the stronger electron donor. Thus in the initial state over 80% of the macrocycles occupy this position. However, the electrochemical oxidation of the benzidine group forces the macrocycle into the second position. The reversibility of this process can be achieved by reducing the benzidine group back to its initial state. Response of the systems shown in Fig. 6.6 and 6.7 can be applied not only in molecular machines but also in sensors and logical elements that can be used in memory devices.

6.3.2 Electronics on the basis of organic molecules or their aggregates chemionics

6.3.2.1 The need for miniaturization of electronic devices

Faster computers with bigger memory capabilities and smaller size and energy consumption are a must for further technological development. More complicated tasks can be handled with computer speed and memories doubling every 3 years and an increase in computers efficiency being accompanied by the shrinking of their sizes. One of the first electronic computers ENIAC occupied several rooms and weighted 30 tons [51], its counterparts in the seventies were of the size of a wardrobe while todays palmtops are more efficient than mainframes of the 1980-ties. However, this miniaturization process cannot

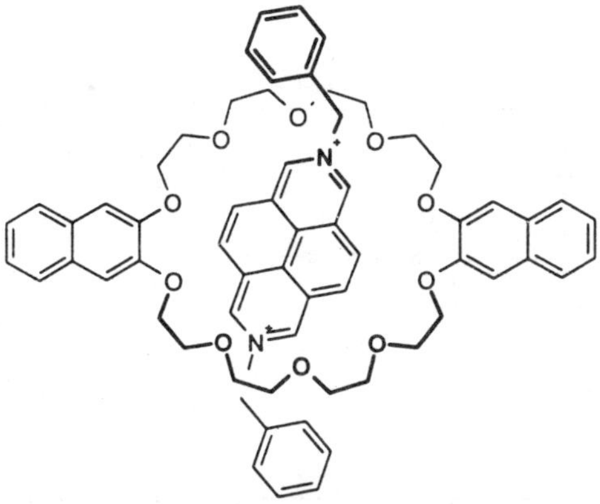

153

proceed much further since, for instance, the walls between memory cells will soon become too thin to protect their content. Therefore a completely different approach 'from the bottom up' has been proposed to obtain much smaller electronic devices with single-molecule or single-supermolecule devices as the ultimate goal. DNA and RNA computers which lie outside the scope of this book, can serve as a good illustration of the concept [52]. Enormous technological challenges must be overcome to realize it. Molecules and supramolecular systems acting as wires, switches and conductors, have been reported. However, as illustrated by the example of a single carbon single-walled nanotube field-effect transistor [53], its incorporation into a device involves other elements obtained using current microtechnology, resulting in a device which offers no gain in size in comparison with current silicon transistors.

The schemes presenting the logical operation of AND, OR and XOR (eXclusive OR) gates are shown in Fig. 6.8. Balzani and coworkers [54] proposed a pseudorotaxane **153** which can reversibly decompose (i. e., unthread) by chemical stimuli thus modeling the XOR logic gate. The complex **153** composed of a macrocyclic electron donor 2,3-dinaphtho-30-crown-10, and an elongated electron acceptor 2,7-dibenzyldiazapyrenium dication, unthreads upon addition of stoichiometric amounts of acids or amines. Rethreading can then be induced by addition of amine. Both processes are accompanied by strong changes in the fluorescence properties of the system enabling monitoring and modeling of the gate.

6.3.2.2 (Supra)molecular wires, conductors, semi- and superconductors, and so forth

According to Lehn [2] "The processing of molecular information via molecular recognition events implies a passage from the molecular to the supramolecular level. By endowing photo-, electron-, or iono-active components

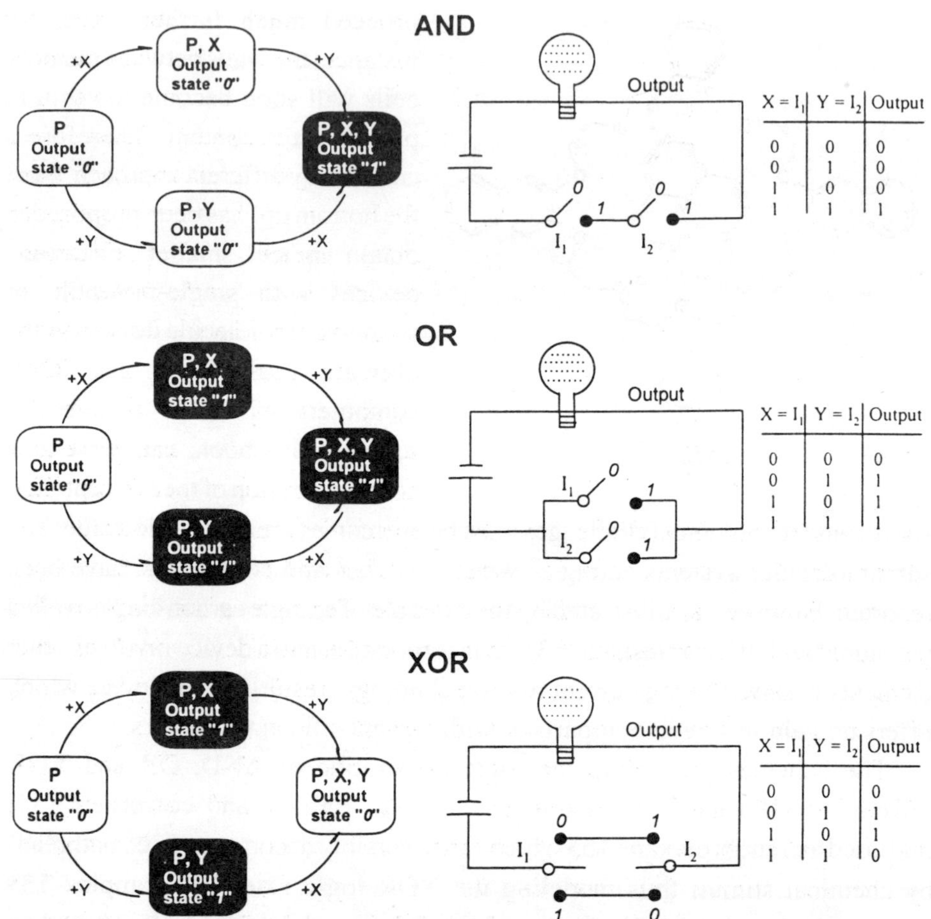

Figure 6.8. Schemes of .AND., .OR. And .XOR. logical gates.

with recognition features, one may be able to design programmed molecular systems that undergo self-assembly into organized and functional (photonic, electronic, ionic) supramolecular devices, and these are in turn likely to reach nanometer dimensions. Thus the chemistry of supramolecular devices, chemionics, is the chemistry of molecular recognition-directed, self-organized, and functional entities of supramolecular nature" [2]. Several exciting systems which could act as molecular wires, conductors, semiconductors, etc. have been proposed. A few of them will be briefly presented below, but basic technological problems involving understanding of the operation of mesoscopic systems, their reproducible production, and the assembly of them into electronic devices are still enormous. Obstacles are huge and very few such systems have been presented in the literature to date. One type of these devices is represented by a room-temperature transistor based on a single carbon nanotube and a kinked single-walled nanotube acting like a rectifying diode [53, 55]. On the basis of molecular modeling [56], the kink is believed to be a defect in the nanotube consisting of the presence of a pentagon and a heptagon on opposite sides of the standard hexagon carbon lattice. As briefly discussed earlier, these devices serve as a good illustration of the problems encountered by miniaturization of electronic devices since they are of the size of a conventional chip. Another type of supramolecular devices is represented by multilayer Langmuir-Blødgett organic rectifiers [57] and monolayer Langmuir-Blødgett photodiode [58].

One of the most promising types of organic metals and superconductors to be used eventually as wires is based on bis(ethylenedithio)tetrathiafulvalene **154** complexes with 7,7,8,8-tetracyanoquinodimethane **155** [59]. Molecular wires have been proposed by Lehn [2] as elements connecting parts of electronic circuitry enabling electron flow between them. An interesting attempt to create insulated molecular wires was reported by the Ito group [60]. The authors have observed that at low temperatures (below 275 K) polyaniline **156** in solution with β-cyclodextrin **11** changes its conformation from coil to rod. The finding was interpreted in terms of formation of polyrotaxane complex with cyclodextrins that form an insulating cover on the rod. Lehn and coworkers inserted viologen molecules acting as wires into micelle walls mimicking charge transfer through the membrane [2] whilst Schoorbeck and coworkers [61] observed efficient intermolecular charge transport in self-assembled fibers of mono- and bithiophene bisurea derivatives. A formation of a metallic wire of single gold atoms was reported by Yanson et al. [62] while DNA-templated assembly and electrode attachment of a thicker conducting silver wire was described by Braun and coworkers [63a]. Carbon nanotubes which have also been proposed to be used as wires are briefly discussed in Sections 7.5. Yet another type of molecule which could be applied as nanowires was reported by Simon [63b]. The system consisting of columnar liquid crystals formed

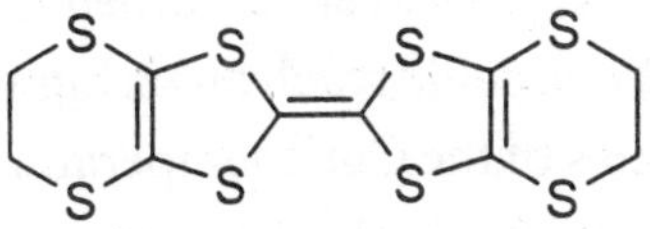

S S S S S S S S

154

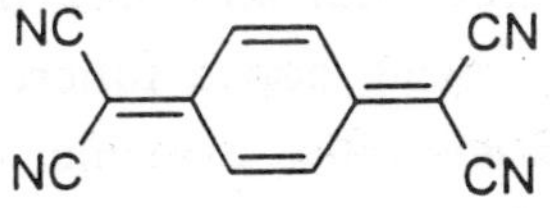

NC CN NC CN

155

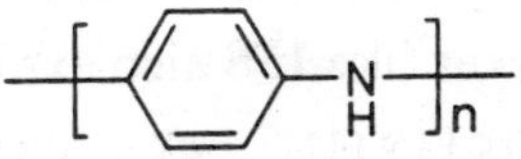

N H n

156

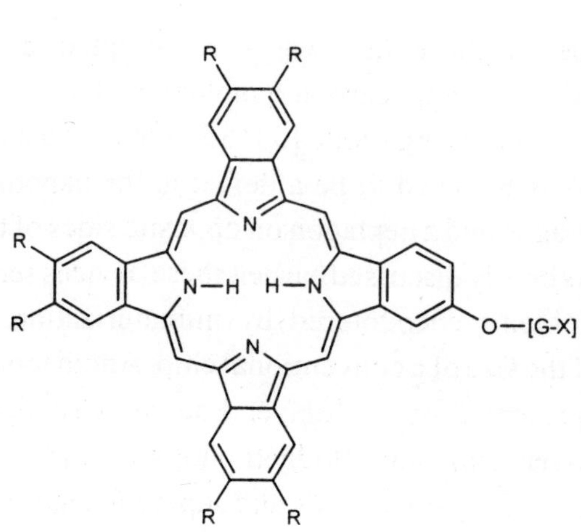

X = 1, R = $O(CH_2CH_2O)_2Me$
X = 2, R = $O(CH_2CH2O)_2Me$
X = 3, R = $O(CH_2CH_2O)_2Me$

157

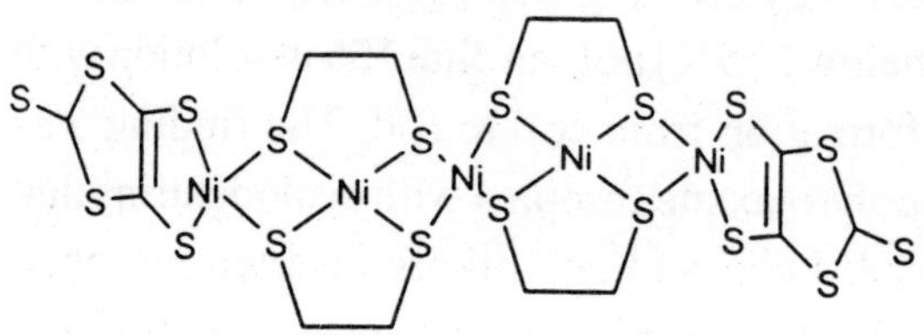

158

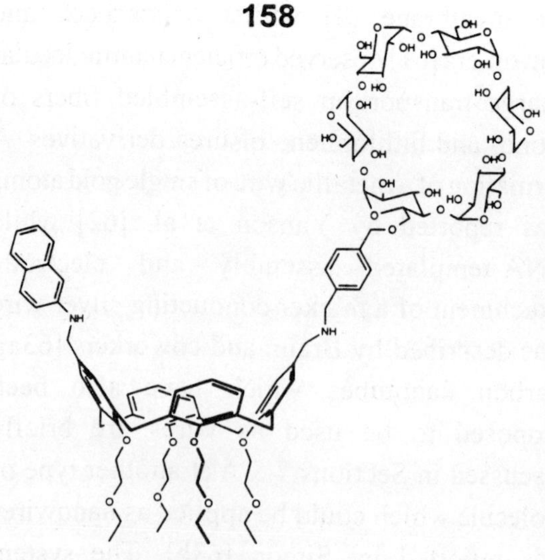

159

from heptaalkyl substituted phthalocyanines **157** exhibited one-dimensional energy migration and charge transport. Thin organic magnetic films developed by J. Miller and coworkers exhibit electrical properties ranging from insulating to semiconducting and may be flexible and transparent [64a]. The properties of such materials may be studied by the computer modelling which is widely used in rational drug design [64b]. A detection in the early 1990s that certain doped fullerenes are superconducting at low temperatures was met with euphoria, since the lack of resistance in such a conductor would significantly cut energy losses. However, it turned out that one could reach much higher values of the Curie temperature, that is, the highest temperature at which the superconductivity takes place, with other materials than by using doped fullerene [65]. Some other examples of unfulfilled fullerene promises are presented in Sect. 7.5.

Polynuclear nickel complexes like **158** also exhibit conductivity or even superconductivity [66]. Such compounds have been found also to be potentially important for the assembly of molecular ferromagnets [67].

6.3.2.3 Sensors and switches

A change in, mainly photochemical or redox, properties upon supramolecular aggregate formation constitutes the basis of operation of sensors or switches. Thus some examples of aggregates which may be applied in such systems were presented earlier in this Chapter.

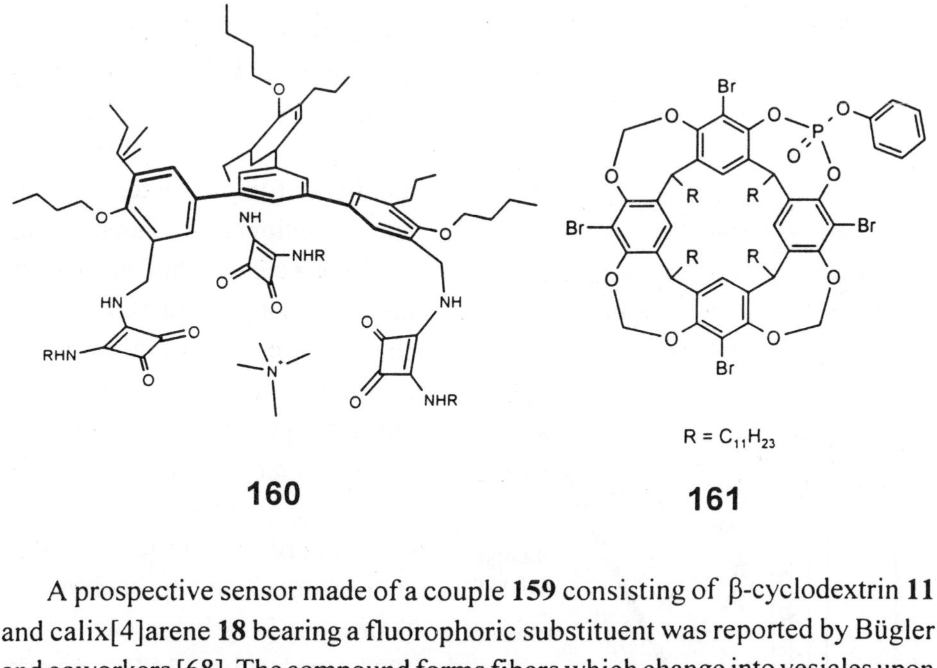

NH NHR HN O O RHN O O NH O NHR O N

160

Br O O P O O O Br R R Br R R O O O O Br

$R = C_{11}H_{23}$

161

A prospective sensor made of a couple **159** consisting of β-cyclodextrin **11** and calix[4]arene **18** bearing a fluorophoric substituent was reported by Bügler and coworkers [68]. The compound forms fibers which change into vesicles upon complexation, forcing the fluorophore out of the cyclodextrin cavity. As a consequence, the intensity of fluorescence is reduced. In another approach to molecular sensors Dermody and coworkers [69] applied β-cyclodextrin-functionalyzed polymeric films to enhance the selectivity of molecular sensors. A fluorescent sensor for barbiturates presented in Fig.6.9 makes use of the selective complexation of these guests due to hydrogen bonds formation [70]. Costa et al. [71] synthesized a molecule **160** acting as a fluorescence sensor for

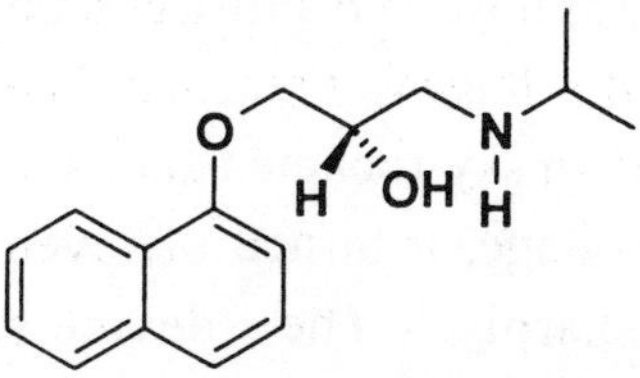

O N H OH H

162

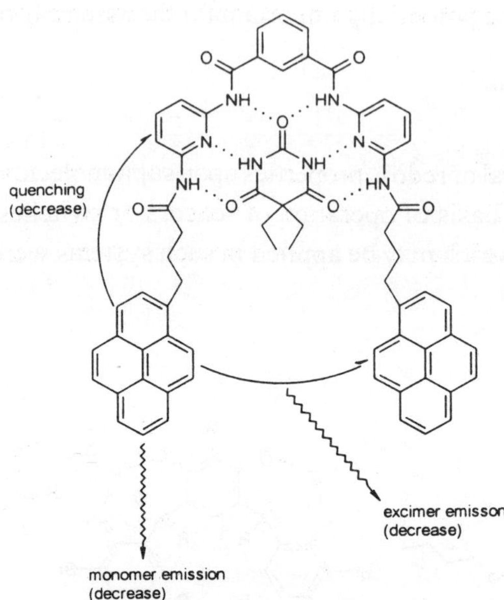

Figure 6.9. The mode of operation of a barbiturate sensor.

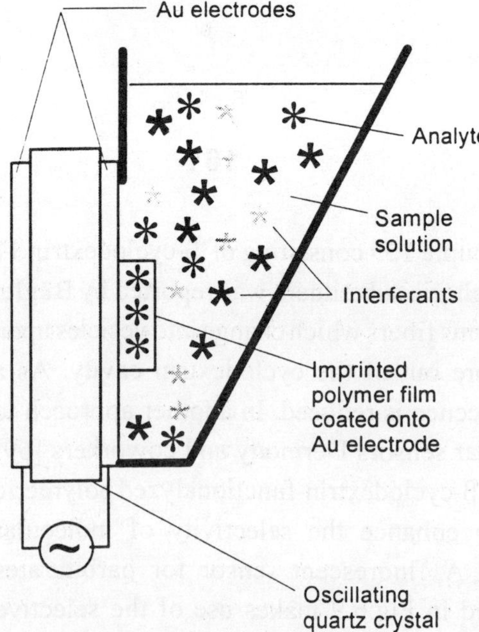

Figure 6.10. The operation mode of microgravimetric detection of S-propanolol using S-imprinted polymer.

choline-containing phospholipids. The selectivity of this sensor is also based on hydrogen bonding interactions. A supramolecular sensor for the detection of alcohols on the basis of cavitands **161** was developed by Dalcanale with coworkers [72].

An enantio-selective sensor for recognizing β-blocking drug S-propranolol **162** was proposed by Haupt and coworkers [73]. The sensing in this device is executed by a polymer film imprinted with the latter enantiomer. As shown in Fig. 6.10, the weight of the film glued to the oscillating quartz crystal changes as a result of the selective complexation of the S-enantiomer of the drug enabling its detection.

A supramolecular electronic switching device with quite nonstandard properties at low temperatures was reported by the Tour group [74]. It consisted of ca. 1000 molecules **163** spanned between two gold electrodes. When a steadily increasing voltage was applied to this system at 60 K no current was observed until a certain threshold voltage was reached. The current then sky-rocketed and, at a higher voltage, it turned off even more sharply. The device's characteristic peak-to-valley ratio (PVR), that is the ratio of the peak 'on' current to the 'off' (leakage) current, was greater than 1000:1, while in standard semiconductor devices based on gallium arsenide the PVR is usually less than 100:1.

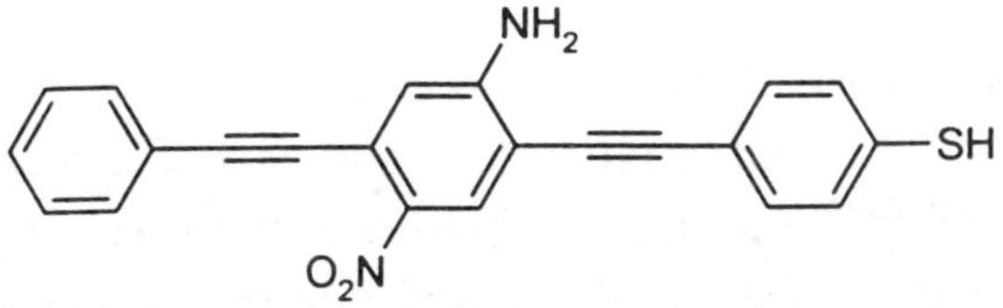

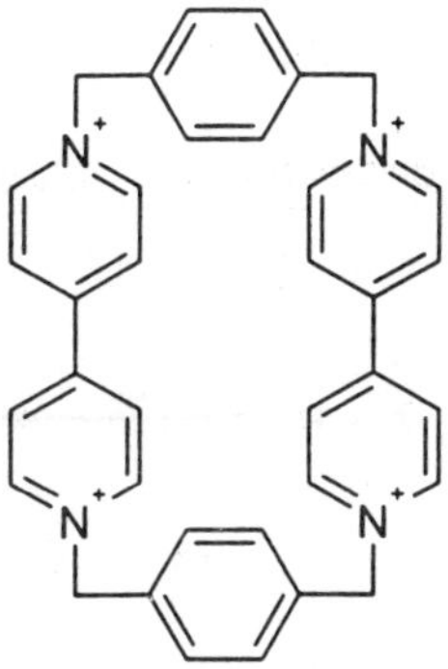

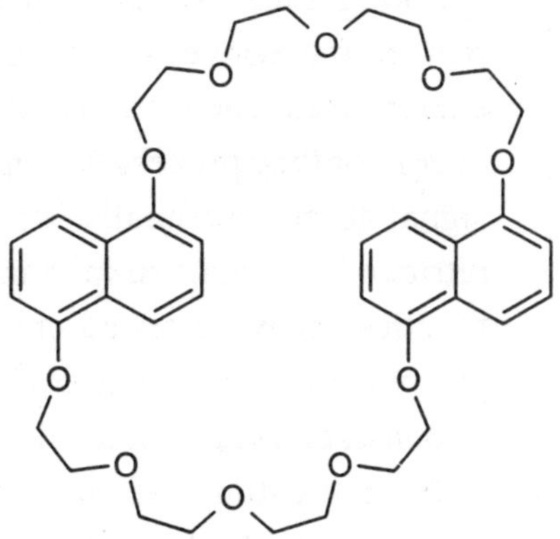

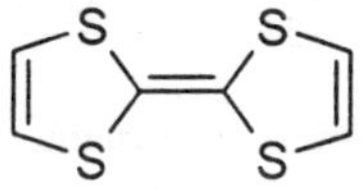

A more complex behaviour exhibits a three-pole supramolecular switch proposed by Ashton and coworkers [75]. The system consists of two macrocycles (the π-accepting tetracationic cyclophane cyclobis(paraquat-*p*-phenylene) **164** and the π-donating polyether 1,5-dinaphtho-38-crown-10 **165**) which can host tetrathiafulvalene **166** guest capable of existence in three stable forms: neutral TTF(0), cation radical $TTF^{+\cdot}$ and dication one TTF^{2+}. Electrochemical experiments showed that depending on the potential range TTF can be: (1) free in the $TTF^{+\cdot}$, (2) complexed with **164** as TTF(0); or (3) included within the cavity of **165** in the TTF^{2+} state. Such a behaviour could provide interesting opportunities for the design of electrochromic displays and other devices.

6.3.2.4 Photochemical devices

Light consists of photons, which can be considered as both energy quanta and information bits, and photochemical processes in Nature perform the two major functions of photosynthesis and vision [76,6]. More generally speaking, these processes either take part in the conversion of the solar energy into chemical or electrical energy or they process the input light signal to be eventually stored and retrieved. Systems mimicking these functions may be called energy and information processing devices. Let us consider a supramolecular system A~B consisting of two weakly interacting parts A and B (we do not agree with the opinion of V. Balzani and F. Scandola that a covalently linked but weakly interacting system may also be called supramolecular, as stated in Ref. 6). Its excitation by light results either in excited states localized on A or B, causes an electron transfer between A and B or leads to the excited states considerably delocalized on both A and B. Analogously, oxidation and reduction in a supramolecular system can be described either as oxidation or reduction of specific components or of the whole species (Fig. 6.11). The following equations describe the most important localized processes of photo excitation, energy and electron

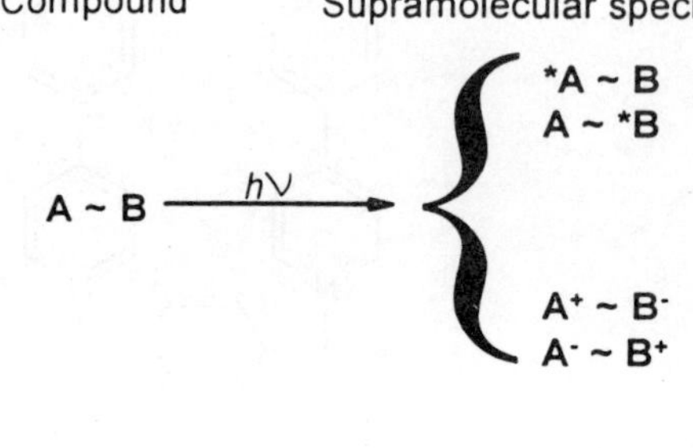

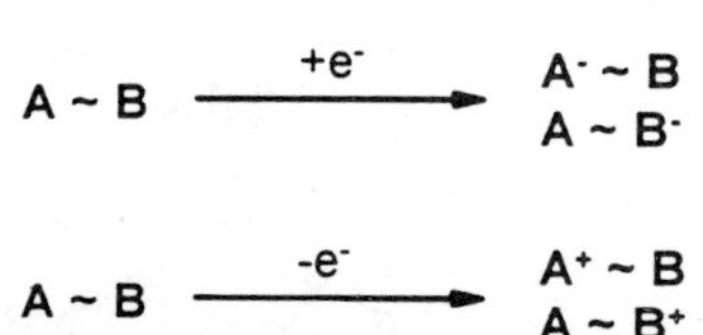

Figure 6.11. Scheme of light excitation (top), oxidation (middle) and reduction (bottom) in supramolecular system A~B.

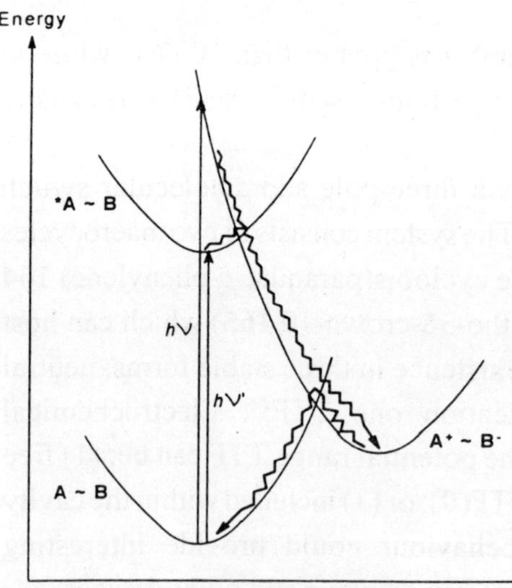

Figure 6.12. Scheme of photoexcitation with the following photoinduced electron transfer and radiative electron transfer.

transfers and charge transfer between A and B (corresponding to so-called optical electron-transfer process), respectively,

$$A{\sim}B + h\nu \Rightarrow {*A}{\sim}B, \quad (6.1)$$
$$*A{\sim}B \Rightarrow A{\sim}{*B}, \quad (6.2)$$
$$*A{\sim}B \Rightarrow A^{+}{\sim}B^{-}, \quad (6.3)$$
$$A{\sim}B + h\nu \Rightarrow A^{+}{\sim}B^{-}. \quad (6.4)$$

Let us note that only one of two photoexcitation (and electron transfer) processes is shown in the equations. A thermal back electron-transfer process may follow photoexcitation (Eq. 6.1), photoinduced electron transfer (Eq. 6.3) and radiative electron transfer (Eq. 6.4)

$$A^{+}{\sim}B^{-} \Rightarrow A{\sim}B + h\nu. \quad (6.5)$$

The relationships among all these processes are schematically shown in Figure 6.12.

Photoinduced bond breaking, light emission, or electron transfer may be used in such simple processes as photochemical synthesis, photodecomposition, photochromism, or photoluminescence, but migration of electronic energy, vectorial electron transfer, on/off switching of receptor reaction and other more complex functions induced by light require the cooperation of several components and can be carried out by supramolecular aggregates. For instance, in Nature in green plants, algae and bacteria, thanks to photosynthesis, water and carbon dioxide are converted into oxygen and carbohydrates producing oxygen in the atmosphere and the organic material supporting the food chain and responsible for the accumulation of fossil fuels

$$H_2O + CO_2 + h\nu = O_2 + 1/6\,(C_6H_{12}O_6) \quad (6.6)$$

The chemical reaction described by Eq. 6.6 is endothermic, it is made possible by light energy quanta $h\nu$. In green plants the incident light is collected by so-called antennae consisting of a large number of chromophoric groups (pigments) which in addition to the absorption of the photon energy transfer it through the whole system to the reaction centers (Fig. 6.13), leading to the charge separation reaction described by Eq. 6.4. The positive hole enables, with the help

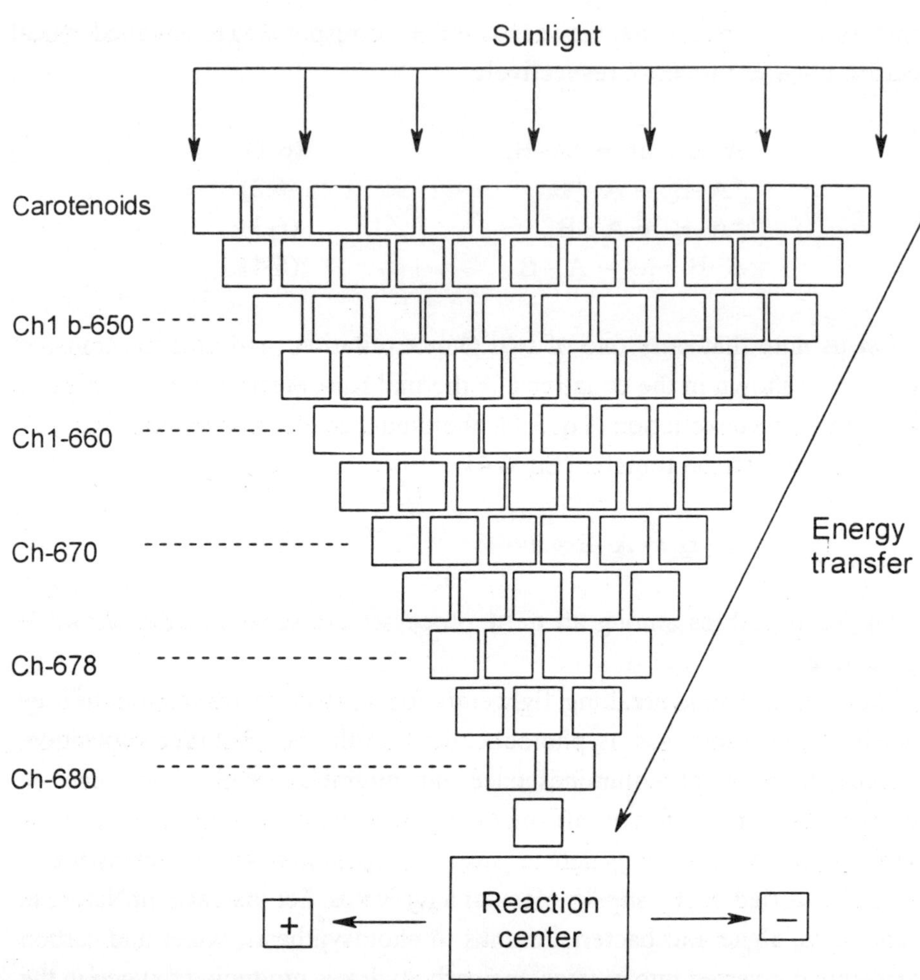

Figure 6.13. Schematic representation of an antenna unit of a natural photosynthetic system. The various pigments are kept together by intermolecular forces.

of an enzyme containing manganese, the oxidation of water to oxygen whilst the negative charge takes part in the reduction of $NADP^+$ **167** to nicotinamide adenine dinucleotide phosphate (NADPH) **168** and in the production of adenosine triphosphate ATP **14**. (The transformation of ATP to ADP catalysed by macrocycles synthesized in the Lehn group was presented in Chapter 1). **167** and **14** are later used for the synthesis of carbohydrates from CO_2. From the point of view of photochemical devices, two processes play the most important role in photosynthesis: photo-induced charge separation and the antenna effect. They can be, and actually are, used in photoelectrochemical cells (solar batteries) for the

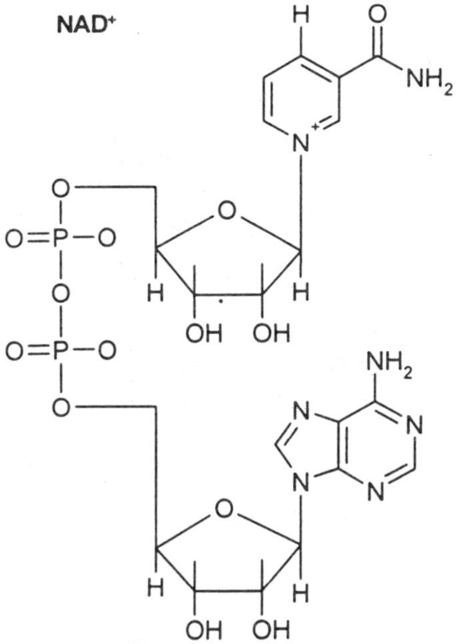

NAD⁺ **167**

NADH **168**

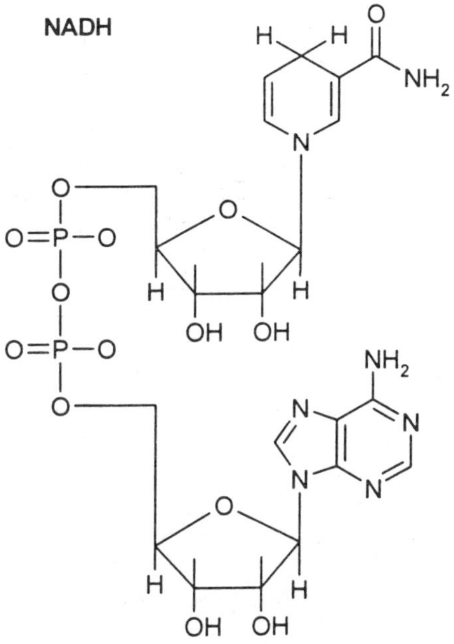

conversion of visible light energy into electricity [77]. The simulation of the light harvesting and the following charge separation processes with Langmuir-Blødgett film mimicking the light harvesting and electron transfer reactions proceeding in course of photosynthesis in the living systems in the lipid bilayer thylakoid membrane was studied by the Fujihara group [78].

Contrary to energy-processing photosynthesis, in vision light acts only as a signal. To be able to react to the absorption of a single photon a highly effcient amplification process operates in living organisms. A recent X-ray analysis of a trans-membrane protein bacteriorhodopsin forming the photochemically active part (i. e. reaction center) of a photoreceptor cell has helped to deepen our understanding of the process of vision [79] which is far from being fully clarified. In the core of vision is the photoisomerisation of the red 11-*cis* isomeric form of the chromophore retinal **169** to the *all*-trans form. Using additional sources of energy coming

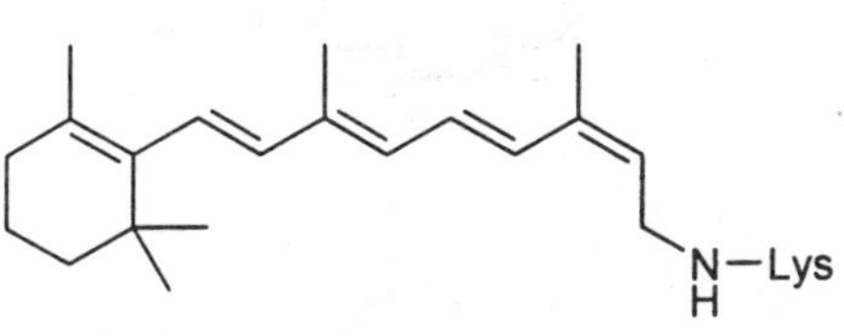

169

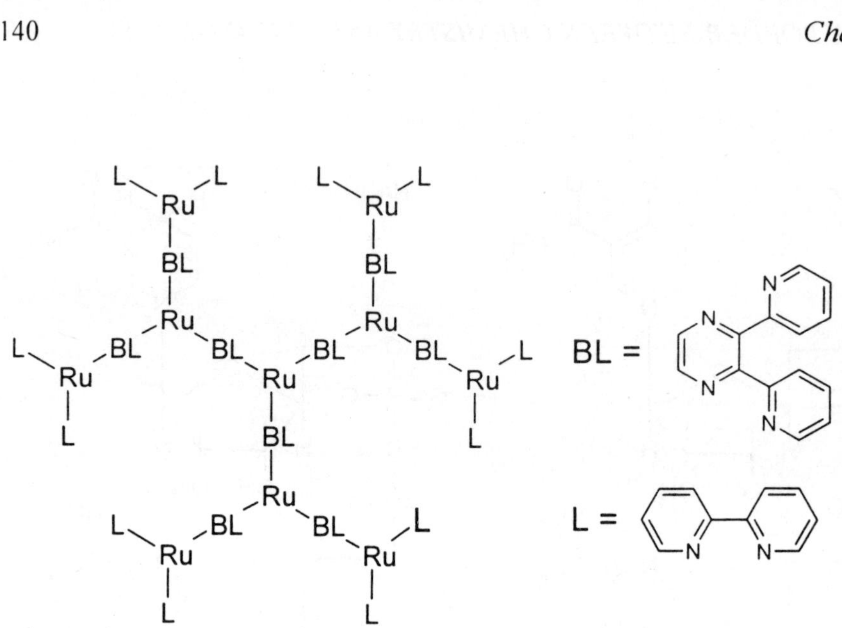

BL =

L =

170

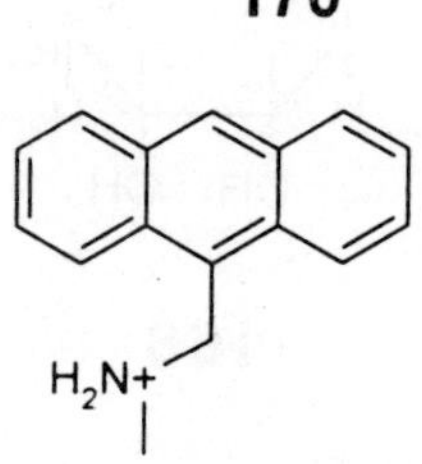

171

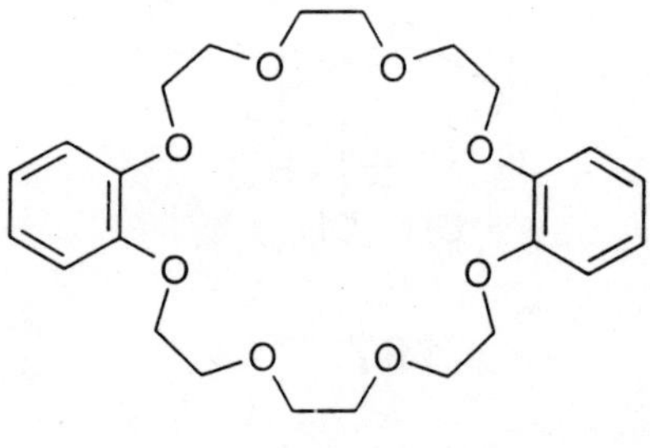

172

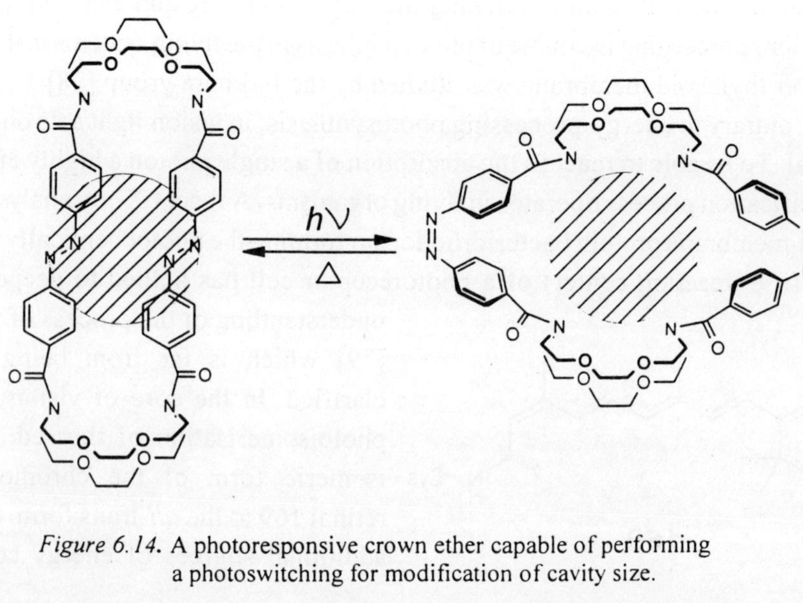

Figure 6.14. A photoresponsive crown ether capable of performing a photoswitching for modification of cavity size.

from metabolism, this process triggers a sequence of multiple chemical reactions, yielding an amplification factor of 500 000 in the conversion of incident photons to chemical mediators of membrane permeability sufficient to produce a measurable neural pulse at the synaptic termination. Photoinduced structural changes may also be used in devices performing a photoswitching function. Shinkai and Manabe [80] proposed such a device the idea of which is schematically presented in Fig. 6.14.

Only a few out of numerous photochemical devices proposed can be presented here. The self-assembled dendritic decanuclear complex **170** (see Section7.6 for information on dendrimers) was designed by Balzani group so that the energy absorbed by all the units in the complex is transferred via efficient intercomponent process from the center to the perifery, where it is re-emitted by the osmium(II)-subunits in form of near-IR luminescence [81]. Therefore, the system represents a typical light-harvesting antenna. An efficient electronic energy transfer between the protonated form of 9-aminomethyloanthracene **171** and aromatic crown ethers like **172** [82] also observed by Balzani and coworkers could be applied in sensors.

6.3.3 Pharmaceutical, cosmetic, and food industries

The quiet revolution taking place in the pharmaceutical industry in about the last 20 years consists not only in creating new drugs on the basis of better understanding of their action but also in changing ways in which drugs are administered. Both these approaches are deeply rooted in Supramolecular Chemistry. By knowing an active site one can model the most effective prospective drug using not only experimental methods (Quantitative Structure Activity Relations QSAR [83]), but also computational approaches (Computer-Assisted Molecular Design, CAMD) [84,85]. This procedure bearing the name 'rational drug design' considerably speeds up the development of new drugs, which can take more than 10 years and 3 billion dollars (a spending of about one million dollars per day). It should be stressed that even if the CAMD approach were to allow one only to exclude some molecules as unsuitable as a new drug or drug precursor, appreciable cost cutting would be achieved by avoiding synthesis and expensive pharmacological trials of these substances.

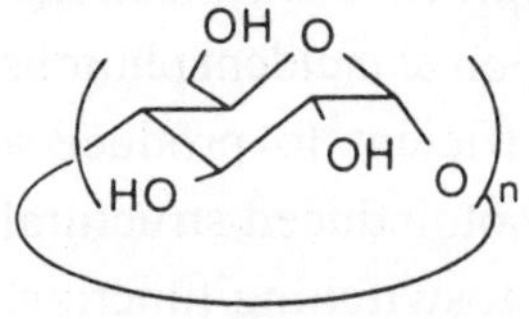

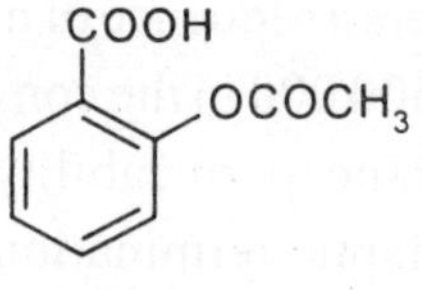

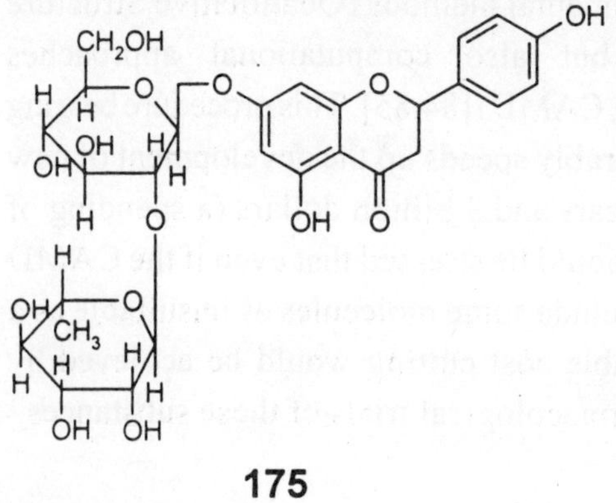

Complexing a drug with a cyclodextrin **173** has opened a new venue in drug administration since its encapsulation in the macrocycle increases solubility of poorly soluble drugs, stabilizes them preventing decomposition by moisture, light etc. [86] (see, however, the aspirin **174** case, in which its use in the form of cyclodextrin complex was prevented by catalytic action of the cyclodextrin on the drug [87]). It also reduces irritation caused by certain drugs and decreases their bad taste or smell. Thus the complexation can considerably increase the drug bioavailability and lead to less frequent drug administration. This is achieved by a slow release of the medicine which also allows for its more uniform content in the organism, enabling its administration once in a few days instead of several times a day [86]. Today more than 10 drugs are marketed in the form of cyclodextrin complexes. More highly-targeted drug delivery and release systems such as the one mimicking mast cell secretory granules found in fatty tissue are studied [88]. Using supramolecular assemblies as building blocks in biomolecular engineering as microsurgery materials is thought to be another prospective application [89]. Menger with coworkers speculated even on exploiting the reversibility of self-assembling of fibers to dissolution of blood clots by using bio-compatible fibrous organics to simulate the clots that can be dissolved chemically or thermally [90]. Non-toxicity of cyclodextrins is a prerequisite of their use in agrochemistry, pharmacological, cosmetic and food industries [8]. In addition to the drug formulations involving cyclodextrins that have reached marketing stage, numerous patents describing further application of analogous complexes as drugs exist. Similarly, cyclodextrin use in food industry and cosmetics is booming. Removal of naringin **175** responsible for the unpleasant bitter taste of grapefruit juice [91] and that of cholesterol to obtain products for a cholesterol-free diet [92] show only a few examples of the cyclodextrins application for removal of undesired food components. On the other hand, one can use the analogous complexes as additions to food or cosmetic products to enhance their stability. Garlic extracts and tea [93] marketed in the form of cyclodextrin complexes as well as stabilized foams and fragrances illustrate this point. It should be stressed that using CDs to store fragrances consisting of several substances may pose some problems, since these substances are unlikely to form complexes of the same stability. Thus some ingredients of the mixture may be lost earlier than other resulting in the change in odour.

6.3.4 Environmental protection [94, 95]

'Green chemistry', 'sustainable chemistry', 'clean chemistry', 'environmentally compatible chemistry' or 'environmentally benign synthesis' denote novel approach that gains in importance in industry. According to Anastas and Warner [94a], "Green chemistry is a set of principles that reduces or eliminates the use or generation of hazardous substances in the design, manufacture and application of chemical products".

The chemical industry is known to be one of the main sources of pollutants, and its development of taking care of environmental concerns is crucial for the survival of future generations. Therefore there is a growing public awareness exerting pressure demanding environmental legislation, and stimulating the drive towards clean technology. It should be stressed that a high level of environmenta protection, such as proper disposal of toxic waste, can be costly. Thus developing clean technologies, that is chemical processes reducing (or better avoiding) waste or toxic emissions, is an essential priority. Such technologies include clean syntheses, recycling, as well as purification methods which should take care of pollutants in air, water and soil [94]. Clean synthesis consists in the use of environmentally friendly reactants [95a] and/or their re-use as well as the recycling of hazardous by-products. An interesting idea with this respect is to use a by-product of one process as a raw material for another one [95b]. Optimization of chemical reactions to maximize the yield of the desired product, thus minimizing polluting wastes on the other, are also aims of environmentally friendly technologies. At present novel environmentally friendly processes which can find industrial applications fall outside the realm of supramolecular chemistry [95c]. However, this domain will undoubtedly play an important role in future. An important aspect of environmental protection is a multidisciplinary approach taking into account the whole complexity of the problem. For instance, as shown by T. Gerngross [96], producing plastics from renewable resources can consume considerably more energy than obtaining them from petroleum. On the basis of patent data and scientific literature this author showed that production of polyhydroxyalkanoates (PHAs) from maize is 19 times more energy demanding, requires 7 times more water and 22% more steam than standard manufacturing if the whole of the costs of PHAs production is calculated. The latter must include the cost of sowing, growing and harvesting of maize as well as those of the transport to the plant, a subsequent processing of starch and the purification of PHAs. 70% of electric power in the USA is obtained by burning coal, gas and petroleum, thus the net effect of PAHs production from maize fermentation results in considerable consumption of fossil fuel reserves. As a consequence, the use of renewable resources does not necessarily lead to a more environmentally friendly technology. However, it should be stressed that the first commercially scaled production unit for polylactide derived solely from annually renewable feedstocks is currently being built by Cargill Dow Polymers and should be opened by the end of 2001 [97]. It is also noteworthy that in the year 2000 about 10% of the polymer production stems from renewable resources and the figure is estimated to be 25% and 90% by 2020 and 2090, respectively [97].

As concerns clean technologies, significant efforts are directed at advancing the possibility of using hydrogen as a fuel, since, when burned in oxygen, this gas does not produce any pollution. It also does not create greenhouse emission and its energy content is more efficient than that of petrol. However, problems associated with the hydrogen storage are one of the main obstacles hampering its application. Storing the gas has been ridiculously impractical so far. If hydrogen were to be used as fuel, a tank for a car would be 3000 times bigger than a conventional petrol tank. Carbon nanotubes (discussed in Section 7.5) may provide a solution of this problem since they have been shown to take up H_2 efficiently [98].

Avoiding the creation of waste and the discharge of pollutants into air, water and soil is obviously a preferable approach to environmental protection. Limiting waste requires a complete change in chemical technology. For instance, by making use of one-pot reactions instead of multistep processes. Such reactions, made possible by preorganization (discussed in Chapter 2), allow one to avoid numerous purification procedures using large amounts of organic solvents.

Synkinesis, that is the creation of supramolecular aggregates, is also more environmentally friendly than standard chemical reactions since it is usually carried out in aqueous solutions. Methods rooted in supramolecular chemistry have been also proposed for removing dangerous polluting agents from the environment after contamination has taken place. Filters capable of extracting polluting agents such as heavy metals may considerably reduce this burden. For instance, selective complexation of cadmium, lead, mercury, and uranium by calixarenes presented in Section 7.2 has been proposed as a highly efficient technique for their removal [99]. Thus this process can be applied for removing lead from domestic water supplies, removing and recycling lead and cadmium from waste discharges from pigment and battery manufacturers, and for cleaning land fill leachates and mining waste streams. Not only calixarenes (Section 7.4) but also cyclodextrins and other selectively complexing agents could be used for air, water and soil purification. However, the stability of these complexes and their disposal also present problems. The best way of solving them would be to recycle heavy metals or other pollutants for their subsequent reuse.

As discussed in Section 4.2.3, the use of microemulsions to destruct half-mustard gas **88** (a warfare agent much less dangerous, but similar in action to mustard gas) was proposed by Menger and coworkers. A treatment of waste waters carrying a heavy load of 'hard' polluting agents with cyclodextrins proved very successful in decreasing the activity of the toxic substances by their complexation with cyclodextrins leading to their partial and temporarily masking. As a result, the variety of yeast and bacteria present in active sludge can cope with the waste remaining in water as they are able to oxidise, hydrolyse or degrade it [100]. Swollen insoluble cyclodextrin polymers have been also shown to remove polychlorinated biphenyls or detergents like lithium dodecylsulphate [101] from waste water.

6.3.5 Microemulsions in cleaning processes [102]

The term 'cleaning' describes several complicated processes, depending on the type of dirt involved, while washing refers specifically to the cleaning process taking place in a water bath containing, amongst others, dissolved amphiphilic compounds called detergents. Dirt is mostly of an organic nature (originating from sebum, food, oils and dyes, and many other biological or industrial sources) and the methods used to remove it depend on the type of dirt involved. Inorganic dust can typically be eliminated using a vacuum cleaner or a brush. Greasy dirt is concentrated within soft fat films by anionic or neutral amphiphiles that form an essential part of practically all washing powders and soaps. The surfaces of such films become polar, thus they can be hydrated forming water soluble fat droplets. To avoid the irreversible precipitation of the droplets caused by Ca^{2+} and Mg^{2+} counter-ions, the latter should be removed by 'builders' such as polyacrylate, sodium phosphate or, most recommendable, with zeolites. Other additives such as carbomethoxyl cellulose prevent dirt redeposition, presumably by adsorption, on the dirt and the fabric, by intensifying surface repulsions.

The procedures effective in removal of soft grease cannot be applied to other types of dirt. For instance, hard organic solids such as stacked multilayers of aromatic dyes (e.g., aniline type of dyes of ink or hemoglobin) on surfaces are not affected by detergents. Their removal is executed by oxidative bleaching or enzymatic degradation. It should be stressed that these kinds of dirt are water soluble before their adsorption onto surfaces. The process of washing pertains to supramolecular chemistry in several aspects: (1) surfaces bind dirt of organic nature in thin layers; (2) neither water-soluble nor water-insoluble dirt material can be removed quantitatively by water alone; (3) the action of detergents consisting in solubilization of soft materials is most effective for dirt compounds with long alkyl chains such as fats, fatty acids, and hydrocarbons; (4) Hydrophobic organic dirt can be also removed by creating structurally ill-defined microemulsions between water and organic compounds; (5) solubilization of aromatic dye molecules can often be achieved only by their destruction by bleaching, or by the use of detergent above its critical micelle concentration (see Chapter 4 for its definition) or by solvents.

There is an interesting development in cleaning products marketed recently [102b]. Cleaning itself is not the problem now and "research and development laboratories are on the hunt for that breakthrough ingredient they hope their customers will adopt to increase sales, boost profits, and make shareholders happy." [102b] For instance, in 1998 Proctor and Gamble introduced cyclodextrin-based odor-removing spray in which the oligosugar molecules capture odourous compounds on carpets and textiles keeping them from floating into the air.

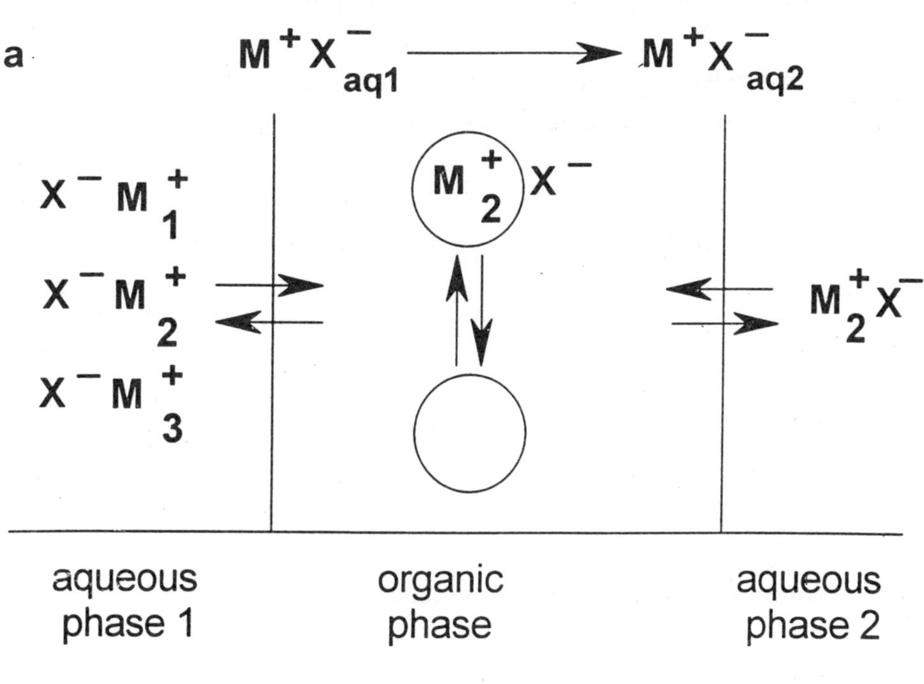

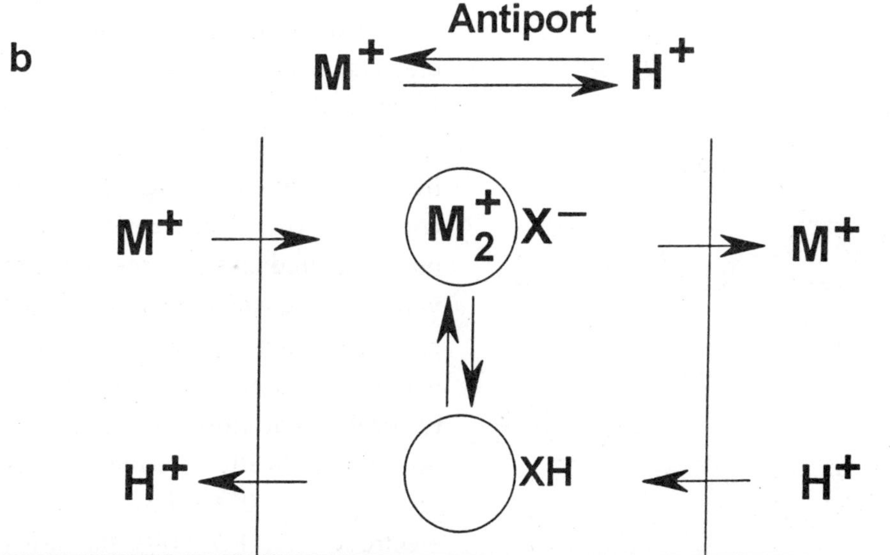

Figure 6.15. The (a) symport and (b) antiport modes of ionophore operation.

6.3.6 Cation extracting systems ionophores [103a]

Strong and selective affinity of crown ethers to metal cations have been applied in so-called liquid membranes (which are completely different from mono- and bilayer membranes discussed in Chapter 4). A scheme of their operation is presented in Fig. 6.15. The device enables either selective extraction of a cation from the mixture (Fig. 6.15a) or, as shown in Fig. 6.15b, transport. The complexation of benzo-15-crown-5 fluorophore with γ-cyclodextrin **68** yields a system **176** with exceptionally high potassium ion sensivity and selectivity in water [103b].

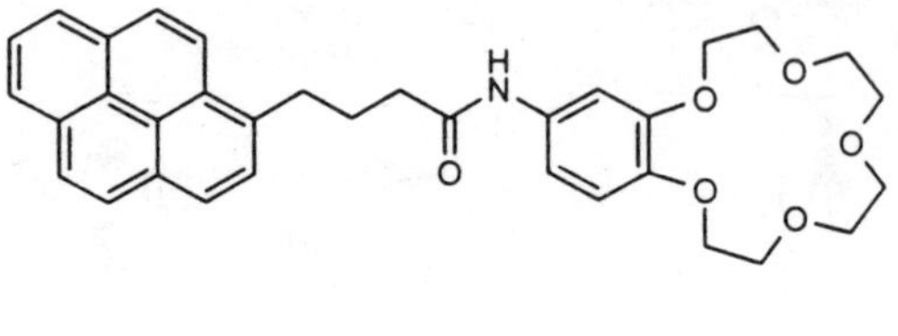

176

6.3.7 Other applications of supramolecular systems

Only few out of many proposed applications of supramolecular systems can be mentioned here. Nanotube 'tweezers' which enable manipulation and characterization of nanoscale clusters and wires developed by Lieber and Kim [104] seem to be one of the first tools necessary for nanoscale manufacturing. The device consists of two electrodes deposited onto a tapered glass micro-pipette to which two conducting multiwalled carbon nanotubes were attached. The nanotubes, forming the arms of the tweezers, bent closer to each other upon the application of a bias voltage ranging from 0 to 8.3 V to the electrodes. At 8.5 volts the arms close, whilst with the voltage removed they relax to their original position.

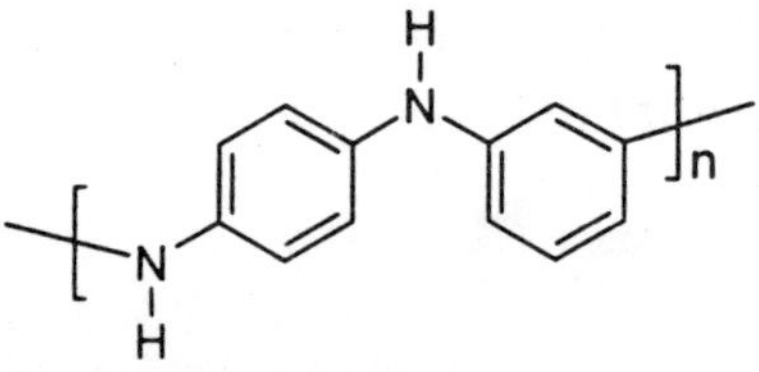

177

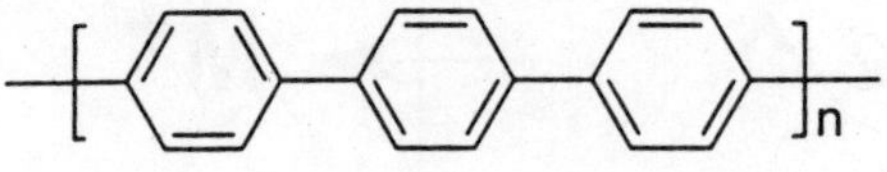

178

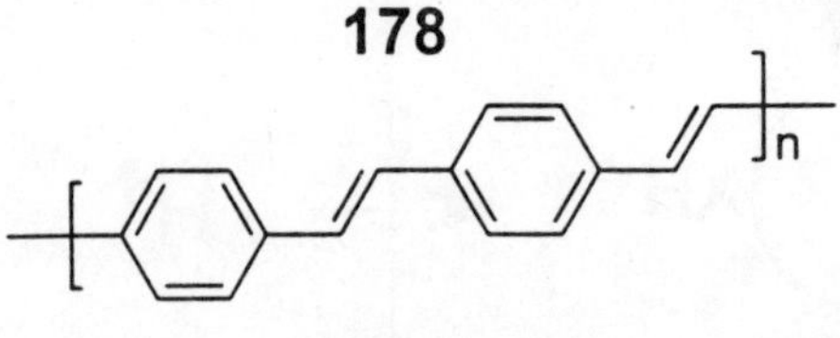

179

Using scanning probe microscope one can follow the manipulation process. For instance, the nanotweezers were shown to be able to grab and pick up clusters or nanowires with diameter of 500 nm. Moreover, one can immediately probe their electrical properties, since the tweezer arms serve as conducting wires. The probing may not be limited to manmade materials but it can also be used to manipulate and modify substructures within cells once the nanowire tips will be insulated.

Various approaches to organic magnets have been explored. Langmuir-Blødgett films of single-molecule nanomagnets has been reported by Coronado, Mingotaud and coworkers [105a]. The macroscopic ferromagnetic behaviour was detected in doped *meta-para* aniline oligomers like **177** [105b].

The application of Langmuir-Blodgett films as rectifiers and/or switches have been also proposed. Peterson [106] investigated two semiconductors polyparaphenylene **178** and polyphenylenevinylene **179** with these purposes in mind. These systems with the chain lengths of at least 20 units could also be used as photovoltaic devices since their electroluminescence should be readily detectable. The doping of such materials may be necessary but at present it is not clear whether they will form Langmuir-Blodgett films when doped.

6.4. Supramolecular Catalysis

6.4.1 Introduction

Modelling enzymatic processes in living organisms, on one hand, and developing effective catalysts for industrial application, on the other, are the most important driving forces of the field. These studies enable one better to understand the operation of living organisms at the molecular level but, to our knowledge, until now no supramolecular catalyst for industrial use has been developed. It should be stressed first that supramolecular catalysis differs considerably from the classical catalysis since both acceleration or inhibition of a chemical reaction can take place in the aggregate under study. Secondly, the amount of catalyzing agent in supramolecular catalysis is much larger. Thus one should rather speak about the acceleration of a chemical reaction than about catalysis *sensu stricte*. Moreover, selectivity of a chemical reaction involving supramolecular aggregate is much more effective than the acceleration of reaction rates. Similarly to the whole of supramolecular chemistry, supramolecular

Catalysis of the 15, 15' double bond scission of beta-carotene

H =

R = 4-*n*-heptylphenyl

Acceleration of a Diels-Alder reaction by a self-assembled dimeric capsule

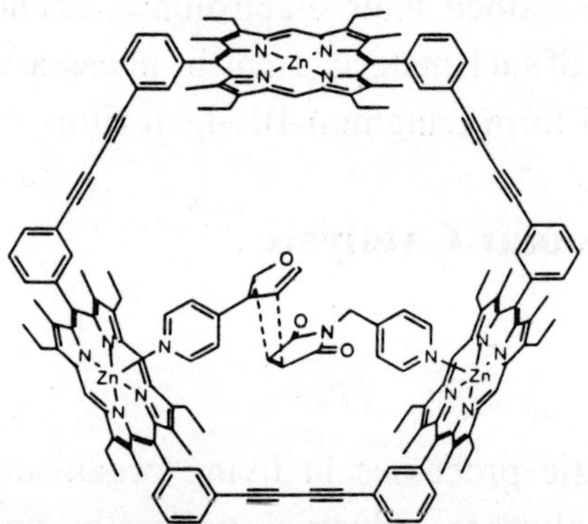

Reversing stereoselectivity of a Diels-Alder reaction by preorganization of reactants within a porphyrinic host

A synthetic self-replicating system by Kiedrowsky *et al.*

Figure 6.16. Examples of host-catalyzed reactions.

catalysis may be divided according to the type of the aggregate involved since it can be based on a catalytic action of a macrocyclic host, that of microemulsions, micelles or vesicles or on the catalytic activity of mesoporous materials. Several host-catalyzed reactions are schematically presented in Fig. 6.16. Two assumptions are made in the transition state theory of chemical reactivity, that are neither exact or rigorous but allow one to understand reactivity at the molecular level. The first assumption defines the transition state complex as a hypothetical unstable transient species in fast equilibrium with the reactants and the second states that the overall reaction rate is determined by the decomposition of the activated complex. Supramolecular catalysis obeys the general rule (Fig. 6.17) stating that catalytical action consists in stabilization of a transition state of the reaction [107] and in the rapid release of the product. In a reaction catalyzed by the formation of a supramolecular transition state complex, substrates are inserted into either an active site of an enzyme, a host cavity, a layer, a micelle, a vesicle or micropores bringing their functional groups into a close contact for relatively long time. In addition, the incorporation can stabilize favourable conformations of the substrates. Functional groups in the active site or cavity such as hydrogen bonding donor or acceptor groups can additionally exert 'true' catalytic action. Contrary to that of substrates and transition state complex, the fit of the product into the active site or catalyst cavity must be worse to enable its release. Otherwise, as sometimes happens with artificial catalysts [108], the reaction is inhibited. As mentioned in Section 5.3.5, enzymes enable high selectivity and reaction yields taking place in living

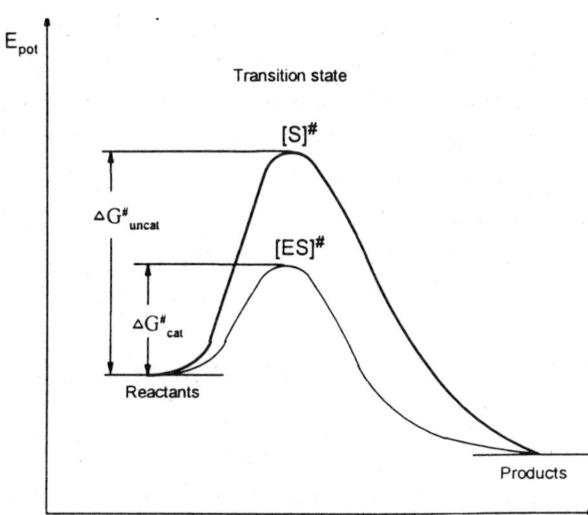

Figure 6.17. Three stages of supramolecular catalysis:
1) Selective binding of reactant(s) based on their recognition by the receptor that may bear reactive group(s);
2) Transformation of the bound species;
3) Release of the products, thus regenerating the catalyst.

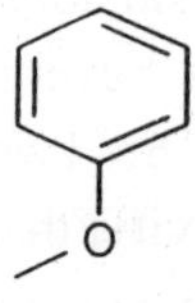

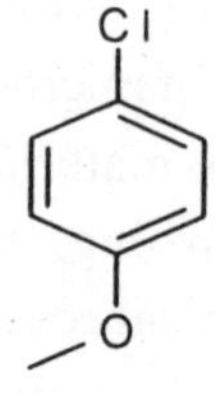

180 **181**

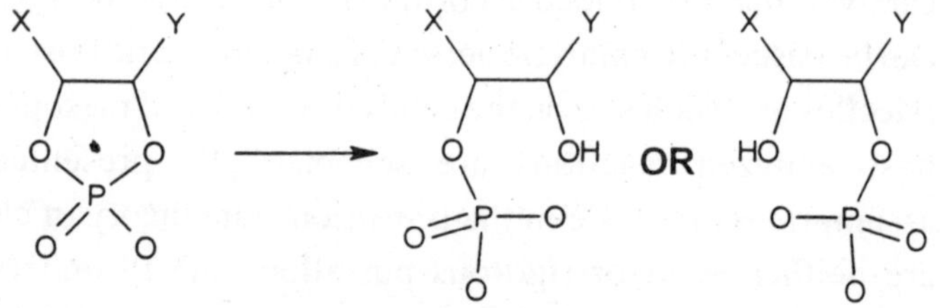

Figure 6.18. Ring-opening of asymmetric phosphodiesters which can give two different products.

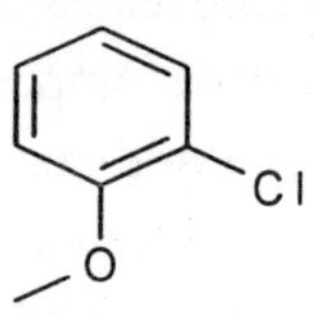

182

organisms; thus together with their mimics, they are the main object of studies in supramolecular catalysis. Native enzymes are organic macromolecules (in a single globular structure) made up of polypeptide chains consisting of several hundreds of amino acid residues. As such they fall outside the scope of this book. Enzymes active sites, as we understand them today, are their relatively small fragments that can be modelled by artificial enzymes for which the names synzymes or chemzymes are sometimes used.

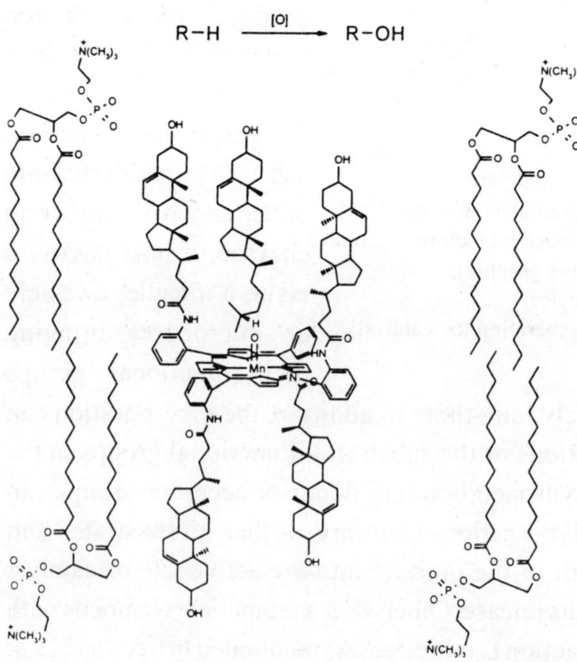

Figure 6.19. Catalytic oxidation of a steroid in the bilayer.

6.4.2 Enzyme mimics

Regio- or stereo-selectivity of enzymes action is much easier to mimic than the enormous yields of reactions they catalyze (accelerations factors of 10^6-10^{12}). A simple example of this kind is provided by chlorination of anisol **180** that produces only *p*-chloroanisol **181** in the presence of α-cyclodextrin **13** while both **181** and **182** are formed without the latter factor [109]. By using suitably

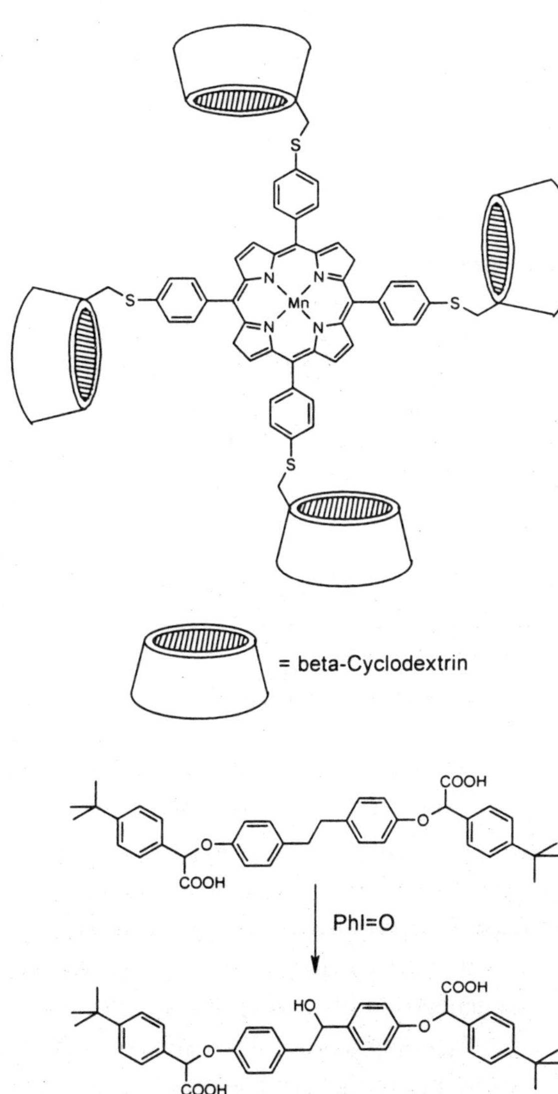

Figure 6.20. Oxidation reaction catalysed by tetra-cyclodextrin-porphyrin (PhI=O denotes iodosobenzene).

designed hosts Breslow [110] and Hamilton groups [111] were able to control the outcome of ring-opening of cyclic phosphodiesters (Fig. 6.18) directing the substrate along one reaction pathway. Several mimics of selective oxidation of inactive carbon centers by cytochrome P-450 have been reported. An impressive example of vesicular catalysis of this type was proposed by Grooves and Neumann [112]. In the process the activation of a CH bond is carried out by the catalyst embedded in the bilayer forming a vesicle wall together with the steroid substrate shown in Fig. 6.19. Another kind of a porphyrin catalyst, mimicking a similar reaction without involvement of aggregate formation, was proposed by Breslow [113]. In the latter case (Fig. 6.20) phenyl rings bearing large *p*-substituents complexed with cyclodextrins pendant on the porphyrins ensure a close contact of oxidation site with the porphyrin core thus catalyzing the reaction. Halogenation of the porphyrin enabled the authors to avoid catalyst destruction during the reaction. Another enzyme carrying oxidation reactions in living organisms is pyruvate oxidase. Mattei and Diederich have recently developed a system mimicking the action of this enzyme [114]. The system, oxidizing 2-naphthyl aldehyde to the

Figure 6.21. Diedrich's model of pyruvate oxidase system.

corresponding methyl ester in methanol (Fig. 6.21) reaches a catalytic turnover number of ca. 100. It binds the aromatic aldehyde inside its cavity, enables the formation of a covalent intermediate by reaction with its thiazolium group, oxidizes the intermediate by intramolecular transfer of a hydride equivalent to the flavin residue and, finally, releases the product by solvolysis. Catalytic turnover is carried out by electrochemical regeneration of flavin. It is one of the few catalytic supramolecular systems described in literature that have achieved a genuine preparative scale. Two latter examples belong to transformation reactions. A catalyst involving a cyclodextrin derivative was also presented for fission reaction by Zhang and Breslow [115]. Cyclodextrins as enzyme mimics have been briefly discussed in Section 5.3.5. (2-C_{60})-*bis*(triphenylphosphine)palladium complex **183** (double bonds in the formula have been omitted for clarity) was found to exhibit catalytic

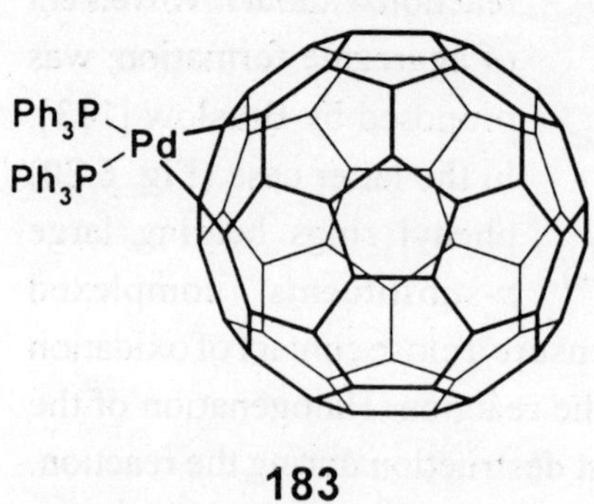

properties both as homogenous and heterogeneous catalyst. In the former case it was assumed to mimic enzyme action [116].

6.4.3 Macrocyclic host molecules, medium-sized aggregates (microemulsions, micelles,vesicles, etc.) and mesoporous materials as catalysts

Numerous macrocyclic hosts have been shown to exert the catalytic activity on the included guest. This activity can be even undesirable. For instance, as mentioned before, cyclodextrins **173** use as aspirin carrier proved unsuccessful in view of a catalytic decomposition of the drug [87]. Two interesting host systems catalyzing the Diels-Alder reaction have been reported. Sanders and coworkers [117] analyzed the influence of the cavity constraints on the exo- or endo-adduct **184** and **185**, respectively, formation involving cyclic porphyrin trimers **186** and **187** with acetylene linkages. At 30° C for the smaller 1,1,2-trimer **186**, the endo 500-fold acceleration was observed

exo *endo*

184 **185**

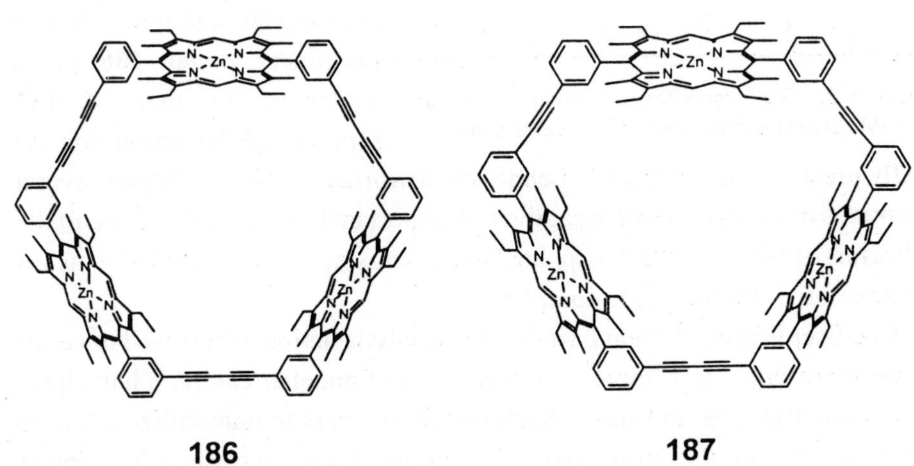

186 **187**

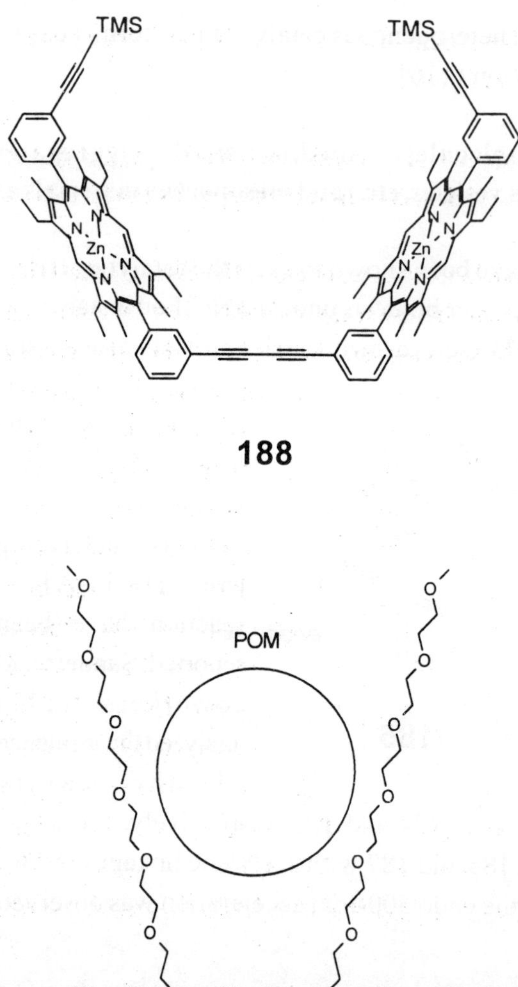

Figure 6.22. The immobilization of polyoxometalate POM catalyst within a polyethyleneglycol layer.

while the exo-adduct was preferred with **187**. Interestingly, at 60° C the former host looses its stereoselectivity. This evokes the question of the catalyst rigidity. As discussed in Chapters 1 and 3, enzymes are known to exhibit a certain flexibility allowing them to adapt themselves to the guest owing to conformational changes described by the induced fit model. However, most syntheses involving supramolecular catalysts are limited to rigid cage compounds. According to Sanders [108] the failure to design an effective catalyst for industrial use may be owing to the neglect of the importance of the host's flexibility in catalytic processes. The effective and exo-selective catalytic action of the linear dimer **188** [117] analogous to that of **187** supports the Sanders notion. As briefly presented in Chapter 1, Lehn with coworkers [118] developed several substituted macrocycles which catalyze an important biochemical process of the hydrolysis of adenosine triphosphate **14** to adenosine diphosphate **15** by means of formation of intermediate complex **16**.

A self-replicating system based on the catalytic action of reversed micelles has been presented in Chapter 1. Other cases of micellar catalysis have been discussed in Ref. 119. The use of semicrystalline fibers to immobilize catalysts and substrates was also proposed [120]. Another approach to the enhancement of catalytic activity by exploiting supramolecular aggregates consists in partial immobilization of polyoxometalate POM catalyst within a polyethylene glycol layer covalently bound to a solid support (Fig. 6.22) [121].

Porous materials such as organic zeolites discussed in Section 8.4 are vigorously studied since they often exhibit catalytic activity [122]. An interesting type of zeolite materials is obtained by tucking chiral amines inside pores of a commercially available zeolite. Such an approach allowed Ramamurthy's group to enhance stereoselectivity of a photochemical reaction [122b].

The catalytic action of alkalides and electrides will be briefly discussed in Section 7.1.3.

6.5 Concluding Remarks

In addition to numerous applications presented in this chapter, several examples of more specific ones are presented in other chapters. Cyclodextrins as enzyme mimics have been briefly discussed in Sect. 5.3.5, hydrogen storing in nanotubes is mentioned in Sect. 7.5, and huge deposits of methane in the form of clathrate hydrates found mainly in sea water that surmount carbon, oil, and gas combined as fuel resource that are presented in Section 8.3.3 illustrate this point. The latter case clearly shows a fragile balance between promise, the enormous energy gain we could obtain, and technological obstacles associated with obtaining methane from the clathrates. As shown by the numerous examples presented in this chapter, a massive effort must be made to bridge this gap. As shown by the proposal of using $C_{60}F_{60}$ **189** as an ideal lubricant, as well as by fullerene superconductivity, discussed in Section 7.5, not all of the numerous suggestions of prospective applications will come to fruition.

Supramolecular aggregates are a rich source of systems which could serve as molecular devices. Chemically driven unthreading and rethreading of a [2]pseudorotaxane **190** [123] provide one example of large amplitude motion that in future can be used in logic devices [124] and computer technologies [125]. DNA based computers fall outside the scope of this book although they operate as supramolecular aggregates [126].

Several other applications of supramolecular systems have been proposed. Amongst those that are thought to be capable of bringing enormous benefits are devices making use of nonlinear optical phenomena [3,127]. Another exciting possibility of application of supramolecular systems includes the use of nematic

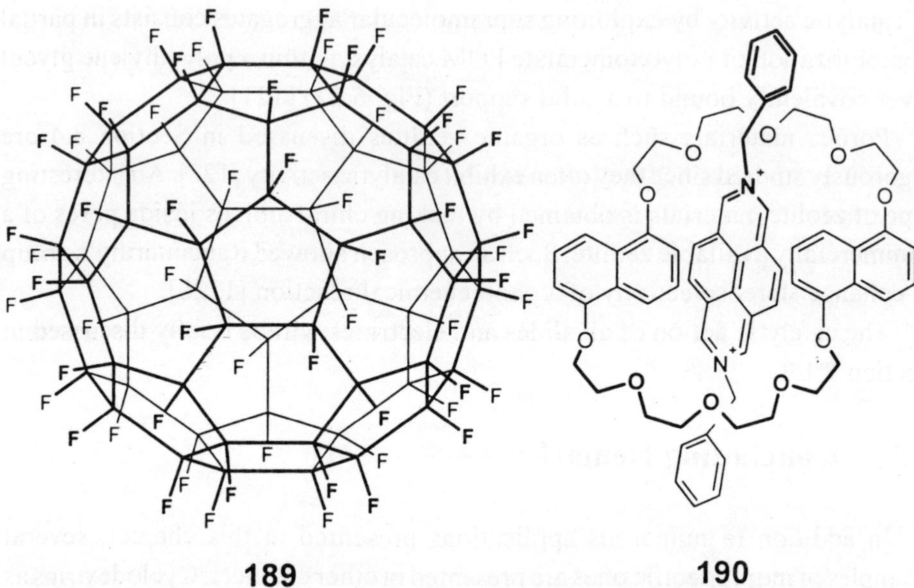

189 190

crystals as media for real-time holography [128]. Even if present day applications of Supramolecular Chemistry are limited and huge obstacles have to be overcome, the prospects are bright and far reaching, promising to change many aspects of our lives.

REFERENCES

1. D. Bradley, Chem. Soc. Rev., 1995, 24, 379.
2. J.-M. Lehn, Angew. Chem. Int. Ed. Engl., 1988, 27, 89.
3. G. H. Wanniere, *Linear and Nonlinear Properties of Molecules*, VCH, Weinheim, 1993; J.-M. Andre, J. Delhalle, Chem. Rev., 1991, 91, 843.
4. *Comprehensive Supramolecular Chemistry*, vol. 10, Pergamon, Oxford, 1996.
5. N. Sacchetti, Int. J. Modern Phys. B, 2000, 14, 2617; B. R. Lehndorff, High TC Superconductors for Magnet and Energy Techn.: Fund. Aspects, 2001, 171, 1.
6. V. Balzani, F. Scandola, *Comprehensive Supramolecular Chemistry*, vol. 10, p. 687, Pergamon, Oxford, 1996.
7. R. F. Service, Science, 1998, 282, 2179; A. H. Tullo, Chem. Eng. News, 2000, June 26, 20.

8. *Comprehensive Supramolecular Chemistry*, vol. 4, Pergamon, Oxford, 1996; J. Szejtli, *Cyclodextrin Technology*, Kluwer Academic Publishers, Dordrecht, 1988.

9. G. R. Desiraju, *Comprehensive Supramolecular Chemistry*, vol. 6, p. 1, Pergamon, Oxford, 1996.

10. V. G. Videnova-Adrabinska, *The Hydrogen Bond as a Design Element of the Crystal Architecture. Crystal Engineering. From Biology to Materials*, Oficyna Wydawnicza Politechniki Wroclawskiej, Wroclaw, Poland, 1994.

11. A. I. Kitaigorodsky, *Molecular Crystals and Molecules*, Academic Press, New York, 1973.

12. Van der Waals and electrostatic interactions are sometimes called together nonbonded interactions. However, in most works this name refers to the van der Waals interaction only.

13. O. Bastiansen, Acta Chem. Scand., 1952, 6, 205; C. P. Brock, K. L. Haller, J. Phys. Chem., 1984, 88, 3570.

14. J. D. Dunitz, Pure Appl. Chem., 1991, 63, 177.

15. (a) Y. Inoue, Y. Takahashi, R. Chujo, Carbohydr. Res., 1985, 144, C9; (b) C. Betzel, W. Saenger, B. E. Hingerty, G. M. Brown, J. Am. Chem. Soc., 1984, 106, 7545; (c) V. Zabel, W. Saenger, S. A. Mason, J. Am. Chem. Soc., 1986, 108, 3664.

16. G. P. Charbonneau, Y. Delugeard, Acta Crystallogr., Sect. B, 1976, 32, 1420. However, there is a possibility that an isomer less stable in solution crystallized.

17. M. G. Usha, R. J. Wittebort, J. Am. Chem. Soc., 1992, 114, 1541.

18. Y. Inoue, T. Okuda, R. Chujo, Carbohydr. Res., 1985, 141, 179; Y. Yamamoto, M. Onda, Y. Takahashi, Y. Inoue, R. Chujo, *ibid.*, 1988, 182, 41.

19. E. B. Brouwer, G. D. Enright, C. I. Ratcliffe, G. A. Facey, J. A. Ripmeester J. Phys. Chem. B, 1999, 103, 10604.

20. O. Ermer, Helv. Chim. Acta, 1991, 74, 1339.

21. F. H. Allen, O. Kennard, R. Taylor, Acc. Chem. Res., 1983, 16, 146.

22. F. H. Allen, G. P. Shields, in *Implications of Molecular and Materials Structure for New Technologies*, J. A. K. Howard, Ed., F. H. Allen, Ed., P. Shields, Eds, Kluwer, Dordrecht, 1999, p. 291.

23. J. Bernstein, M. C. Etter, L. Leiserowitz, in *Structure Correlation*, J. D. Dunitz, H. B. Bürgi, Eds., VCH, Weinheim, 1994.

24. J. D. Watson, F. H. C. Crick, Proc. Roy. Soc. Lond., 1954, A223, 80.

25. G. A. Geffrey, W. Saenger, *Hydrogen Bonding in Biological Structures*, Springer, Berlin, 1991.

26. (a) C. T. Seto, G. Whitesides, J. Am. Chem. Soc., 1990, 112, 6409; (b) J. A.Zerkowski, C. T. Seto, D. A. Wierda, G. Whitesides, *ibid.*, 1990, 112, 9025; (c) C. T. Seto, G. Whitesides, *ibid.*, 1991, 113, 712; (d) G. Whitesides, J. P. Mathias, C. T. Seto, Science, 1991, 254, 1312.

27. M. Tadokoro, K, Isobe, H. Uekusa, Y. Ohashi, J. Toyoda, K. Tashiro, K. Nakasui, Angew. Chem. Int. Ed. Engl., 1999, 38, 95.

28. E. S. Shubina, N. L. Belkova, A. N. Krylov, E. V. Vorontsov, L. M. Epstein, D. G. Gusev, M. Niedermann, H. Berke, J. Am. Chem. Soc., 1996, 118, 1105.

29. D. S. Reddy, B. S. Goud, K. Panneerselvam, G. R. Desiraju, Chem. Commun., 1993, 663.

30. J. J. Wolff, F. Gredel, T. Oeser, H. Irngartinger, H. Pritzkow, Chem. Eur. J., 1999, 5, 29.

31. M. Mascal, J. Hansen, P. S. Fallon, A. J. Blake, B. R. Heywood, M. H. Moore, J. P. Turkenburg, Chem. Eur. J., 1999, 5, 381.

32. D. B. Amabilino, J. F. Stoddart, Chem. Rev., 1995, 95, 2725.

33. Van der Waals interactions are composed of dispersive attraction changing as r^{-6} with distance, and repulsion changing as r^{-12} prevailing when the atoms are close.

34. G. R. Desiraju, A. Gavezzotti, J. Chem. Soc. Chem. Commun., 1989, 621.

35. (a) P. N. W. Baxter, J.-M. Lehn, G. Baum, D. Fenske, Chem. Eur. J., 1999, 5, 102; (b) P. N. W. Baxter, J.-M. Lehn, B. O. Kneisel, G. Baum, D. Fenske, *ibid.*, 113.

36. Ref. 10, p. 61.

37. M. Gdaniec, M. J. Milewska, T. Polonski, Angew. Chem. Int. Ed. Engl., 1999, 38, 392.

38. T. Klaus, R. Joerger, E. Olsson, C. G. Granqvist, Proc. Natl. Acad. Sci. USA, 1999, 96, 13611.

39. *Feynman Lectures on Computation*, R. P. Feynman, R. W. Allen, Ed., A. J. G. Hey, Ed., Perseus Books Group, July 2000.

40. K. E. Drexler, *Nanosystems: Molecular Machinery, Manufacturing and Computation*, Wiley, New York, 1992.

41. K. E. Drexler, *Engines of creation*, Anchor Books, 1987.

42. www.chemweb.com/alchem9/spotlight/sp_991022_nanotech.html.

43. R. M. Baum, Chem. Eng. News, 1993, Nov. 22, p. 8.

44. J. F. Stoddart, Angew. Chem. Int. Ed. Engl., 1991, 30, 70.

45. W. Krätschmer, K. Fostiropoulos, D. Huffmann, Chem. Phys. Lett., 1990, 170, 167.

46. J. H. Schön, C. H. Kloc, B. Batlogg, Nature, 2000, 408, 549.

47. R. Taylor, A. G. Avent, T. J. Dennis, J. P. Hare, H. W. Kroto, D. M. R. Walton, J. H. Holloway, E. G. Hope, G. J. Langley, Nature, 1992, 355, 27.

48. (a) J.-P. Sauvage, Acc. Chem. Res., 1998, 31, 611; (b) J.-C. Chambron, S. Chardon-Noblat, A. Harriman, V. Heitz, J.-P. Sauvage, Pure Appl. Chem., 1993, 65, 2343.

49. (a) S. Shinkai, M. Ishihara, K. Ueda, O. Manabe, J. Chem. Soc. Chem. Commun., 1984, 727; (b) R. A. Bissell, E. Cordova, A. E. Kaifer, J. F. Stoddart, Nature, 1994, 369, 133.

50. J.-P. Sauvage, Bull. Pol. Acad. Sci. Chem., 1998, 46, 289.

51. ENIAC - Electronic Numerical Integrator and Computer was built for American Army by J. P. Eckert and J. W. Mauchley at the Pennsylvania State University in 1946. It weighed about 30 tons. consisted of 18 thousand vacuum lamps and semiconducting diodes. - Gazeta Wyborcza, 18 marca 2000, Supermaket p. 3.
52. Q. H. Liu. L. M. Wang, A. G. Frutos. A. E. Condon. R. M. Corn. L. M. L. Smith. Nature. 2000. 403. 175; D. Faulhammer. A. R. Cukras. R. J. Lipton. L. F. Landweber. Proc. Natl. Acad. Sci. USA, 2000, 97, 1385.
53. S. J. Tans. A. R. M. Verschueren, C. Dekker, Nature, 1998. 393. 49.
54. A. Credi. V. Balzani. S. J. Langford, J. F. Stoddart, J. Am. Chem. Soc.. 1997. 119. 2679.
55. S. J. Tans. A. R. M. Verschueren, C. Dekker, Nature, 1998, 393, 49. C. Dekker, *ibid*, 1999. 402, 273.
56. (a) S. Wilson. *Chemistry by Computer*, Plenum Press. New York. 1986. Chapters 5. 7; (b) H. Dodziuk, *Modern Conformational Analysis. Elucidating Novel Exciting Molecular Structures*, VCH Publishers, New York, 1995, Chapters 3, 11.
57. E. E. Polymeropoulos, D. Möbius, H. Kuhn, Thin Solid Films, 1980. 68. 173; M. Sugi, K. Sakai, M. Saito, Y. Kawabata, S. Yizima, Thin Solid Films, 1985. 132, 69; C. M. Fischer, M. Burghard, S. Roth, K. V. Klitzing, Europhys. Lett., 1994, 28, 129.
58. M. Fujihira, K. Nishiyama, H. Yamada, Thin Solid Films, 1985. 132. 77.
59. J. M. Williams, A. J. Schluttz, U. Ceiser, K. D. Carlson. A. M. Kini. H. H. Wang. W.-K. Kwok. M. H. Whangbo, J. E. Schirber, Science. 1991, 252, 1501.
60. K.-I. Yoshida, T. Shimomura, K. Ito, R. Hayakawa, Langmuir, 1999. 15. 910.
61. F. S. Schoonbeek, J. H. van Esch, B. Wegewijs, D. B. A. Rep. M. P. de Haas. T. M. Klapwojk, R. M. Kellogg, B. L. Feringa, Angew. Chem. Int. Ed. Engl., 1999, 38, 1393.
62. A. I. Yanson, G. R. Bollinger, H. E. van den Brom, N. Agrait, J. M. van Ruitenbeek, Nature, 1998. 395. 783.
63. (a) E. Braun, Y. Eichen, U. Sivan, G. Ben-Yoseph, Nature. 1998. 391. 775; (b). J. Simon, in *Nanostructures Based Molecular Materials*, W. Göpel, Ch. Ziegler. VCH, Weinheim, 1992, p. 267.
64. (a) J. S. Miller, A. J. Epstein, MRS Bull., 2000, 25, 21; A. P. Alivisatos. P. F. Barbara, A. W. Castleman, J. Chang, D. A. Dixon, M. L. Klein, G. L. McLendon. J. S. Miller, M. A. Ratner. P. J. Rossky. S. I. Stupp, M. E. Thompson, Adv. Mater., 1998. 10. 1297; (b) K. I. Pokhodnaya, A. J. Epstein, J. S. Gray, Adv. Mater., 2000. 12, 410.
65. J. G. Bednorz, K. A. Müller, Z. Phys. B, 1986, 64, 189.
66. T. Sheng. X. Wu, W. Zhang, Q. Wang, X. Gao. P. Lin. J. Chem. Soc. Chem. Commun., 1998. 263.
67. A. T. Coomber. D. Beljonne. R. H. Friend. J. L. Bredas. A. Charlton. N. Robertson. A. E. Underhill. M. Kurmoo. P. Day. Nature. 1996. 380. 144.

68. J. Bügler, N. A. J. M. Sommerdijk, A. J. W. G. Visser, A. van Hoek, R. J. M. Nolte, J. F. J. Engbersen, D. N. Reinhoudt, J. Am. Chem. Soc., 1999, 121, 28.

69. D. L. Dermody, R. F. Peez, D. E. Bergbreiter, R. M. Crooks, Langmuir, 1999, 15, 885.

70. I. Aoki, T. Harada, T. Sakaki, Y. Kawahara, S. Shinkai, J. Chem. Soc. Chem. Commun., 1992, 1341.

71. S. Tomas, R. Prohens, G. Deslongchamps, P. Ballester, A. Costa, Angew. Chem. Int. Ed. Engl., 1999, 38, 2208.

72. R. Pinalli, F. F. Nachtigall, F. Ugizolli, E. Dalcanale, Angew. Chem. Int. Ed. Engl., 1999, 38, 2377.

73. K. Haupt, K. Noworyta, W. Kutner, Anal. Commun., 1999, 36, 391.

74. J. Chen, M. A. Reed, A. M. Rawlett, J. M. Tour, Science, 1999, 286, 1550.

75. P. R. Ashton, V. Balzani, J. Becher, A. Credi, M. C. T. Fyfe, G. Mattersteig, S. Menzer, M. R. Nielsen, F. M. Raymo, J. F. Stoddart, M. Venturi, D. J. Williams, J. Am. Chem. Soc., 1999, 121, 3951.

76. V. Balzani, A. Credi, F. Scandola, in *Transition Metals in Supramolecular Chemistry*, L. Fabrizzi, A. Poggi, Kluwer, 1994, 1; V. Balzani, F. Scandola, *Supramolecular Photochemistry*, Ellis Horwood, 1991, 355.

77. N. Vlakopoulos, P. Liska, J. Augustynski, M. Grätzel, J. Am. Chem. Soc., 1988, 110, 1216; M. Grätzel, Comments Inorg. Chem., 1991, 12, 93; M. Grätzel, K. Kalyanasundaram, in *Photosentitization and Photocatalysis Using Inorganic and Organometallic Compounds*, K. Kalyanasundaram, M. Grätzel, Eds., Kluwer, Dordrecht, 1993, 247.

78. M. Fujihira, in *Nanostructures Based Molecular Materials*, W. Göpel, Ch. Ziegler, VCH, Wienheim, 1992, p. 27.

79. F. Hucho, Angew. Chem. Int. Ed. Engl., 1998, 37, 1518; E. Pebay-Peyroula, G. Rummel, J. P. Rosenbusch, E. M. Landau, Science, 1997, 277, 1676.

80. S. Shinkai, O. Manabe, Top. Curr. Chem., 1984, 121, 67.

81. (a) G. Denti, S. Campagna, S. Serroni, M. Ciano, V. Balzani, J. Am. Chem. Soc. 1992, 114, 2944; (b) A. Juris, V. Balzani, S. Campagna, G. Denti, S. Serroni, G. Frei, H. U. Güdel, Inorg. Chem., 1994, 33, 1491.

82. M. Montalti, R. Ballardini, L. Prodi, V. Balzani, J. Chem. Soc. Chem. Commun., 1996, 2011.

83. S. P. Gupta, Chem. Rev., 1987, 87, 1183; 1989, 89, 1765.

84. R. A Prentis, Y. Lis, S. R. Walker, Br. J. Clin. Pharmac., 1988, 25, 387; W.J. Egan, K.M. Merz, Jr., J.J. Baldwin, J. Med. Chem., 2000, 43, 3867.

85. Ref. 56b, p. 238 - 240.

86. Ref. 56b, p. 217.

87. O. S. Tee, B. K. Takasaki, Can. J. Chem., 1985, 63, 3540.

88. P. F. Kiser, G. Wilson, D. Needham, Nature, 1998, 394, 459; P. Berressem, Chem. Brit., 1999, 35, 29.

89. F. Giulieri, M.-P. Krafft, J. G. Rjess, Angew. Chem. Int. Ed. Engl., 1994, 33, 1514.

90. F. M. Menger, S. S. Lee, X. Tao, Adv. Matter., 1995, 7, 669.

91. M. Puri, S. S. Marwaha, R. M. Kothari, J. F. Kennedy, Crit. Rev. Biotech., 1996, 16, 145.

92. H. M. Suh, J. Ahn, H. J. Kwon, Asian-Austr. J. An. Sci., 2001, 14, 548.

93. Several Japanese patents have been cited in H. Hashimoto, *Comprehensive Supramolecular Chemistry*, vol. 3, Pergamon, Oxford, 1996, p. 483.

94. P. T. Anastas, J. C. Warner, *Green Chemistry*, Oxford University Press, Oxford, 1998; *Frontiers in Benign Chemical Syntheses and Processes*, P. T. Anastas, T. C. Williamson, Eds, Oxford University Press, Oxford, 1998.

95. (a) C. Bolm, O. Beckmann, O. A. D. G. Dabard, Angew. Chem. Int. Ed. Engl., 1999, 38, 907; (b) *Pollution: Causes, Effects and Control*, R. M. Harrison, Ed., The Royal Society of Chemistry, London, 1996; (c) J. H. Clark, D. J. Macqarrie, Chem. Soc. Rev., 1996, 25, 303.

96. T. Gerngross, Chem. Eng. News, 1999, Sept. 13, p. 39.

97. American BioIndustry yearly conference Bio2000, Bostom, USA, 27 March, 2000.

98. P. Chen, X. Wu, J. Lin, K. L. Tan, Science, 199, 285, 91; A. C. Dillon, K. M. Jones, T. A. Bekkedahl, C. H. Klang, D. S. Bethune, M. J. Heben, Nature, 1997, 386, 377.

99. P. Morrell, Industrial Environmental Management, 1999, 10, 4.

100. J. Szejtli, J. Mater. Chem., 1997, 7, 575.

101. G. Wenz, K. Engelskirchen, H. Fischer, H. C. Nicolaisen, S. Harris (Henkel), Ger. Pat., DE 4009621 A1, 1990.

102. (a) J.-H. Fuhrhop, J. Köning, *Membranes and Molecular Assemblies. The Synkinetic Approach*, The Royal Chemical Society, Cambridge, 1994, p.44 - 46; (b) McCoy, Chem. Eng. News, 2000, 78, 37.

103. (a) T. Hayashita, *Comprehensive Supramolecular Chemistry*, vol. 6, Pergamon, Oxford, 1996, p. 671; (b) A. Yamauchi, T. Hayashita, S. Nishizawa, M. Watanabe, N. Teramae, J. Am. Chem. Soc., 1999, 121, 2319.

104. P. Kim, C. M. Lieber, Science, 1999, 286, 2148.

105. (a) M. Clemente-Leon, H. Soyer, E. Coronado, C. Mingotaud, C. J. Gomez-Garcia, P. Delhaes, Angew. Chem. Int. Ed. Engl., 1998, 37, 2842; O. Kahn, Chem. Brit., 1999, 35, 25; (b) M. M. Wienk, R. A. J. Janssen, J. Am. Chem. Soc., 1997, 119, 4492.

106. I. R. Peterson in *Nanostructures Based Molecular Materials*, W. Göpel, Ch. Ziegler, VCH, Wienheim, 1992, p. 195.

107. L. Pauling, Chem. Eng. News, 1946, 24, 1375; L. Pauling, Nature, 1948, 161, 707.

108. J. K. M. Sanders, Chem. Eur. J., 1998, 4, 1378.

109. R. Breslow, P. Campbell, J. Am. Chem. Soc., 1969, 91, 3085; R. Breslow, H. Kohn, B. Siegel, Tetrahedron Lett., 1976, 1645.

110. R. Breslow, C. Schmuck, J. Am. Chem. Soc., 1996, 118, 6601.

111. S. Liu, Z. Luo, A. D. Hamilton, Angew. Chem. Int. Ed. Engl., 1997, 109, 2678.

112. J. T. Grooves, R. Neumann, J. Org. Chem., 1988, 53, 3891.

113. R. Breslow, B. Gabriele, J. Yang, Tetrahedron Lett., 1998, 39, 2887; R. Breslow, Y. Huang, X. Zhang, J. Yang, Proc. Natl. Acad. Sci. USA, 1997, 94, 11156.

114. P. Mattei, F. Diederich, Helv. Chim. Acta, 1997, 80, 1555.

115. B. L. Zhang, R. Breslow, J. Am. Chem. Soc., 1997, 119, 1676.

116. E. Sulman, V. Matveeva, N. Semagina, I. Yanov, V. Bashilov, V. Sokolov, in *Recent Advances in the Chemistry and Physics of Fullerenes and Related Materials*, K. M. Kadish, R. S. Ruoff, Eds., 1998, 6, 1186.

117. Z. Clyde-Watson, A. Vidal-Ferran, L. J. Twyman, C. J. Walter, D. W. J. McCallien, S. Fanni, N. Bampos, R. S. Wylie, J. K. M. Sanders, New J. Chem., 1998, 22, 493.

118. W. M. Hosseini, A. J. Blacker, J.-M. Lehn, J. Am. Chem. Soc., 1990, 112, 3896.

119. L. S. Romsted, C. A. Bunton, J. H. Yao, Curr. Opin. Colloid In., 1997, 2, 622.

120. J.-H. Fuhrhop, W. Helfrich, Chem. Rev., 1993, 93, 1565.

121. R. Neumann, M. Cohen, Angew. Chem. Int. Ed. Engl., 1997, 36, 1738.

122. (a) Y. Aoyama, Top. Curr. Chem., 1998, 198, 131; (b) V. Ramamurthy, A. Joy, S. Uppili, J. Am. Chem. Soc., 2000, 122, 728; A. Joy, J. R. Scheffer, V. Ramamurthy, Org. Lett., 2000, 2, 119.

123. M.-V. Diaz, N. Spencer, J. F. Stoddart, Angew. Chem. Int. Ed. Engl., 1997, 36, 1904.

124. A. P. da Silva, H. Q. N. Gunaratne, C. P. McCoy, Nature, 1993, 364, 42.

125. P. Ball, L. Garwin, Nature, 1992, 355, 761; D. Bradley, Science, 1993, 259, 890.

126. R. A. Bissell, A. P. de Silva, H. Q. N. Gunaratne, P. M. L. Lynch, G. E. M. Maguire, C. P. McCoy, K. R. A. S. Sandanayake, Top. Cur. Chem., 1993, 168, 223; N. N. Li, P. M. Jacoby, Chem. Eng. News, 2000, March 6, 60.

127. H. C. Visser, D. N. Reinhoudt, F. de Jong, Chem. Soc. Rev., 1994, 23, 75.

128. A. Miniewicz, S. Bartkiewicz, A. Januszko, J. Parka, J. Incl. Phenom. Macrocycl. Chem., 1999, 35, 317; J. M. Kang, J. Rebek, Nature, 1997, 385, 50; Z. Clyde-Watson, A. Vidal-Ferran, L. J. Twyman, C. J. Walter, D. W. J. McCallien, S. Fanni, N. Bampos, R. S. Wylie, J. K. M. Sanders, New J. Chem., 1998, 22, 493.

Chapter 7

THE MOST INTERESTING MACROCYCLIC LIGANDS WHICH ARE HOSTS FOR INCLUSION COMPLEXES

7.1 Crown Ethers and Coronands, Cryptates and Cryptands [1]

7.1.1 Introduction

As noted in Chapter 3, Pedersen synthesized dibenzo-18-crown-6 **44** which marked the beginning of inclusion (or host-guest) chemistry as a minor product obtained as a result of the presence of catechol as an impurity in the reaction mixture [2]. However, in agreement with Pasteur saying "Dans les champs de l'observation le hasard ne favorise que les ésprits préparés". ("In the field of observation luck favours only prepared minds"), this discovery was by no means accidental. At that time Pedersen was searching for a compound **191** which would serve as Ca^{2+}-complexing agent. He synthesized and patented it [3]. The spin-off of this research was the discovery of the crown ethers, their ability to complex alkali metals and, finally, the Nobel Prize in 1987. This was totally new since only neutral

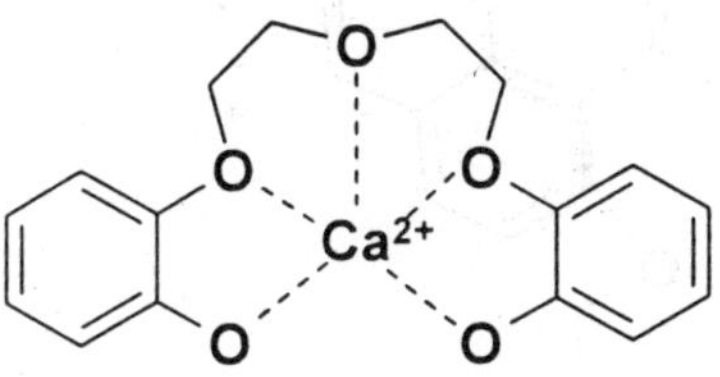

191

192

193

194

195

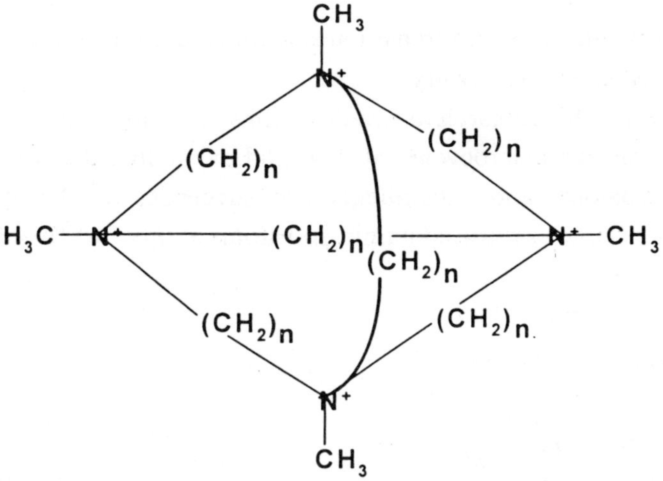

CH_3 N^+ $(CH_2)_n$ $(CH_2)_n$ H_3C N^+ $(CH_2)_n$ $(CH_2)_n$ N^+ CH_3 $(CH_2)_n$ $(CH_2)_n$ N^+ CH_3

196

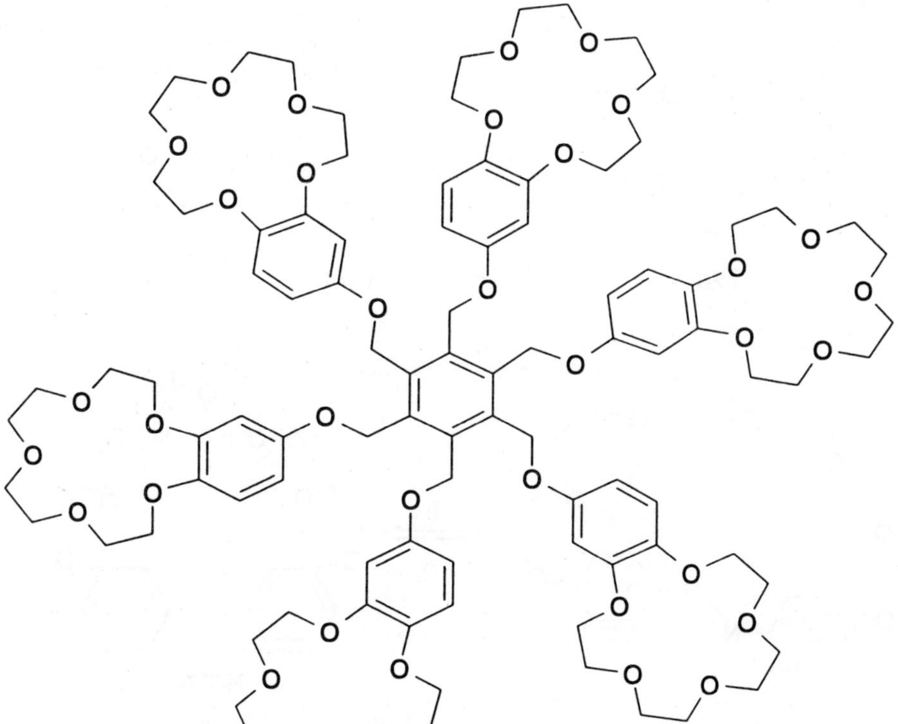

O O O O O O O O O O O O O O O O O O O O O O O O O O O O O O O O O O O O O O O O O O

197

neutral complexing agents for transition and coinage metals were known at the time of the Pedersen's discovery.

He continued this research in two directions. The first one consisted in the synthesis of various macrocycles of this kind expanding the ring's size and varying heteroatoms and substituents. He succeeded in the synthesis of macrocycles as large as those containing 60 atoms at times when only very few

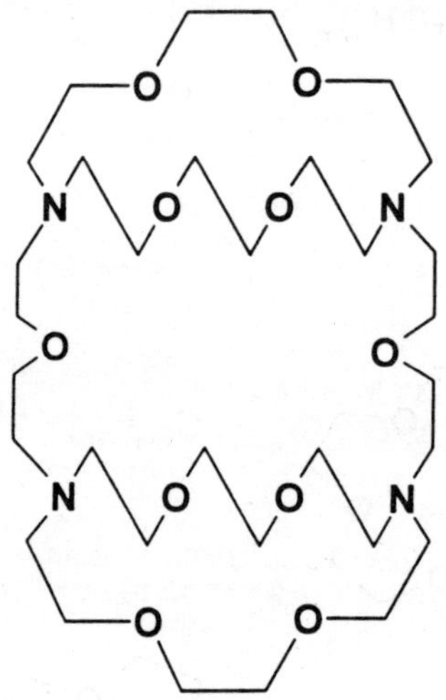

198

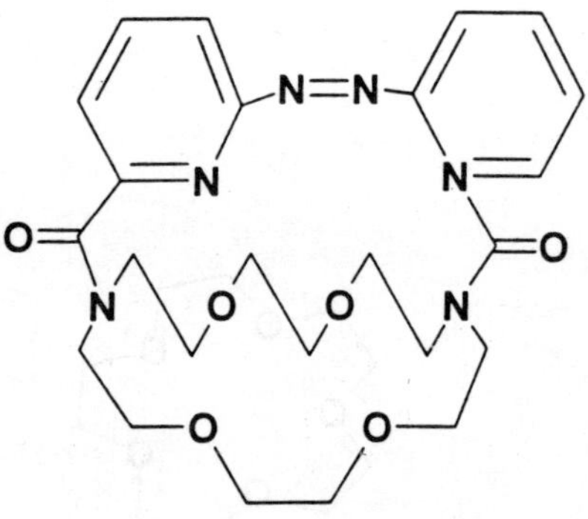

199

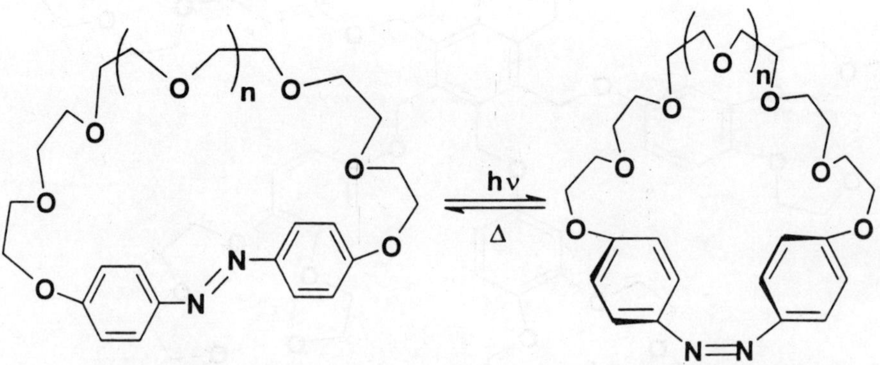

(E)-form

200a

(Z)-form

200b

such large macrocycles with well-defined structures were known. He also showed that nitrogen and sulfur atoms containing crown ethers could be easily obtained [4]. The studies of the crown ethers complexation behaviour determined the second direction of the Pedersen efforts. He observed that crown ethers selectively bound not only potassium, sodium, and other alkali ions but were also capable of complexing alkaline earth metal cations and alkylammonium ions. As discussed in Section 3.2, at the very beginning Pedersen recognized that systematic names of crown ethers were not 'user friendly' and he proposed the very convenient system for their designation. The analogous system for cryptands was proposed by Lehn [5]. Both systems were briefly presented in this Section. As briefly described in Chapter 3, Pedersen and Lehn led foundations of this burgeoning research field. Syntheses of thousands of such molecules varying in size and substituents and containing not only oxygen and nitrogen atoms but also sulfur atoms or aromatic rings built into the polyether macrocycle like in **192** [6] followed. Their modifications include numerous polycyclic systems exemplified by Cram's chiral ethers **193-195** [7], symmetrical N4 cage **196** [8] capable of tetrahedral recognition discussed in Chapter 3 and aesthetically pleasing **197** [9]. Bis-crown macrocycles may exhibit 'face to face' arrangement of the ether rings **198** [10] or isomers interconverting by photochemically induced change of configuration at N=N bond **199** influencing their complexing ability [11]. In the latter case, photoinduced isomerisation of 2,2'-azopyridine bridge dramatically changes cordination behaviour of the ligand towards heavy metal ions Cu(II), Ni(II), Co(II) and Hg(II). **200** exemplifies photoresponsive systems providing photochemical control of complexation.

7.1.2 Crown ethers and cryptands syntheses

As a matter of fact, the synthesis of crown ethers was patented in Britain as early as the mid 1950s by Stewart, Waddan and Borrows [12]. Cyclooligomerization of ethylene oxide in the presence of alkyl aluminum, zinc or magnesium compounds produced dioxane and other cyclic materials with cyclic tetramer **201** prevailing among the macrocyclic oligomers. Similarly to the famous C_{60} case which was reported without any appreciation of its importance [13] prior to the Nobel Prize inning report [14], the authors of the British patent (not interested in the complexing ability of the macrocycles) failed to recognize

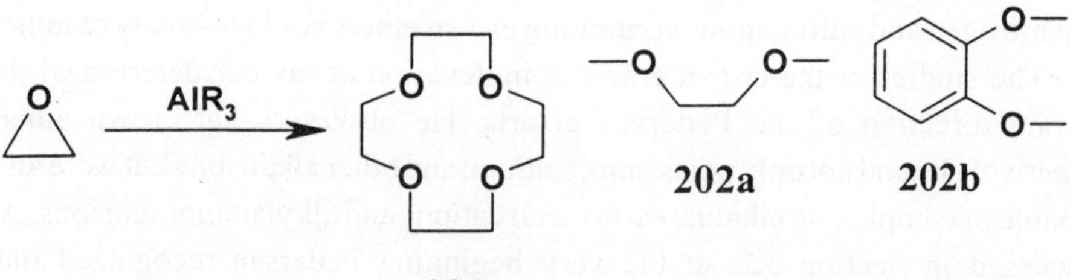

Figure 7.1.1. Cyclooligomerization of ethylene oxide leading to crown ether.

the potential of their finding.

The ethylenedioxy unit **202a** (Fig. 7.1.1) is the most common building block of most crown ethers and cryptands. This is owed to: (a) the flexibility of this system which is less rigid that 1,2-phenylenedioxy unit **202b**; (b) **202a** can easily assume the *gauche* conformation allowing for appropriate alignment of the acceptor groups of the donor; and (c) the availability of this unit in common industrial chemicals. On the other hand, acetal and propyleneoxy units are less suitable as crown ethers fragments since the former is acid labile whilst the latter must be converted from its more stable all-*anti* conformation to the one with oxygen atoms pointing inwards the macrocycle.

There is no general synthetic strategy allowing one to obtain any of the highly diversified members of crown ethers or cryptand families. Crown ethers can be formed by reacting linear polyethers that are electrophilic at one end and nucleophilic at the other. In such a reaction either the macrocycle can be formed or linear oligomeric (or polymeric) species are obtained (Fig. 7.1.2). One can shift the equilibrium between these processes by carrying out the reaction under *high dilution conditions* or by making use of the template effect. In the former case there is more probability that the ends of the reacting molecule will meet forming the macrocycle (first order reaction) prior to the meeting of two molecules resulting in the formation of the linear molecule (second order reaction). High dilution prohibits the latter reaction but it does not enhance the rate of the former one. In the case of template effect, adding of metal ion forces polyether conformation suitable for the macrocyclic ring closure. A typical reaction scheme involving the templating action of a sodium ion, *i.e.* preorganization, is given in Fig. 7.1.3. The importance of template

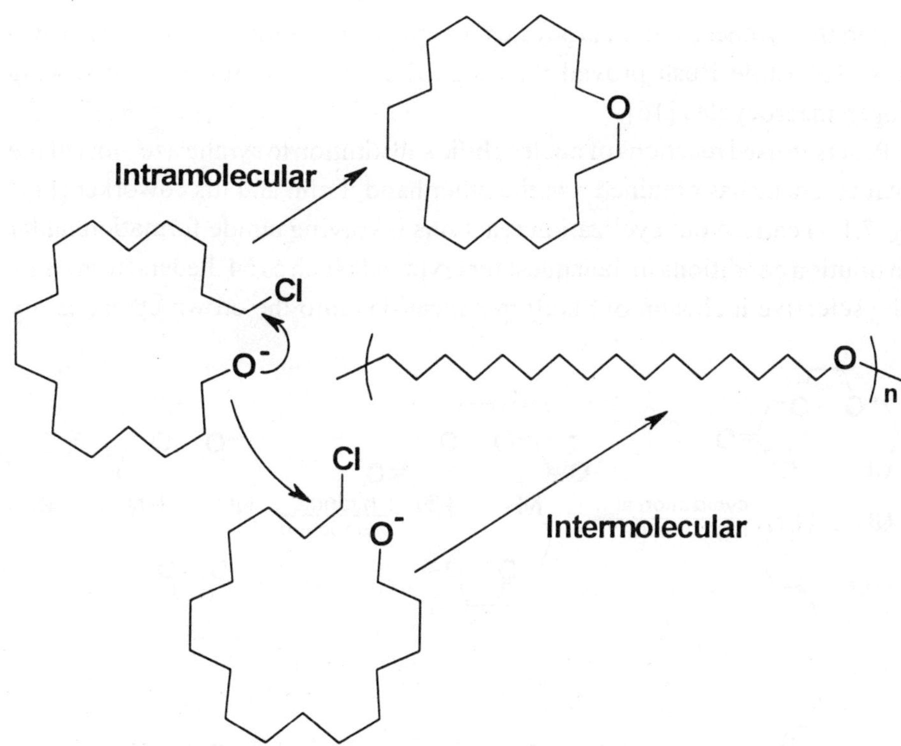

Figure 7.1.2. The formation of macrocycle and the completing oligomerization reaction.

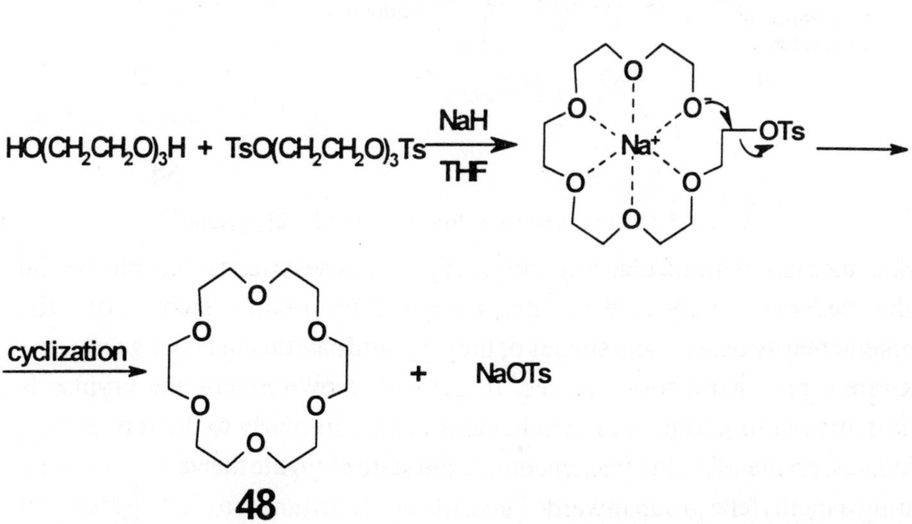

$$HO(CH_2CH_2O)_3H + TsO(CH_2CH_2O)_3Ts \xrightarrow[THF]{NaH} [Na^+ \text{ template complex}] \longrightarrow$$

$$\xrightarrow{\text{cyclization}} \mathbf{48} + NaOTs$$

Figure 7.1.3. The formation of 18-crown-6 making use of Na^+ template.

effect in the syntheses of macrocycles was proved by Greene in case of crown ethers [15] while Bush proved their significance in the reactions involving nitrogen macrocycles [16].

Pedersen used reactions of nucleophilic substitution to synthesize most of the crown ethers he has obtained. On the other hand, Lehn and his coworkers [17] (Fig. 7.1.4) carried out cyclization reactions involving amide formation under high dilution conditions in their quest for cryptands such as **54**. Pedersen analysis of the selective inclusion of alkali metal cations into the crown ethers cavity

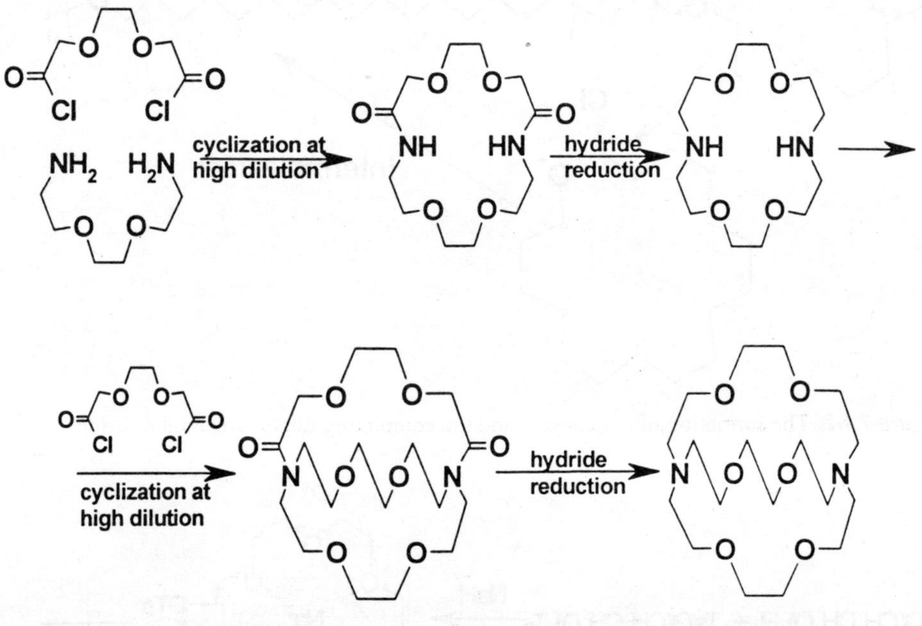

cyclization at high dilution

hydride reduction

cyclization at high dilution

hydride reduction

54

Figure 7.1.4. The scheme of formation of [2.2.2]cryptand.

marked the start of molecular recognition studies. As described in Chapters 2 and 3, the Pedersen analysis was later extended by Lehn's studies of the complementarity of sizes and shapes of the cryptand cavities and their guests, and by Cram's preorganization studies. In general, crown ethers and cryptands exhibit analogous complexation behaviour. Thus, similarly to the former host molecules, cryptands in the free, uncomplexed state elongate the vacant cavity by rotating a methylene group inward. Thus, the N...N distance in [2.2.2]-cryptand **54** across the cavity is extended to almost 70 pm [18] whilst, in the complexed

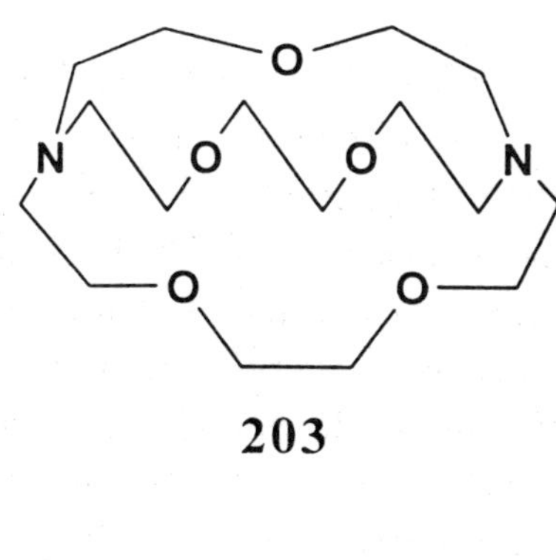

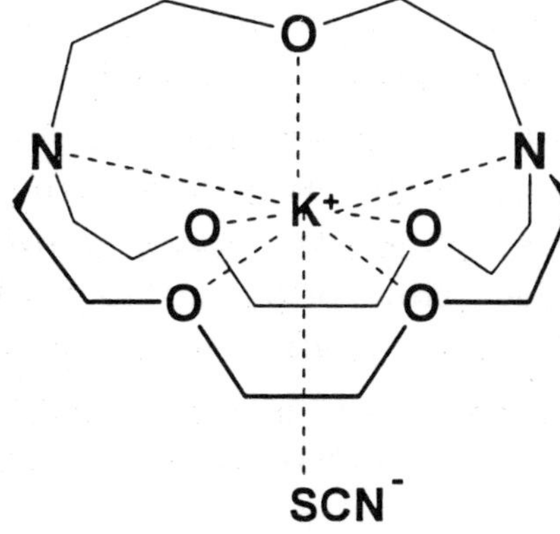

Figure 7.1.5. The structure of the complex of [2.2.1]cryptand with KSCN.

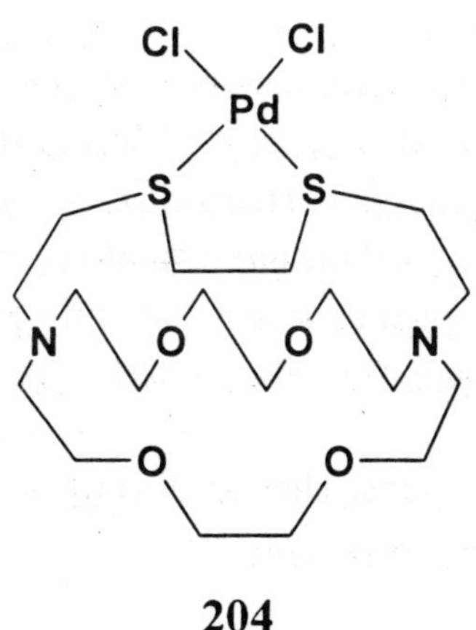

form with a spherical K^+ cation of ca. 26.6 pm in diameter, its internal size is ca. 28 pm [19]. The average structure of the complex exhibits an approximate three-fold symmetry axis. As concerns the smaller [2.2.1]-cryptand **203**, it is too small to include the K^+ cation. Therefore in the latter molecule (Fig. 7.1.5) there is the cation binding on the 18-membered ring surface accompanied by its solvation from above by the oxygen atom of the short chain while the thiocyanate anion solvates the cation from the bottom [20]. The structure of this complex resembles that of the analogous lariat complex. By an unfavourable match of the host and guest sizes, as is the case for the palladium dichloride complex **204**, a so-called exclusive complex is built [21]. The use of crown ethers as liquid membranes was presented in Sect. 6.3.6. Their use as sensors and switches have been also proposed [22]. However, the most exquisite application of these compounds as components of reducing reagents will be described in the next section.

7.1.3 Alkalides and Electrides

Alkali metal cations are ubiquitous in chemistry since these atoms can easily lose their single valence electron. The corresponding anions and isolated electrons are much more exotic species although Sir Humphry Davy note in 1808 about blue and bronze coloured potassium ammonia solutions indicates that this

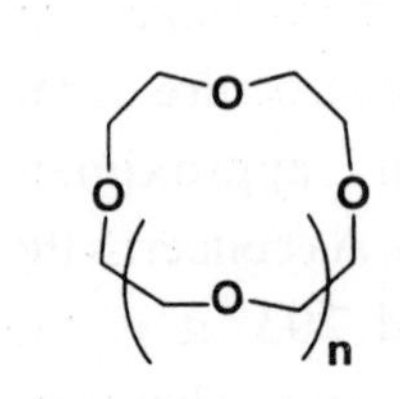

205a 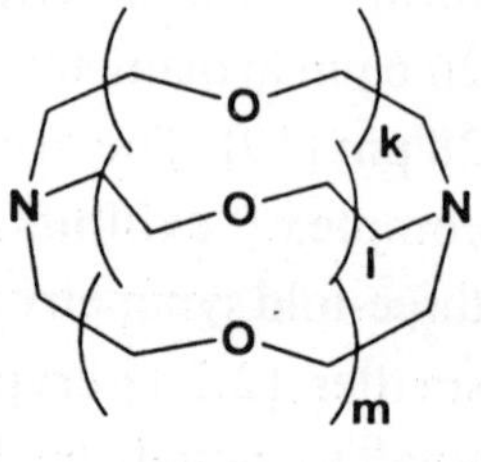 206a

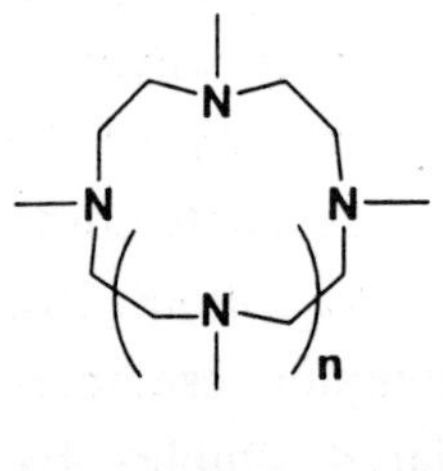

205b 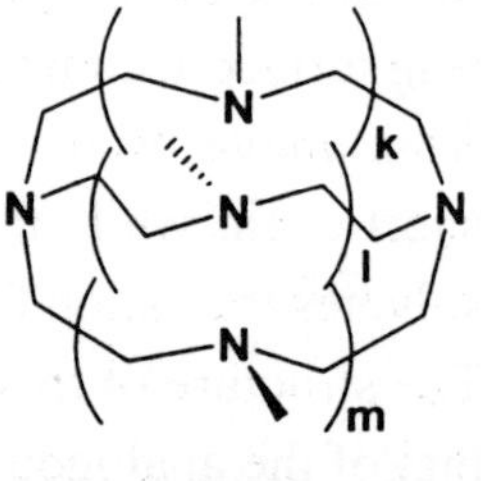 206b

was the first observation of solvated electrons. A strong tendency of crown ethers **205a**, cryptands **206a**, and their aza-analogues **205b** and **206b**, respectively, to complex alkali metal cations has opened new exciting possibilities allowing one to obtain not only ionic crystals having the corresponding anions but also crystals in which ionic sites are occupied by a single electron [23]. Such compounds are called alkalides and electrides, respectively. They consist of an alkali cation buried inside a crown or cryptand (or their aza analogue) and the corresponding anion obtained as a result of the disproportionation of an alkali metal into the cation and anion. The reaction of alkalide formation with one alkali metal is described by the equation:

$$2M(s)+nL \rightleftharpoons M^+(L_n)M^-(s) \tag{1}$$

The equilibrium (1) is very sensitive to several factors. The complexing ability of L, the electron affinity of M and the lattice energy of the resulting salt drive the alkalide formation while the lattice energies of the solid metal M(s) and the complexant L, the ionization energy of M, and the unfavourable entropy of formation of the well ordered crystalline product oppose it. Not only alkalides with the cation and anion of the same element but also mixed ones such as $K^+C(222)Na^-$ have also been obtained [24].

Even more exotic than alkalides are electrides. They are formed instead of alkalides when an excess of a strong complexant is present:

$$M(s)+nL \rightleftharpoons M^+(L_n)e^-(s) \tag{2}$$

Carefully purified solvents, specialized rigorously clean glassware, vacuum technique ($p < 10^{-5}$ torr) and low temperatures (usually < 230 K) are necessary for synthesizing alkalides and electrides from cryptands (or **205a**, **206a** or **206b**) and the corresponding metal(s). Their stability and reactivities should not be confused. They all react vigorously with air and moisture, but when kept under vacuum $Na^+C(222)Na^-$ is stable for hours at room temperature [25a]. Even more stable are crystalline salts of Na^- and K^- with aza[2.2.2]cryptand **207** [25b]. They are stable in a vacuum at room temperatures for months. On the other hand, $Li^+C(222)e^-$ rapidly decomposes above 230K [23b]. The identity of these unusual salts was proved by X-ray analysis of more than 30 alkalides involving Na^- through Cs^- [23b]. The only other metal anions of proven structure known today are auride Au^- [26] and nonastannide Sn_9^{4-} [27] ones. The crystals of both alkalides and electrides consist of large, closely packed cations having 800-1000 pm in diameter. The anions in the former salts usually fit into the 'empty' cavities between the closely packed cations. In electrides, the electrons are trapped in these voids.

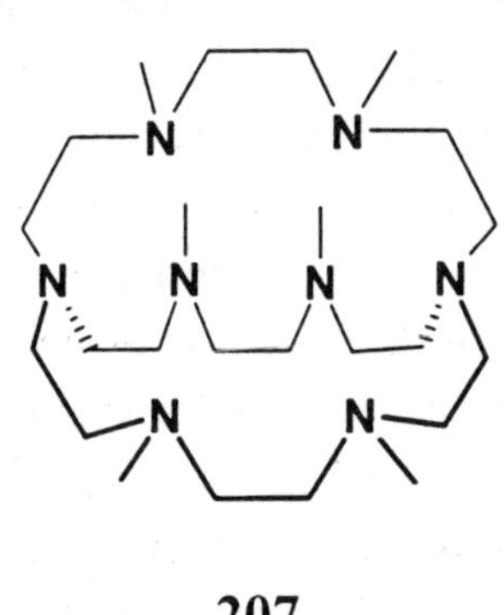

207

An interesting comparison have been carried out between the X-ray structures of $Na^+C(222)Na^-$ and the iodine salt $Na^+C(222)I^-$ [28]. Sodium cation is known to have a very small radius. Interestingly, the radius of the corresponding anion is much larger, very close to that of I^-. As a result there is no significant difference between Na^+-N, Na^+-O, C-N and C-O bonding distances in the latter complexes. Similarly, all corresponding nonbonded distances are very close in both salts. On the other hand, C1-C2 bond lengths in the latter cryptates having sodium and iodine anions differ by 15 pm and large differences (from 15 to 50 pm) have been found between the distances between the negative ion (Na^- or I^-) and Na^+, N or O atoms. Considerable differences in bond angles (up to 15°) and those in torsional angles (up to 37°) reflect the flexibility of the host [2.2.2]cryptand **54** enabling the accomodation of various guests inside its cage.

Interestingly, a considerable number of trapped electrons are present in most alkalides. However, their concentration (up to several percent) can be significantly reduced by applying an excess of metal during the synthesis. Na^-

ions have $1s^2 2s^2 p^6 3s^2$ configuration. As such they should be diamagnetic in pure sodides, but even a small admixture of electrons in the crystals dramatically changes their magnetic and optical properties.

The X-ray structures of five electrides, representing even greater novelty, have cation packing analogous to those in the corresponding alkalide structures. However, the anionic sites in the electrides are X-ray empty with no detectable electron densities. Anions, that is electrons, in electrides can be considered to be 'stoichiometric F-centers' since they are trapped at anions vacancies similarly to the entrapment of electrons in electron doped alkali halide crystals [29]. As mentioned above, alkalides are diamagnetic. The magnetic behavior of electrides is much more complex since the interaction of loosely bound electrons with one another depends on the structure of the solid and the nature of complexant.

NMR is one of the most easy and effective methods in alkalides studies since M^+ and M^- signal positions differ considerably. For instance, Na^+ and Na^- signals of $Na^+C(222)Na^-$ in ethylamine appear at 10.4 ± 0.5 ppm and 62.8 ± 0.2 ppm, respectively, while the signal of uncomplexed Na^+ should lie lower than at -10 ppm [30]. Moreover, the signal of Na^+ of the complex is much broader pointing to a restricted motion of the cation in the cryptand cage.

Most of alkalides and electrides behave as semiconductors characterized by band gaps of few tenth of eV to 2.5 eV. Alkalides and electrides can form not only crystals but also thin films suitable for optical studies. They strongly absorb in the red and infrared regions. A single broad absorption peak at about 1200-1800 nm typical for electrides is analogous to the corresponding band of solvated and trapped electrons in liquids and glasses [31]. The analogous bands in transmission spectra endow the electride with dark blue colour similar to that observed by Davy almost two hundred years ago.

Alkalides and electrides are the strongest known reducing agents in a given solvent due to the presence of M^- or e^-_{solv} in their solution [32]. For instance, only a catalytic amount of 18-crown-6 **48** is necessary to carry out a reduction of alkynes to alkanes at sufficient reaction times [33]. Similarly to reduction of aromatic derivatives in the presence of alkalides and electrides [34], the latter reaction results in a more complete reduction than in the classical Birch reaction.

Highly reactive nanoscale metal particles (3 to 15 nm in diameter) are formed by reducing metal salts to metals by solvated elecrons or alkali metal anions that may also find use in organometallic synthesis [35].

7.1.4 Miscellaneous molecules involving crown ethers, cryptands and related moieties

A plethora of crown ether- or cryptand type molecules have been reported. Some of them are depicted below to show their diversity and versatility. Lariat azaether containing cyclen macrocycle **208** [36] and azaethers involving triazole **209** [37], furane or pyrrole containing macrocycles which can complex two copper **210** [38] or one barium cations **211** [39], spiro-linked crown ethers **212** [40] and cage compounds **213** [41] which, in addition to two alkali cations, could host another guest are few examples of sometimes somewhat exotic crown ethers. Notably, Cram introduced the name *corand* for crown ethers involving ether and pyridine binding sites and sterically confining groups such as **44** or **193** and the name *coraplexes* for their inclusion complexes [42]. Tricyclic molecules **214** [43], **215** [44], **216** [45], **217** [46], **218** [47] represent a small selection of diverse cryptands. **219a** is a pseudocryptand capable of the simultaneous binding of boron and alkali metal cations **219b** [48]. Macrotricyclic cryptand capable of spherical recognition **53** was discussed in Section 3.1. Diprotonated [1.1.1]cryptand **220a** capable of *in-out* conformational isomerism **220b–d** is presented in Fig. 7.1.6. Its parent compound **220a** exhibits extraordinary acid-base and proton transfer properties. On the basis of the NMR study, **220a** was found to be thermodynamically an exceptionally strong base ($pK_a = 17.8$ for the first internal protonation) on one hand and, in view of remarkably slow proton transfer in the protonated species, kinetically it is a remarkably sluggish base, on the other [49]. Crystallographic studies of the cryptand and its mono- and diprotonated forms confirmed that the protons are situated inside the molecular cavity [50].

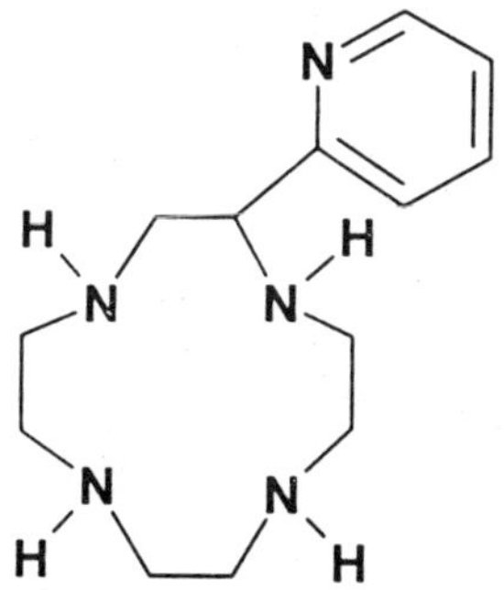

208

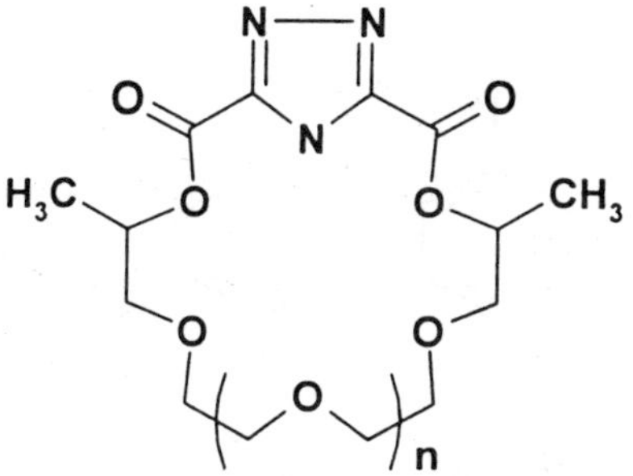

209

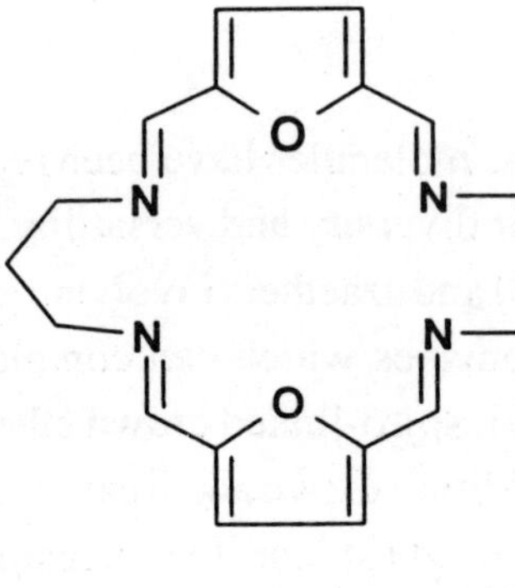

210

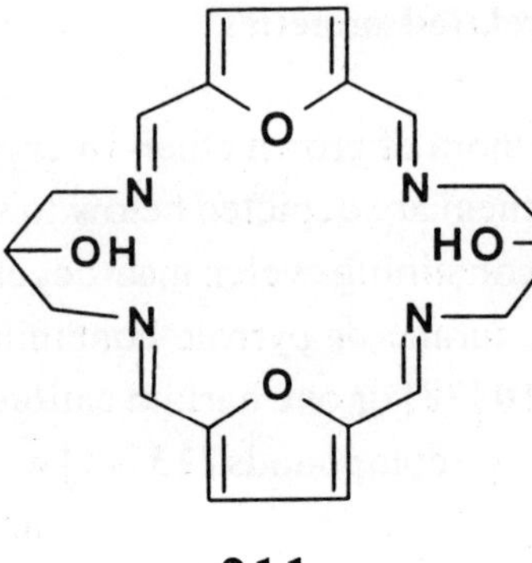

211

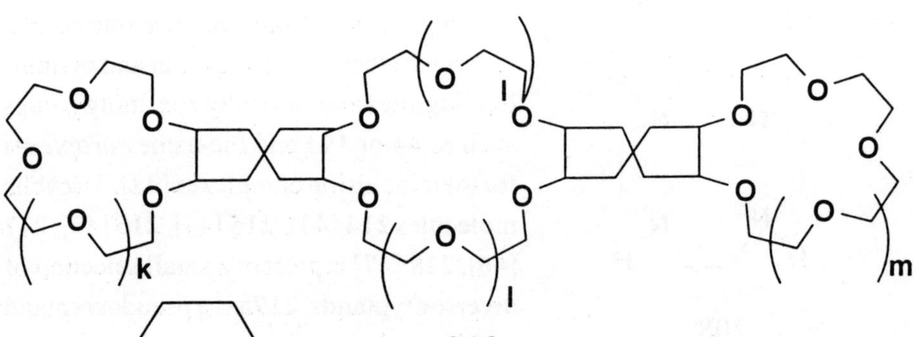

212

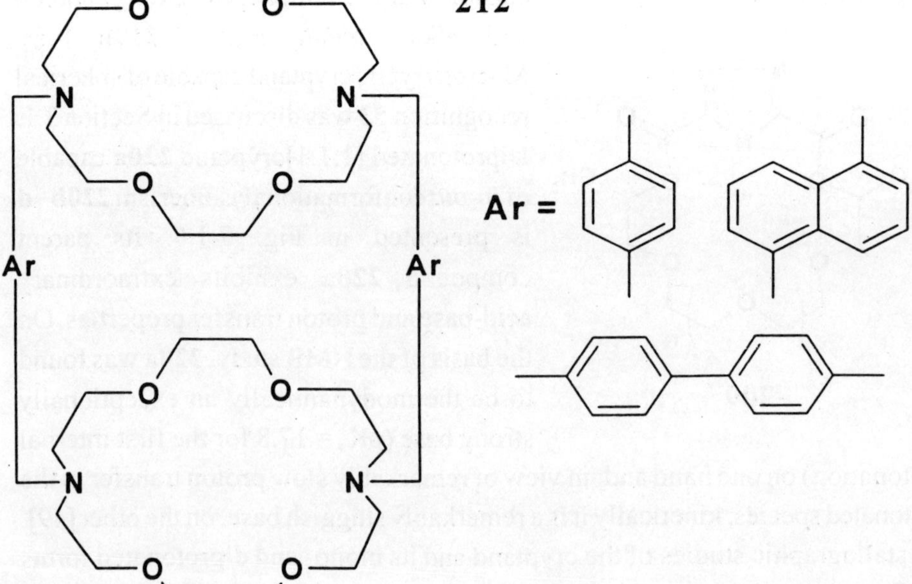

Ar =

213

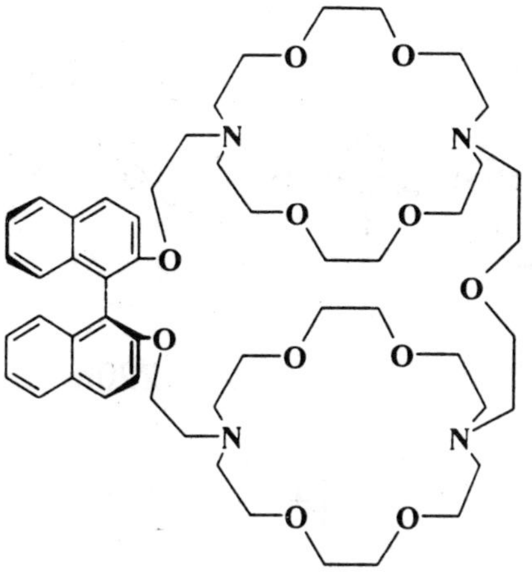

214

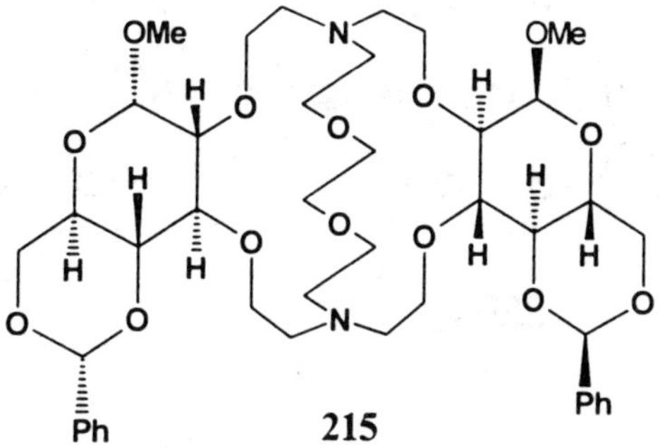

215

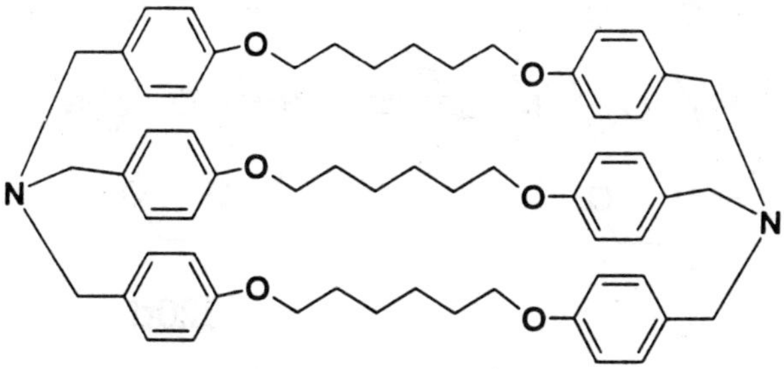

216

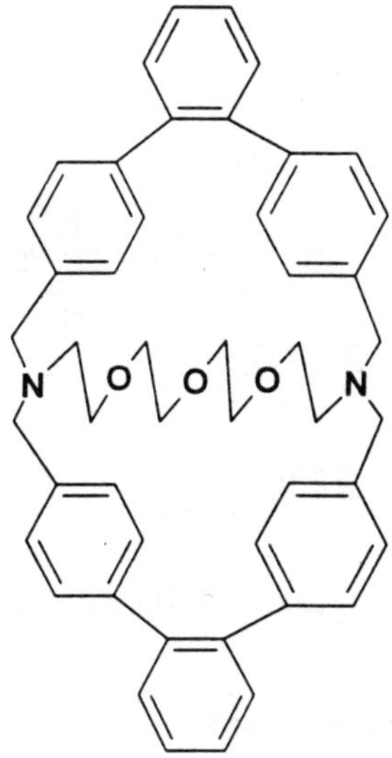

217

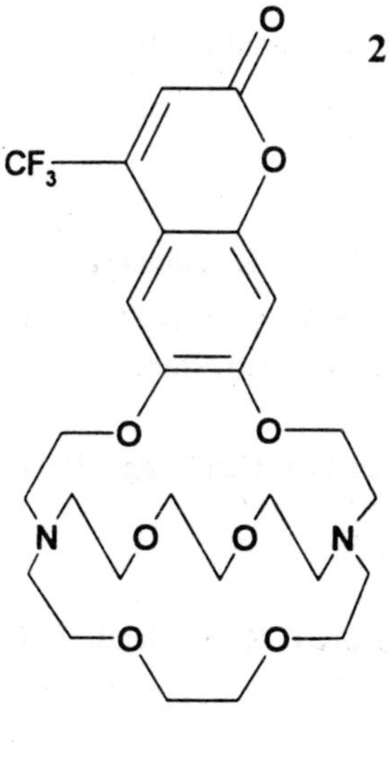

218

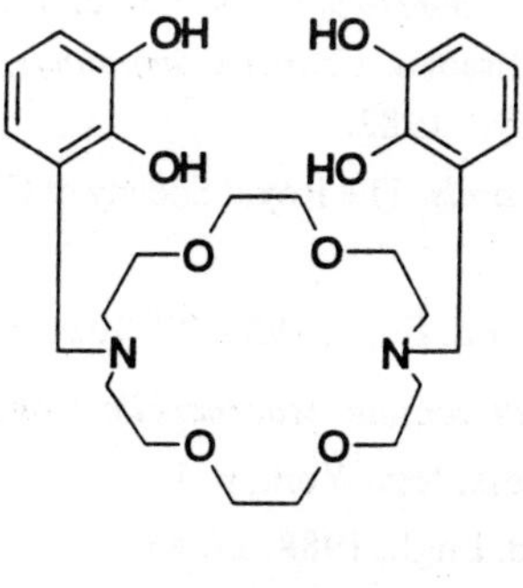

219a

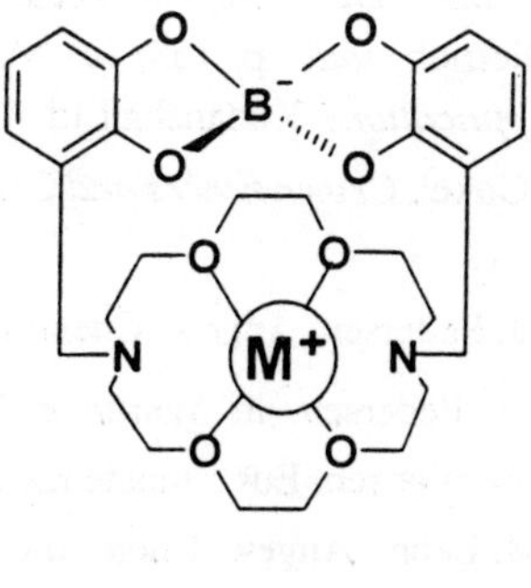

219b

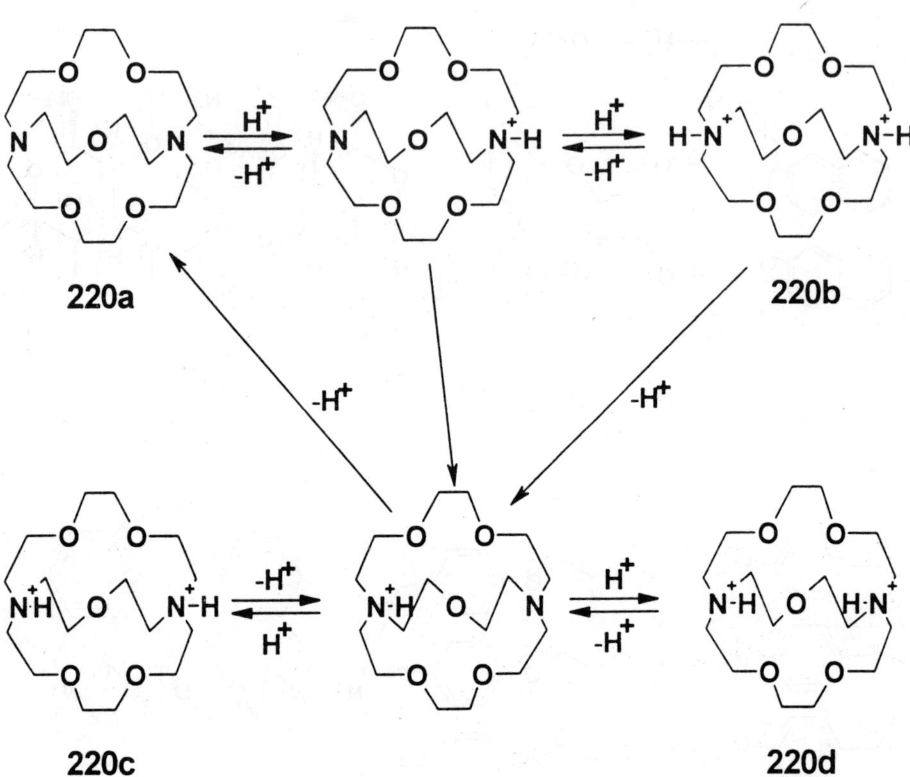

Figure 7.1.6. The protonation-deprotonation reactions and in-out isomerism of [2.2.1]cryptand.

References

1. (a) J. S. Bradshaw, R. M. Izatt, A. V. Bordunov, C. Y. Zhu, J. K. Hathaway, in *Comprehensive Supramolecular Chemistry*, G. W. Gokel, Ed., 1996, vol. 1, p. 35; (b) B. Dietrich, *ibid.*, p. 153; (c) M. Hiraoka, *Crown Compounds, Their Characteristics and Applications*, Kodansha Ltd, Tokyo, 1982.
2. G. Gokel, *Crown Ethers and Cryptands*, The Royal Society of Chemistry. Cambridge, 1994, p. 4.
3. C. J. Pedersen, Angew. Chem. Int. Ed. Engl., 1988, 27, 1021.
4. C. J. Pedersen, in *Synthetic Multidentate Macrocyclic Compounds*, R. M. Izatt, J. J. Christiansen, Eds., Academic Press, New York, p. 1.
5. J.-M. Lehn, Angew. Chem. Int. Ed. Engl., 1988, 27, 89.
6. J. D. Lamb, R. M. Izatt, C. S. Swain, J. S. Bradshaw, J. J. Christensen, J. Am. Chem. Soc.,

1980, 102, 479.

7. D. J. Cram. Angew. Chem. Int. Ed. Engl., 1988, 27, 1009.
8. F. P. Schmidtchen, Chem. Ber., 1980, 113, 864.
9. E. Weber, unpublished results cited in E. Weber, F. Vögtle, Top. Curr. Chem., 1981, 98, 1.
10. J.-M. Lehn, J. Simon, J. Wagner, Angew. Chem. Int. Ed. Engl., 1973, 7, 578.
11. S. Shinkai, O. Manabe, Top. Curr. Chem, 1984, 121, 67.
12. see Ref. 1, p. 3.
13. E. A. Rolfing, D. M. Cox, A. Kaldor, J. Chem. Phys., 1984, 81, 3322.
14. H. W. Kroto, J. R. Heath, S. C. O'Brien, R. F. Curl, R. E. Smalley, Nature, 1985, 318, 162.
15. R. N. Greene, Tetrahedron Lett., 1972, 1793
16. D. H. Busch, Acc. Chem. Res., 1978, 11, 392
17. B. Dietrich, J.-M. Lehn, J.-P. Sauvage, Tetrahedron Lett, 1969, 34, 2885; J.-M. Lehn, Acc. Chem. Res., 1978, 11, 49.
18. B. Metz, D. Moras, R. Weiss, J. Chem. Soc./Perkin Trans. 2, 1976, 423.
19. D. Moras, B. Metz, R. Weiss, Acta Crystallogr. Sect. B, 1973, 29, 383; J.-M. Lehn, Struct. Bonding, 1973, 16, 1.
20. F. Mathieu, B. Metz, D. Moras, R. Weiss, J. Am. Chem. Soc., 1978, 100, 4412.
21. R. Louis, J.C. Thierry, R. Weiss, Acta Crystallogr., Sect. B, 1974, 30, 753.
22. Ref. 1, section 5.5.
23. (a) J. L. Dye, J. E. Jackson, P. Cauliez, in *Organic Synthesis for Materials and Life Sciences*, Z. Yoshida, Ed., VCH, New York, 1992, 243; (b) M. J. Wagner, J. L. Dye, in *Comprehensive Supramolecular Chemistry*, J.-M. Lehn, Ed., Kluwer, Dordrecht, v. 1, p. 477.
24. A. Ellaboudy, M. L. Tinkham, B. van Eck, J. L. Dye, P. B. Smith, J. Phys. Chem., 1984, 88, 3852.
25. (a) J. M. Ceraso, J. L. Dye, J. Am. Chem. Soc., 1973, 95, 4432; (b) J. Kim, A. S. Ichimura, R. H. Huang, M. Redko, R. C. Phillips, J. E. Jackson, J. L. Dye, J. Am. Chem. Soc., 1999, 121, 10666.
26. T. H. Teherani, W. J. Peer, J. J. Lagowski, A. J. Bard, J. Am. Chem. Soc., 1978, 100, 7768.
27. J. D. Corbett, P. A. Edwards, J. Am. Chem. Soc., 1979, 99, 3313.
28. F. J. Tehan, B. L. Barnett, J. L. Dye, J. Am. Chem. Soc., 1974, 96, 7203.
29. J. J. Markham, *F Centers in Alkali Halides*, Academic Press, New York, 1966.
30. J. M. Ceraso, J. L. Dye, J. Chem. Phys., 1974, 61, 1585.
31. J. E. Hendrickson, W. P. Pratt, C.-T. Kuo, Q. Xie, J. L. Dye, J. Phys. Chem., 1996, 100, 3395

32. Z. Jedlinski, Acc. Chem. Res., 1998, 31, 55.

33. D. J. Mathre, W, C. Guida, Tetrahedron Lett, 1980, 21, 4773.

34. R. R. Dewald, S. R. Jones, B. S. Schwartz, J. Chem. Soc. Chem. Commun., 1980, 272 and other papers cited in Ref. 1b.

35. K.-L. Tsai, J. L. Dye, J. Am. Chem. Soc., 1991, 113, 1650.

36. E. Kimura, Y. Katake, T. Koike, M. Shionoya, M. Shiro, Inorg. Chem., 1990, 29, 4991; E. Kimura, T. Koike, M. Takehashi, Chem. Commun., 1985, 385.

37. J. S. Bradshaw, D. A. Chamberlin, P. E. Harrison, B. E. Wilson, G. Arena, N. K. Dalley, J. D. Lamb, R. M. Izatt, F. G. Morin, D. M. Grant, J. Org. Chem., 1985, 50, 3065.

38. M. G. B. Drew, P. C. Yates, J. Trocha-Grimshaw, A. Lavery, K. P. McKillop, S. M. Nelson, J. Nelson, J. Chem. Soc. Dalton Trans., 1988, 347; S. M. Nelson, C. V. Knox, M. McCaan, M. G. B. Drew, *ibid.*, 1981, 1669; H. Adams, N. A. Bailey, D. E. Fenton, R. J. Good, R. Moody, C. O. R. de Barbarin, *ibid.*, 1987, 207.

39. M. G. B. Drew, F. S. Esho, S. M. Nelson, J. Chem. Soc. Dalton Trans, 1983, 1653.

40. E. Weber, J. Org. Chem., 1982, 47, 3478.

41. F. Vögtle, H. Puff, E. Friedrichs, W. M. Müller, Angew. Chem., 1982, 94, 443.

42. D. J. Cram, J. M. Cram, *Container Molecules and Their Guests*. The Royal Society of Chemistry, Cambridge, United Kingdom, 1994, p. 7.

43. B. Dietrich, J.-M. Lehn, J. Simon, Angew. Chem., 1974, 86, 443.

44. P. Bako, L. Fenichel, L. Toeke, Liebigs Ann. Chem., 1990, 12, 1161.

45. N. Wester, F. Vögtle, Chem. Ber., 1980, 113, 1487.

46. L. Rossa, F. F. Vögtle, Liebigs Ann. Chem., 1981, 3, 459.

47. P. Buet, F. Kastenholz, E. Grell, G. Kaeb, A. Haefner, F. W. Schneider, J. Phys. Chem. A, 1999, 103, 5871.

48. E. Graf, M. W. Hosseini, R. Ruppert, N. Kyritsakas, A. De Cian, J. Fischer, C. Estournes, F. Taulelle, Angew. Chem. Int. Ed. Engl, 1995, 34, 1115; A. H. van Oijen, N. P. M. Huck, J. A. W. Kruitzer, C. Erkelens, J. H. van Boom, R. M. J. Liskamp. J. Org. Chem., 1994, 59, 2399; T. Nabeshima, T. Inaba, T. Sagae, N. Furukawa, Tetrahedron Lett., 1990, 31, 3919.

49. P. B. Smith, J. L. Dye, J. Cheney, J.-M. Lehn, J. Am. Chem. Soc., 1981, 103, 6044.

50. H. J. Brügge, D. Carboo, K. von Deuten, A. Knöchel, W. Dreissig, J. Am. Chem. Soc., 1986, 108, 107; *Crown Compounds. Toward Future Applications*. S. R. Cooper, Ed., VCH Publ., New York, 1992.

7.2 Calixarenes [1], Hemispherands, and Spherands [2]

7.2.1 Calixarenes syntheses

The name *calixarenes* was proposed by Gutsche for cyclic oligomers of general formula **221** since the word '*calix*', coming from Greek and Latin, meaning beaker reminds us of the most freqently assumed conformation of the tetramer bearing the name calix[4]arene **18**. The molecules of this type, readily obtained by treatment of *p*-alkylphenol with formaldehyde and base (Fig. 7.2.1)

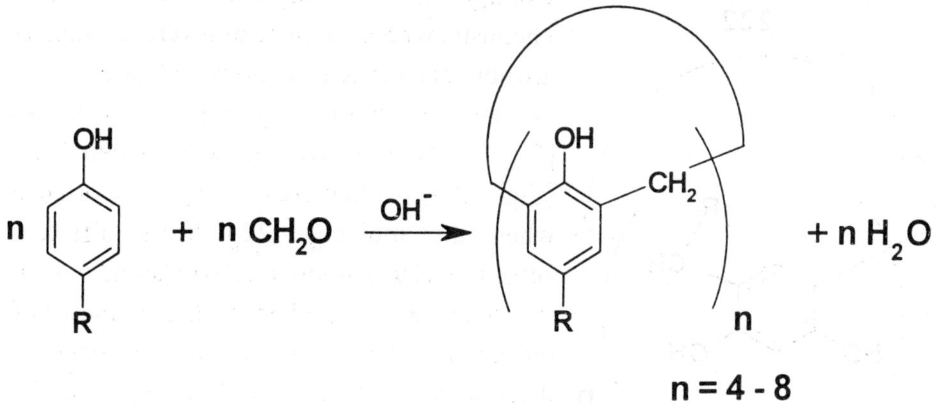

n (OH-phenyl-R) + n CH_2O $\xrightarrow{OH^-}$ [calixarene unit with OH, CH_2, R]$_n$ + n H_2O

n = 4 - 8

221

Figure 7.2.1. The one-pot synthesis of calix[n]arenes.

[1c], have been known for some time, but the interest in this domain was limited to the formation and properties of phenol-formaldehyde condensates. Pedersen's discovery of the complexation ability of crown ethers stimulated the development of calixarene research changing its perspectives from the classical organic chemistry to the supramolecular one. Böhmer [1c] divides these compounds into three groups. The first one **222** consists of *n*-metacyclophanes bearing no substituents on the phenyl rings. Calixarenes **221** (R = *t*-Bu) form the second, the most popular, group while the analogous products of resorcinol condensation **223** bear the name *resorcarenes*.

Contrary to most other synthetic macrocycles, calixarenes are easy available since one can synthesize them using inexpensive chemicals in a 'one-pot'

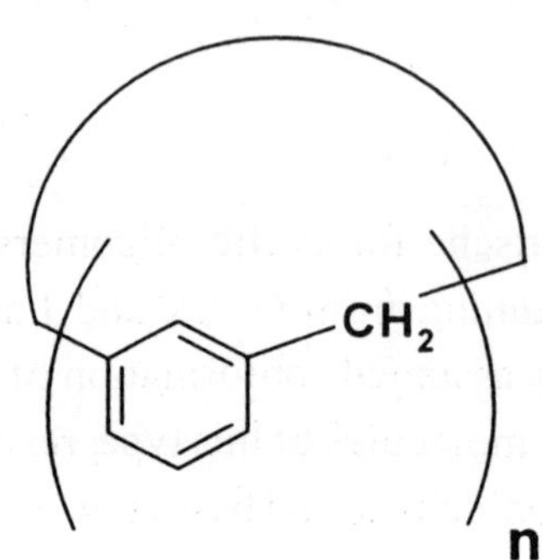

222

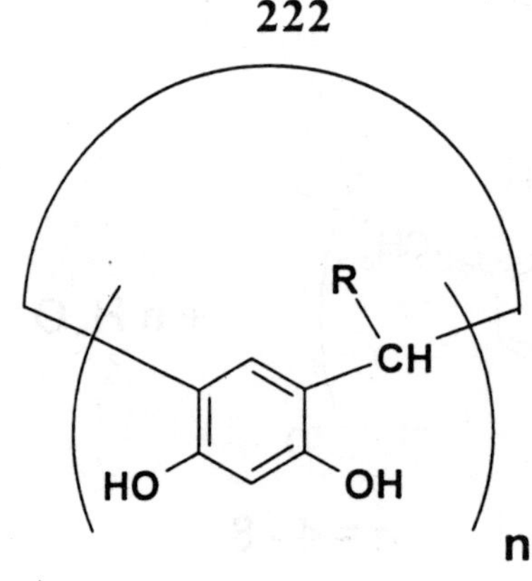

223

reaction. This reaction (Fig. 7.2.1) enables one to obtain tetra-, hexa- or octamer **221** in ca. 50%, 85%, and 63% yields for $n = 4$, 6 and 8, respectively [1c]. Such high yields are quite astonishing since in these reactions 8, 12 or 16 new covalent bonds are formed in a defined manner. Moreover, the formation of these cyclic oligomers is significantly favoured in comparison to their linear analogues. No other example of this kind is known in polymer chemistry where mixtures of several oligomers are obtained in such reactions. Calixarenes **221** can be also obtained by stepwise syntheses (Fig. 7.2.2) [3] or by fragment condensation (Fig. 7.2.3) [4] that are of special importance when the molecule **224** bears different substituents R. Although calix[n]arenes with n = 4, 6, and 8 are best known, the syntheses of the macrocycles with $n = 5$ [5] and 7 [6] and of those with n up to 20 have been reported [7].

In view of the greater reactivity of resorcinol, its cyclic tetramers are synthesized by the acid catalized condensation with less reactive aldehydes (Fig. 7.2.4) [1c]. However, in contrast with calixarenes **221**, only hexamer resorcarene has been reported [8]. In view of different relative configurations at the CHR bridges, **225** can exist in form of four diastereomers [1c].

By bridging calixarenes **226** [9], forming their dimers like **227** with one or several bridges [10] or by combining them with crown ethers **228**[11], calixarenes with several novel architectures and complexation behaviour have been obtained. Some of such systems have been proposed as prospective sensors.

Hydrogen bonded capsule consisting of calixarene dimers will be briefly presented in Section 8.3.2.

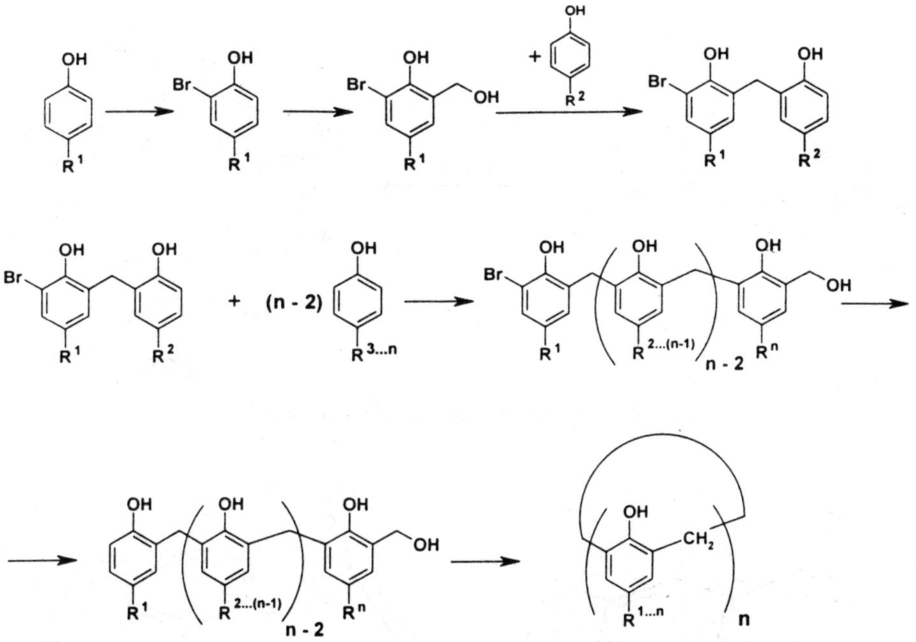

Figure 7.2.2. Stepwise synthesis of calixarenes bearing different substituents.

$R^1 = R^2$ or $R^3 = R^4$

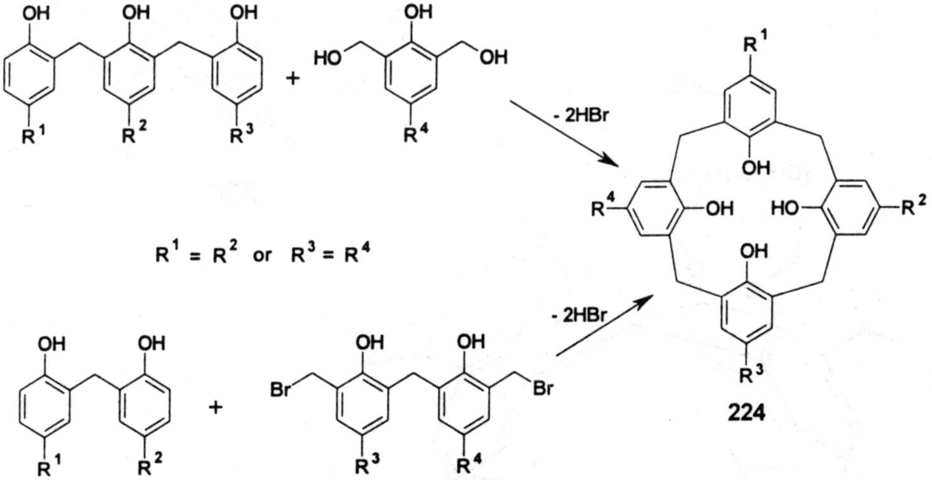

Figure 7.2.3. Two synthetic approaches to calix[4]arenes with different substituents.

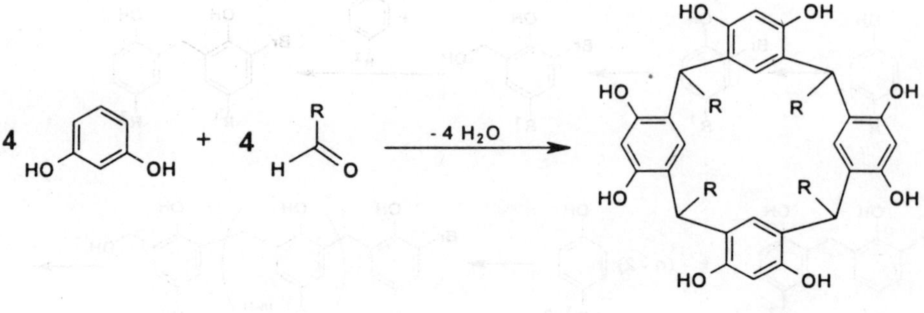

4 resorcinol (HO, OH) + 4 RCHO (H, R, O) —(- 4 H_2O)→ 225

Figure 7.2.4. One-pot synthesis of resorcarene.

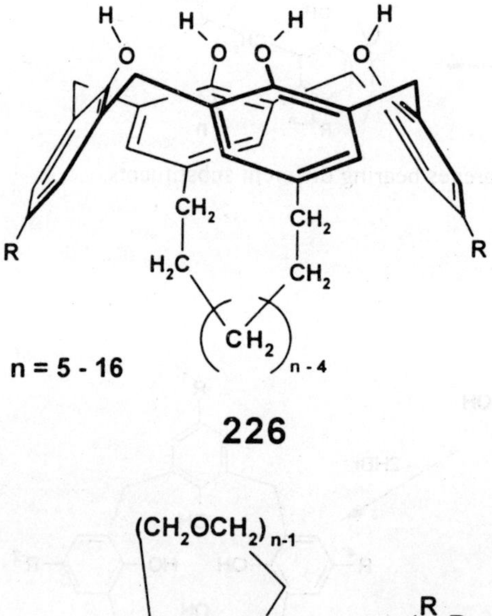

226: n = 5 - 16; CH_2, H_2C, CH_2, CH_2, $(CH_2)_{n-4}$

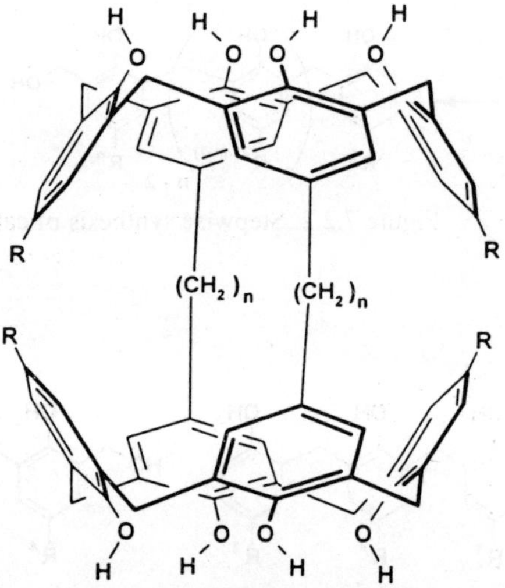

227: $(CH_2)_n$, $(CH_2)_n$

228: $(CH_2OCH_2)_{n-1}$

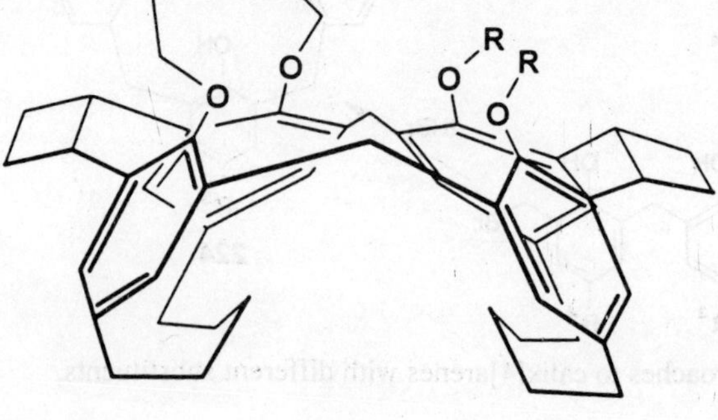

7.2.2 Calixarene conformations

The calix[4]arenes are known to assume one of four conformations: cone (from which its name was coined), partial cone, 1,2-alternate and 1,3-alternate shown in Fig. 7.2.5. The parent calix[4]- and [5]-arenes exhibit only cone conformation. However, at elevated temperatures they undergo ring inversion between two cones shown in Fig. 7.2.6. Other conformations are realized only for *O*-substituted calix[4]arenes.

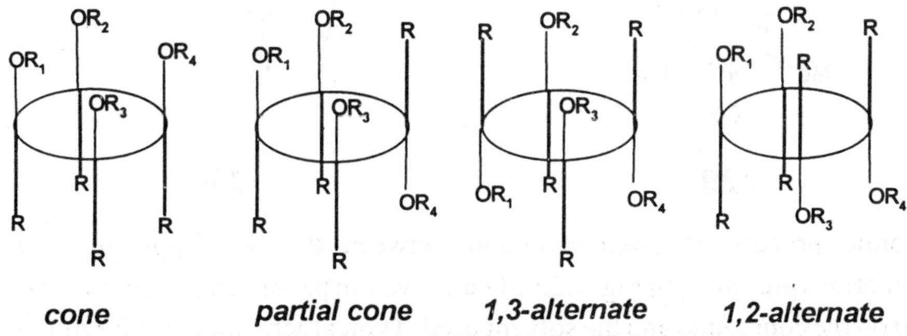

Figure 7.2.5. The conformations of calix[4]arenes.

All crystal structures of calix[4]arenes with free OH groups exhibit the cone conformation stabilized by intramolecular hydrogen bonds [12]. However, the cone may suffer considerable deformations owing to the steric hindrance [13].

In solution, calix[4]arene with *t*-butyl groups **18** shows only four sets of signals for the hydroxy, aromatic, methylene and *t*-butyl protons in the room-temperature ^{1}H NMR spectrum [14]. The CH_2 protons in the molecule are nonequivalent. Their signal splits at temperatures lower than 60° yielding a pair of doublets with the coupling constant of 12–14 Hz typical of the geminal protons [15]. Thus at higher temperatures this signal is an average resulting from

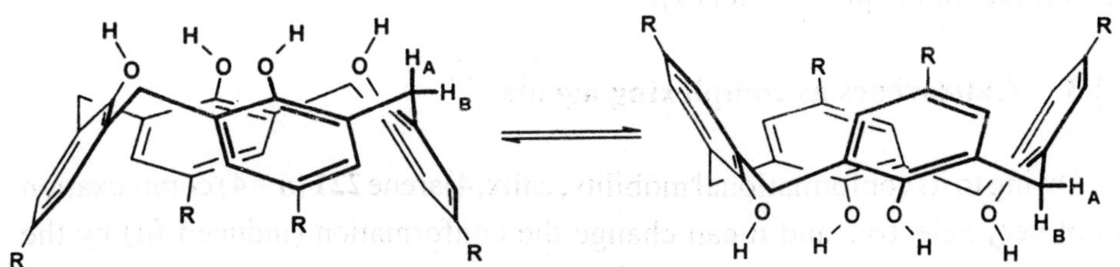

Figure 7.2.6. Calix[4]arene inversion between two conformations.

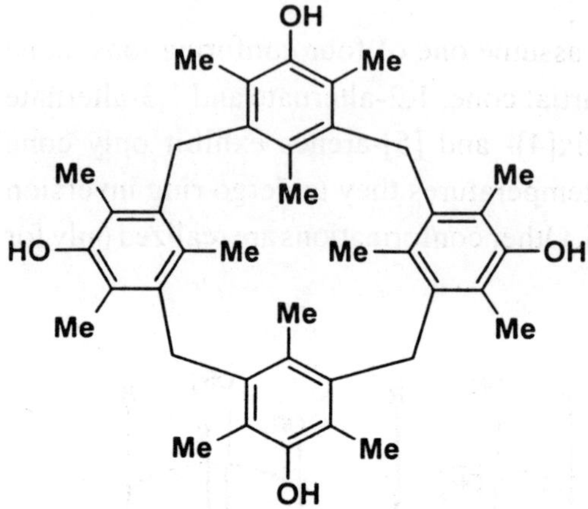

229 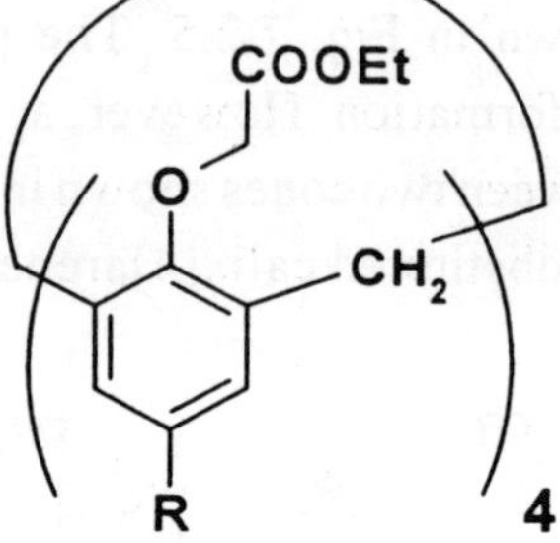 **230**

a dynamic process of interconversion between the two opposite cone conformations presented in Fig. 7.2.6. The activation parameters of this process depend on the compound and the solvent used. Typical activation barrier $\Delta G^{\#}$ is equal to 14 – 16 kcal/mol [16] with $\Delta H^{\#}$ lying between 9 and 17.4 kcal/mol while prevailing majority of $\Delta S^{\#}$ values is small and negative. The low-temperature freezing of dynamic equilibrium, usually yielding only one conformation, has been also observed for higher calixarenes [15, 17].

Hydrogen bonds keeping most calixarene structures in the cone conformation can also be characterized by the position of the ν_{OH} band in IR spectra and the hydroxyl proton signal in the NMR spectra; both are shifted with respect to the corresponding values for the free phenolic OH [1c].

229 provides an example of the 1,3-alternate conformation [18]. For calixarenes with larger substituents, such as **230**, by disturbing the inversion process Iwamoto and Schinkai even succeded in isolating all four possible conformers in the pure form [19].

7.2.3 Calixarenes as complexing agents

Owing to its conformational mobility, calix[4]arene **221** ($n = 4$) complexation is not very selective and it can change the conformation (induced fit) by the

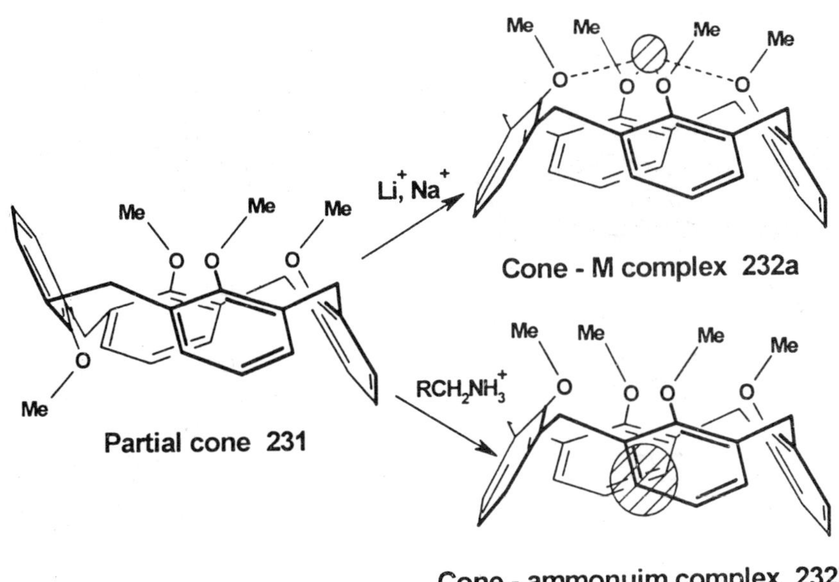

Figure 7.2.7. Cone-partial cone-cone isomerization driven by complexation with cations.

complexation. For instance, tetra-*O*-methylated calixarene 231 adopts the partial-cone conformation in water [20]. In the solid state [21] it includes Li^+, Na^+ and ammonium cations. As a complex it assumes the cone conformation with alkali metals and ammonium group residing in different parts of **232** (Fig. 7.2.7) [1b]. Calix[4]arenes **221** ($n = 4$) form either 1:1 or 1:2 complexes with aromatic guests [22]. In the former case of so-called *endo* complexes, the guest enters the cup-like cavity while in the latter one two calixarene moieties form a capsule hosting the guest. Although toluene (that forms the 1:1 complex) has been found to enter the cavity with methyl group first, the group was proved to rotate freely inside the cavity of the calix [23]. Interestingly, a removal of alkyl groups from the upper rim results in a dramatic drop in complexation ability [24]. Calix[6]- and [8]arenes do not form *endo* complexes since they are too mobile to form a well-defined cavity. However, calix[5]arenes are known to form complexes with aromatic molecules [25].

Calixarene derivatives involving crown ethers like **233** exhibit complexation of alkali or ammonium cations typical of crown ethers that is especially effective in the partial cone conformation [26]. By a proper functionalization selective synthetic receptors for specific hosts have been developed. For instance, two 2,4-diaminotriazine groups at diametrical positions of the upper rim of a

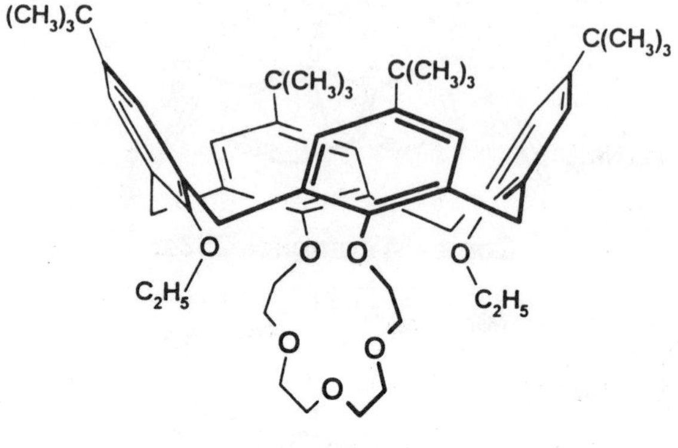

233

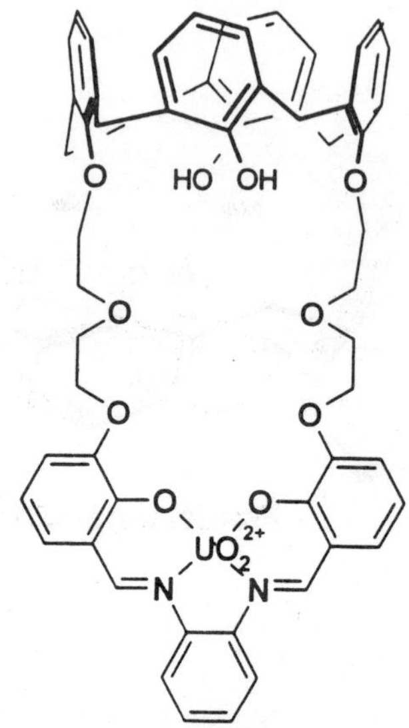

235

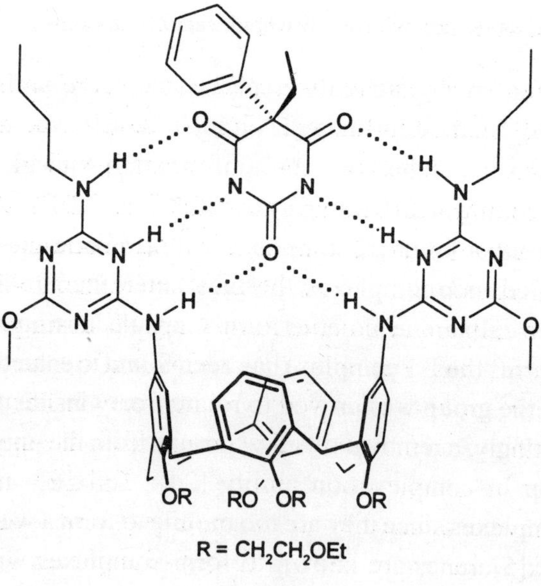

234

236

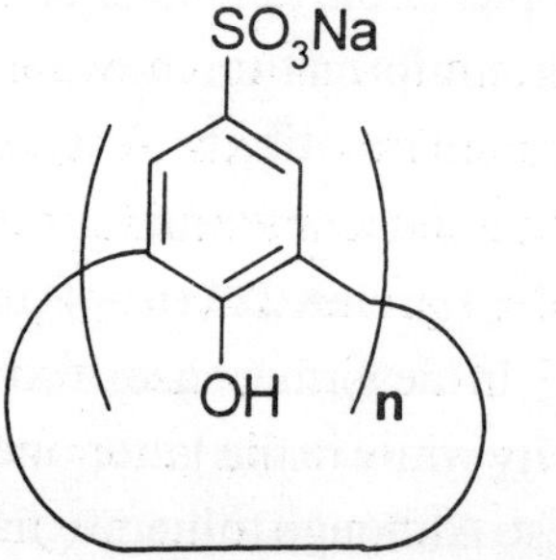

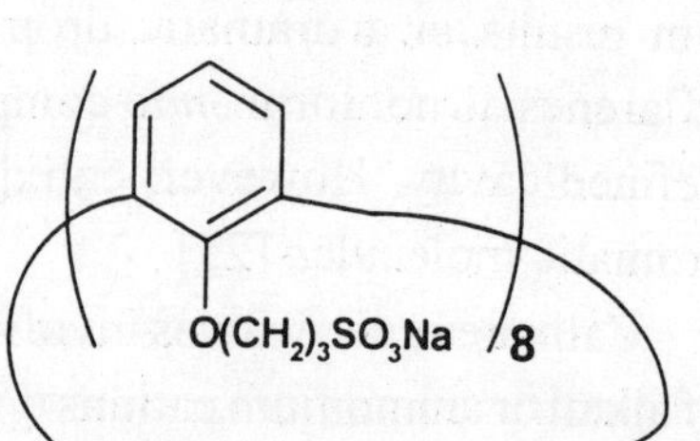

237

calix[4]arene resulted in selective receptor for barbiturates **234** [27]. Several calixarenes-based synthetic receptors for anions have been obtained [28]. A simultaneous complexation of anions and cations have been claimed by the Reinhoudt group by the receptor **235** [29]. The syntheses of highly soluble in water sulphonated calixarenes like **236** extended the calixarenes ability to host neutral molecules and ions [30]. Introduction of the same group on a tether at a lower rim **237** [31] allowed Williamson and Verhoeven to obtain a 1:2 fullerene complex in water exhibiting strong electronic interactions of charge transfer type (CT) between the host and guest.

Exciting motives obtained by hydrogen bonding of calixarenes bearing 2-pyridone [32b], carbonyl and pyridyl [32c], urea [32d], melamine and cyanuric acid [32e] and other acceptor and donor groups have been recently reviewed by Böhmer and Shivanyuk [32a].

7.2.4 Spherands, hemispherands, and other similar macrocycles capable of inclusion complex formation [33]

As discussed in Section 3.3, when complexed 18-crown-6 assumes a circular conformation with CH bonds pointing outside the cavity **48a** (Fig. 3.1), while in the free ligand the structure is squashed with the bonds pointing inside. Thus considerable free energy costs of preorganization must be overcome during the

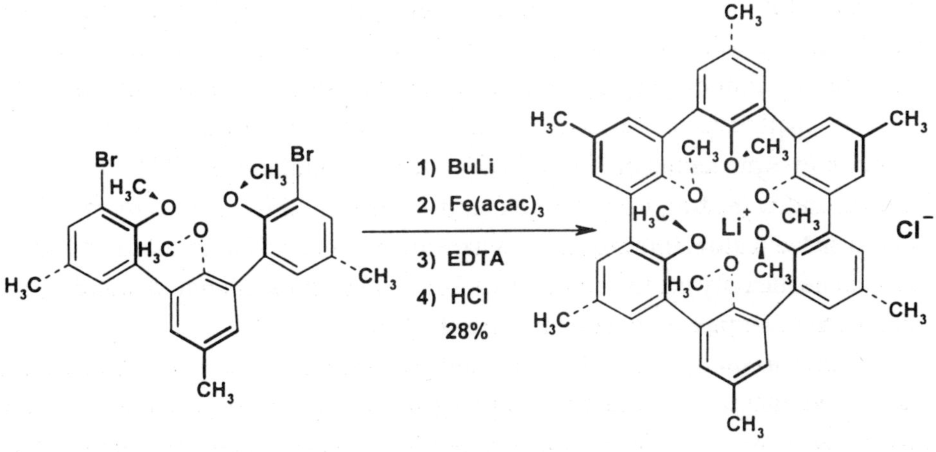

1) BuLi
2) Fe(acac)$_3$
3) EDTA
4) HCl
28%

70 · LiCl

Figure 7.2.8. The template synthesis leading to spherand after demetallation.

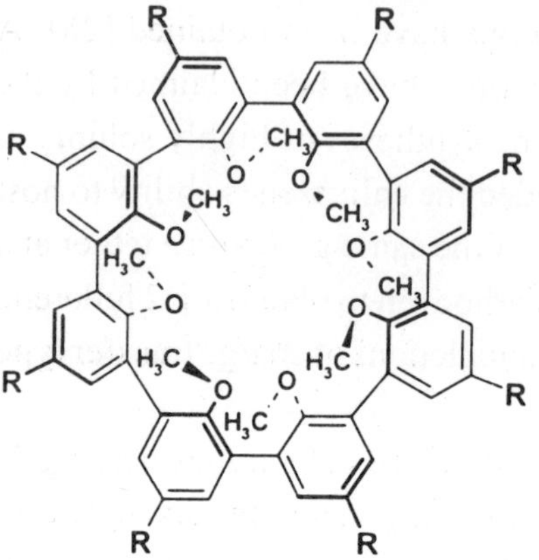

R = H, CH_3

238

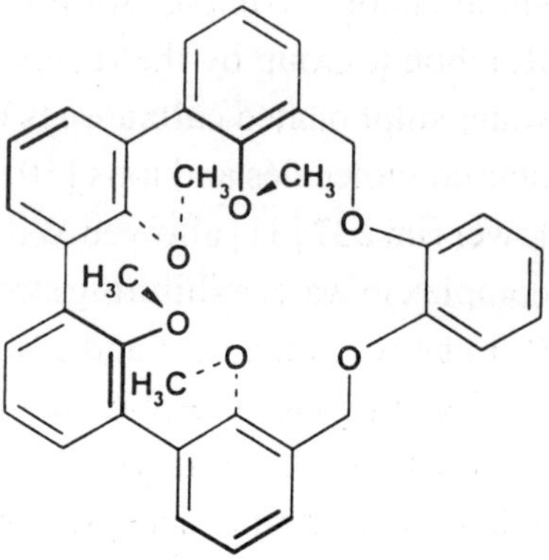

239

complexation. These costs are smaller in the case of calixarenes exhibiting much less conformational mobility in the free state. Cram's spherands [34] such as **70** were developed as the preorganized host molecules which do not change conformation upon the complex formation. (Moreover, contrary to crown ethers **205a** that must be desolvated during the complexation oxygen atoms in **70** are shielded from solvation by its aryl and methyl groups.) It should be noted, however, that increasing the number of anisyl groups to eight in **238** leads to the conformation with two methyl groups of the OCH_3 moieties turned inward, considerably diminishing the energy gain owing to the host preorganization [35].

The synthesis of the latter molecule in form of its Li^+ complex called *spheraplex* is presented in Fig. 7.2.8 [34]. The demetallation of **70** LiCl was then carried out by heating it in 4:1 methanol-water at 125°C. The cavity of **70** can also host Na^+ cation. Interestingly, as a result of induced fit introduced in Section 2.1, the host cavity of the complex of 70 shrinks in the lithium complex while that of the complex with sodium expands [36].

Hemispherands like **66** [37] and **239** [38] are macrocycles in which at least half of the ligating heteroatoms is preorganized prior to complexation (another part being crown ether) while in *cryptahemispherands* like **240** [39] there are two crown type bridges. The latter are very strong binders of alkali metal ions.

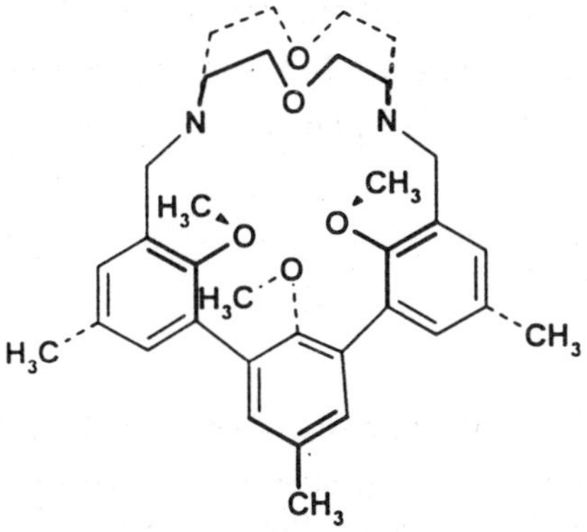

240

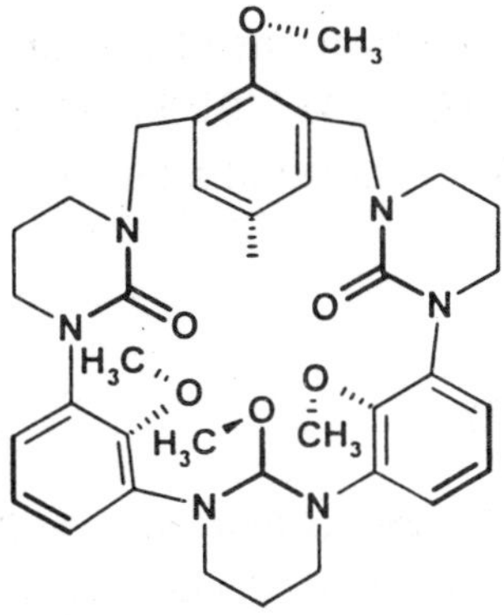

241

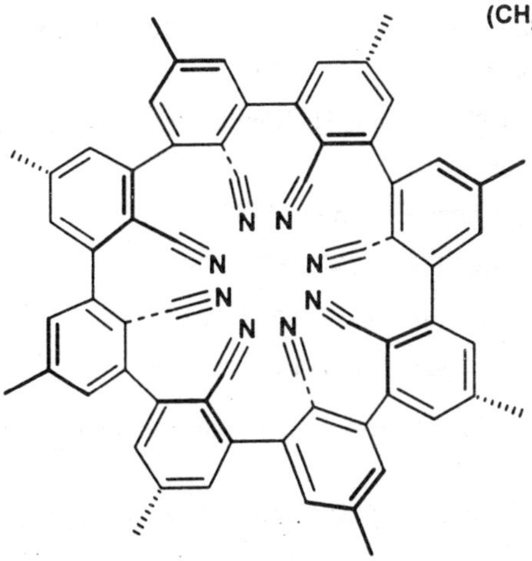

242

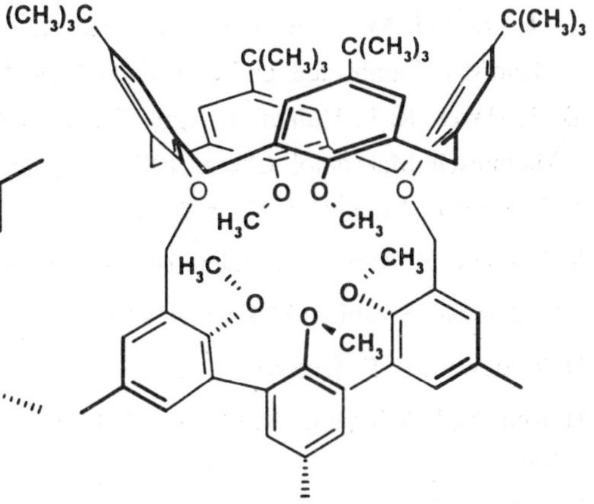

243

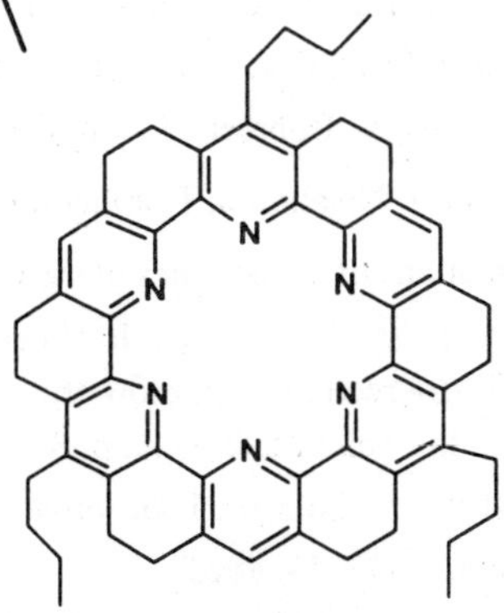

244

Numerous modifications of calixarenes, *spherands* and similar systems have been synthesized: **241** [40], **242** [41], *calixspherands* like **243** [42] and *torands* like **244** [43] to name but a few pertaining interesting ligands.

References

1. (a) C. D. Gutsche, *Calixarenes*, The Royal Society of Chemistry, Cambridge, United Kingdom, 1989; C. D. Gutsche, *Calixarenes revisited*, The Royal Society of Chemistry, Cambridge, United Kingdom, 1998; (b) S. Shinkai, Tetrahedron, 1993, 49, 8933; (c) V. Böhmer. Angew. Chem. Int. Ed. Engl., 1995, 34, 713; (d) M. A. McKervey, M.-J. Schwing-Weil, F. Arnaud-Neu, in *Comprehensive Supramolecular Chemistry*, G. W. Gokel, Ed., 1996, v. 1, p. 537.
2. D. J. Cram, J. M. Cram, *Container Molecules and Their Guests*, The Royal Society of Chemistry, Cambridge, United Kingdom, 1994.
3. B. T. Hayes, R. F. Hunter, J. Appl. Chem., 1958, 8, 743; H. Kämmerer, G. Happel, B. Matthiesch, Makromol. Chem., 1981, 182, 1685.
4. V. Böhmer, L. Merkel, U. Kunz, J. Chem. Soc. Chem. Commun., 1987, 896.
5. K. Iwamoto, K. Araki, S. Shinkai, Bull. Chem. Soc. Japan, 1994, 67, 1499.
6. Y. Nakamoto, S. Ishida, Makromol. Chem. Rapid. Commun., 1982, 3, 705.
7. D. R. Stewart, C. D. Gutsche, J. Am. Chem. Soc., 1999, 121, 4136.
8. H. Konishi, K. Ohata, O. Morikawa, K. Kobayashi, J. Chem. Soc. Chem. Commun., 1995, 309.
9. F. Arnaud-Neu, V. Böhmer, L. Guerra, M. A. McKervey, E. F. Paulus, A. Rodriguez, M.-J. Schwing-Weil, M. Tabatabai, J. Phys. Org. Chem., 1992, 5, 471.
10. H. Goldmann, W. Vogt, E. Paulus, V. Böhmer, J. Am. Chem. Soc., 1988, 110, 6811.
11. Y. Okada, M. Mizutani, F. Ishii, J. Nishimura, Tetrahedron Lett., 1999, 40, 1353.
12. M. Perrin, D. Öhler, Conformations of Calixarenes in the Crystalline State in *Calixarenes. A Versatile Class of Macrocyclic Compounds*, J. Vicens, V. Böhmer, Eds., Kluwer, Dordrecht, 1991, p. 65; G. D. Andreetti, F. Ugozzoli, Inclusion Properties and Host-Guest Interactions of Calixarenes in the Solid State, *ibid.*, p. 87.
13. See, for instance, G. D. Andreetti, V. Böhmer, J. G. Jordan, M. Tabatabai, F. Ugozzoli, W. Vogt, A. Wolff, Z. Org. Chem., 1993, 58, 4023.
14. C. D. Gutsche, Acc. Chem. Res., 1983, 16, 161; T. Harada, F. Ohseto, S. Shinkai, Tetrahedron, 1994, 50, 13377.
15. C. D. Gutsche, L. J. Bauer, Tetrahedron Lett., 1981, 22, 4763.

16. K. Araki. S. Shinkai. T. Matsuda, Chem. Lett., 1989. 581.

17. C. D. Gutsche, Single-step Synthesis and Properties of Calixarenes, in *Calixarenes. A Versatile Class of Macrocyclic Compounds*, J. Vicens, V. Böhmer, Eds., Kluwer, Dordrecht, 1991, p. 3.

18. S. Pappalardo, G. Ferguson, J. F. Gallagher, J. Org. Chem., 1992, 57, 7102.

19. K. Iwamoto, S. Shinkai, J. Org. Chem., 1992, 57, 7066.

20. T. Nagasaki, K. Sisido, T. Arimura, S. Shinkai, Tetrahedron, 1992, 48, 797.

21. K. Araki, H. Shimizu, S. Shinkai, Chem. Lett., in press.

22. A. Pochini, R. Ungaro, in *Comprehensive Supramolecular Chemistry*, v. 2, p. 103.

23. R. Caciuffo, O. Francescangeli, S. Melone, M. Prager, F. Ugozzoli, G. D. Andreetti, G. Amoretti, G. Coddens, H. Blank, Phys. B, 1992, 180/181, 691.

24. R. Ungaro, A. Pochini, G. D. Andreetti, V. Sangermano, J, Chem. Soc. Perkin Trans. 2, 1984, 1979.

25. M. Perrin, S. Lecocq, J. Incl. Phenom. Mol. Recogn. Chem., 1991, 11, 171.

26. E. Ghidini, F. Ugozzoli, R. Ungaro, S. Harkema, A. A. El-Fadl, D. N. Reinhoudt, J. Am. Chem. Soc., 1990, 112, 6979.

27. J.-D. van Loon, J. F. Heida, W. Verboom, D. N. Reinhoudt, Rec. Trav. Chim. Pays-Bas, 1992, 111, 353.

28. Y. Morzehrin, D. M. Rudkevich, W. Verboom, D. N. Reinhoudt, J. Org. Chem., 1993, 58, 7602.

29. D. M. Rudkevich, Z. Brzozka, M. Palys, H. C. Visser, W. Verboom, D. N. Reinhoudt, Angew. Chem. Int. Ed. Engl., 1994, 33, 467.

30. S. Shinkai, K. Araki, T. Matsuda, N. Nishinyama, H. Ikeda, I. Takasu, M. Iwamoto, J. Am. Chem. Soc., 1990, 112, 9053.

31. R. M. Williamson, J. K. Verhoeven, Rec. Trav. Chim. Pays-Bas, 1992, 111, 531.

32. (a) V. Böhmer, A. Shivanyuk, in *Calixarenes in Action*, L. Mandolini, R. Ungaro, Eds., Imperial College Press, Singapore, 2000, p. 203; (b) J.-D. van Loon, R. G. Janssen, W. Verboom, D. N. Reinhoudt, Tetrahedron Lett., 1992, 33, 5125; (c) K. Koh, K. Araki, S. Shinkai, Tetrahedron Lett., 1993, 35, 8255; (d) O. Mogk, E. F. Paulus, V. Böhmer, I. Thorndorf, W. Vogt, J. Chem. Soc. Chem. Commun., 1996, , 2533; R. K. Castellano, J. Rebek, Jr., J. Am. Chem. Soc., 1998, 120, 3657; (e) P. Timmerman, R. Vreekamp, R. Hults, W. Verboom, D. N. Reinhoudt, K. Rissanen, K. A. Udachin, J. Ripmeester, Chem. Eur. J., 1997, 3, 1823.

33. E. F. Maverick, D. J. Cram, in *Comprehensive Supramolecular Chemistry*, G. W. Gokel, Ed., 1996, v. 1, p. 213.

34. D. J. Cram, T. Kaneda, R. C. Helgeson, S. B. Brown, C. B. Knobler, E. F. Maverick, K. N. Trueblood, J. Am. Chem. Soc., 1985, 107, 3645.

35. D. J. Cram, R .A. Carmack, M. P. de Grandpre, G. M. Lein, J. Goldberg, C. B. Knobler, E. F. Maverick, K. N. Trueblood , J. Am. Chem. Soc., 1987, 109, 7068.

36. K. N. Trueblood, E. F. Maverick, C. B. Knobler, Acta Crystallogr., 1991, B47, 389.

37. K. Koenig, M. G. Lein, P. Stackler, T. Kaneda, D. J. Cram, J. Am. Chem. Soc., 1979, 101, 3553.

38. J. A. Tucker, C. B. Knobler, I. Goldberg, D. J. Cram, J. Org. Chem., 1989, 54, 5460.

39. D. J. Cram, S. P. Ho, J. Am. Chem. Soc., 1986, 108, 2998.

40. D. J. Cram, J. B. Dicker, C. B. Knobler, K. N. Trueblood, J. Am. Chem. Soc., 1982, 104, 6828.

41. K. Paek, C. B. Knobler, E. F. Maverick, D. J. Cram, J. Am. Chem. Soc., 1989, 111, 8662.

42. P. J. Dijkstra, J. A. J. Brunink, K. E. Bugge, D. N. Reinhoudt, S. Harkema, R. Ungaro, F. Ugozzoli, E. Ghidini, J. Am. Chem. Soc., 1989, 111, 7567.

43. T. W. Bell, A. Firestone, R. Ludwig, J. Chem. Soc. Chem. Commun., 1989, 1902.

7.3 Carcerands, Hemicarcerands and Novel 'Molecular Flasks' Enabling Preparation and Stabilization of Short-lived Species [1]

Bowl-shaped molecules like **245** and **246** inspired Cram to propose cage molecules like **247** [2] that could host incarcerated smaller guests, hence the name *carcerands*. The latter molecule remained hypothetical but the Cram group

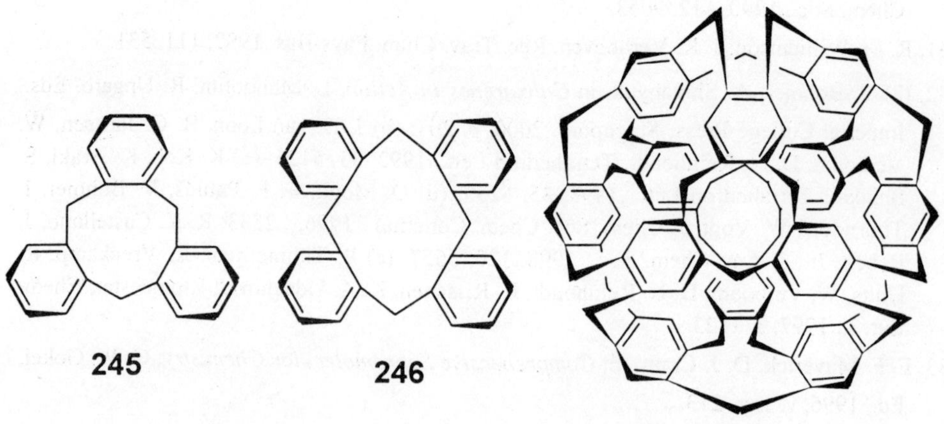

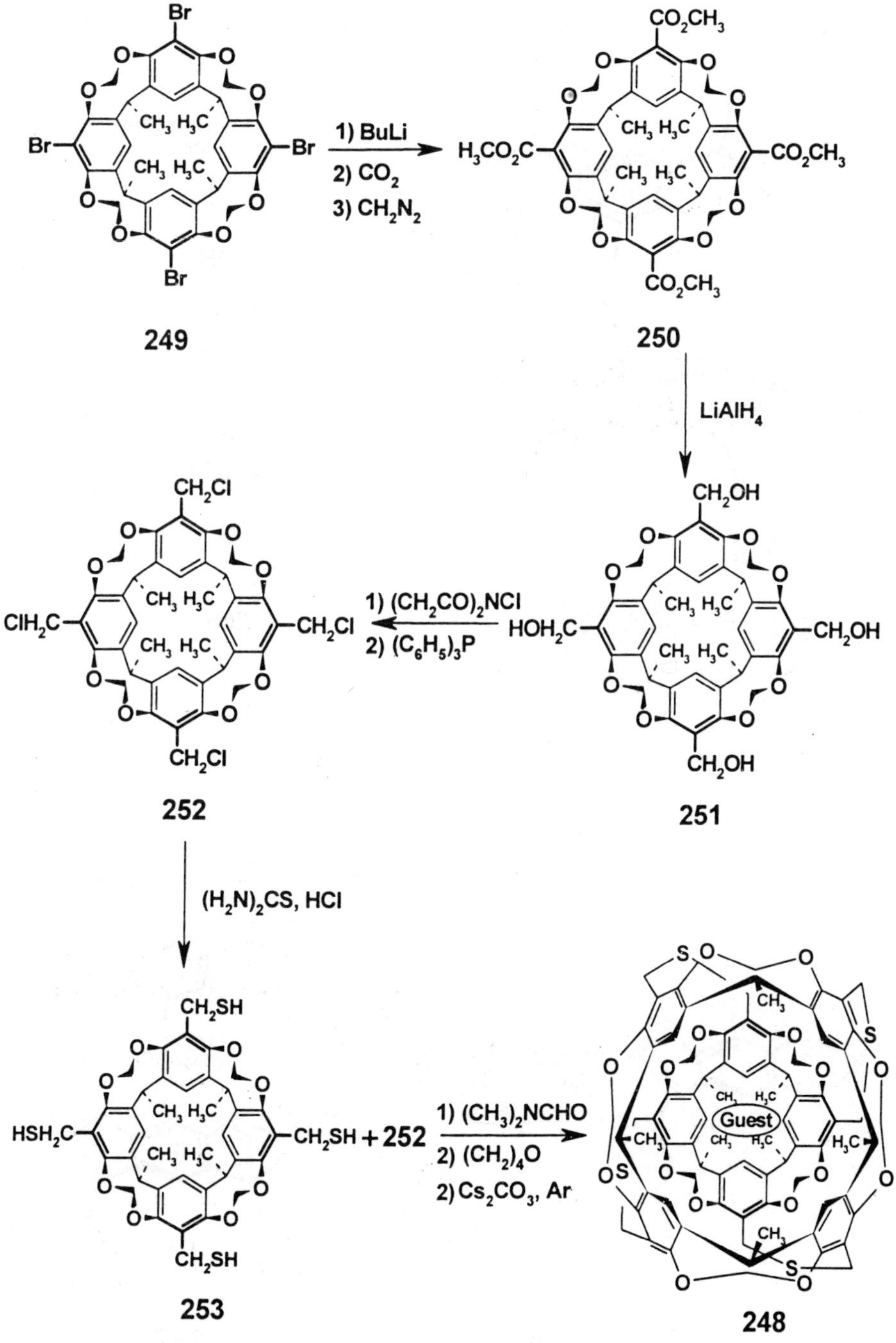

Figure 7.3.1. The synthesis of the first carcerand.

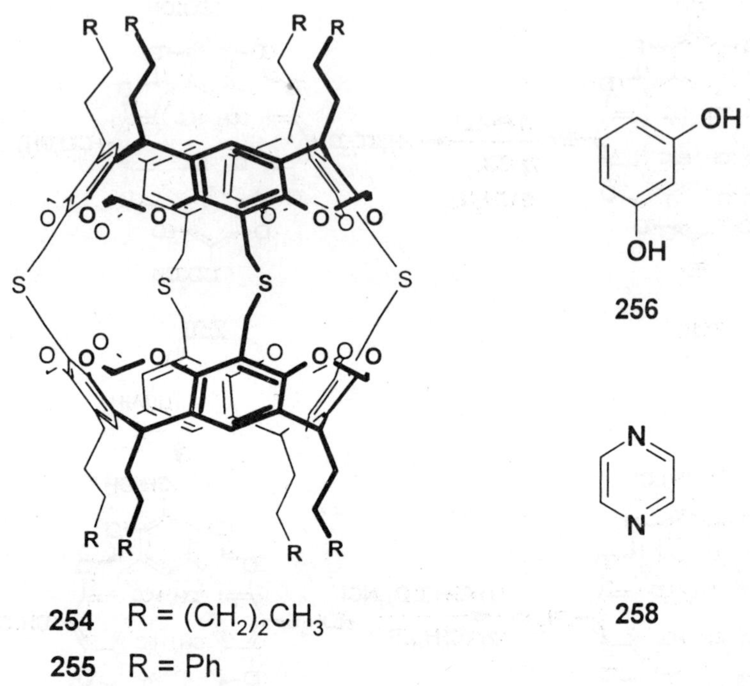

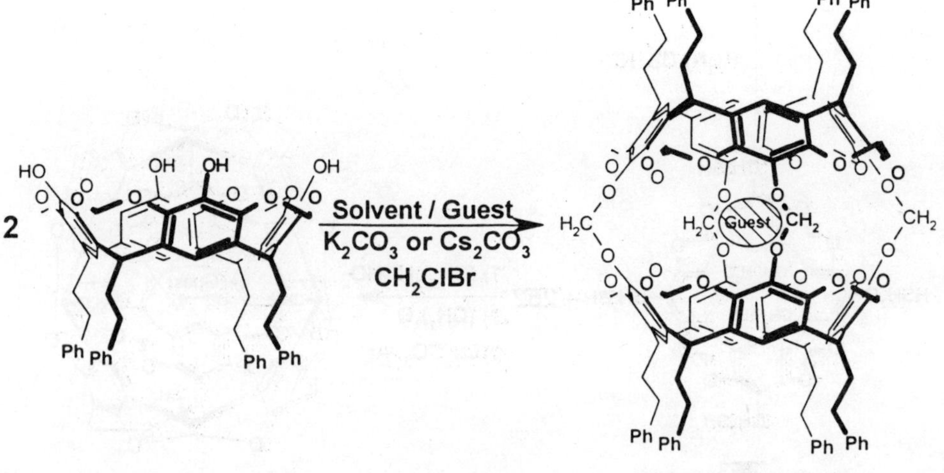

Figure 7.3.2. The formation of **257** using a guest molecule as template.

succeeded in synthetizing **248** from **249** by a five-step synthesis with the total yield of 4.7%. The reaction products insolubility prevented their purification but they were proved to exist in the form of complexes of **248**, called *carceplexes*, with various guests [3]. By changing eight methyl groups by either eight $(CH_2)_4CH_3$ or eight $CH_2CH_2C_6H_5$ groups, the solubility of **254** and **255** was achieved owing to the conformational flexibility of these substituents that have provided the carceplexes with additional solvation sites. The synthesis of **254** from $CH_3(CH_2)_4CHO$ and resorcinol **256** was carried out in eight steps. During the last step of the latter synthesis, the shell closure, one molecule of the reaction solvent became the guest since it was incarcerated into the cage [4]. The guest can be later released from the carceplex by heating. The solvent guest molecule acts as template in the last reaction in Figure 7.3.1. Sherman [5] described highly efficient template syntheses of carcerands **257** (Figure 7.3.2) with the highest yield of 75% obtained with pyrazine guest **258**.

The incorporation of a guest into the carcerand cage influences its properties. For instance, all guest proton NMR signals are shifted upfield by 1 - 4 ppm from their standard positions owing to the large shielding effect of the eight aryl groups of the cage. The signals of the protons in the $O-CH_2-O$ bridges sometimes exhibit splitting owed to the restricted rotation of the guest inside the cage [6] once more showing dynamic character of inclusion complexes discussed in Section 3.4. Aromatic solvent-induced proton shifts are, somewhat surprisingly, larger for the incarcerated guest than for the free guest [7]. The explanation of this observation lies beyond the scope of this book. On the basis of IR spectra of incarcerated amides, Cram and coworkers proposed also a new type of diastereoisomerism in such carceplexes [7]. Another effect forced by the guest incarceration was the increase of the barrier to internal rotation around the nitrogen-carbon amide bond [7]. Dynamic ^1H NMR spectra of Me_2NCOMe and Me_2NCHO in $C_6D_5NO_2$ gave coalescence temperatures of 63°C and 120°C, respectively, and $\Delta G^{\dagger}_{336} = 75$ kJ/mol and $\Delta G^{\dagger}_{393} = 85$ kJ/mol, respectively, for the two methyl signals. The coalescence temperatures for the complexes of **255** with the latter amides are 190°C and 140°C, respectively and the corresponding barriers equal to 85 and 79 kJ/mol, respectively.

The above observations and many other similar ones prompted Cram to define inside carceplex volume as a specific inner phase intermediate between the polar solvent and vacuum.

The portals enabling guests to escape from the cage can be enlarged either by lengthening of the bridges linking two bowl-shaped parts of a carcerand or by omitting one of these bridges. If the portals are so big that the guest molecule can escape at high temperatures but it stays inside at temperatures which allow one to isolate, purify, and characterize the complex, then, according to Cram, such hosts are called *hemicarcerands* and their complexes bear the name *hemicarceplexes*. As presented in Chapter 1, the synthesis of hemicarcerand **5**

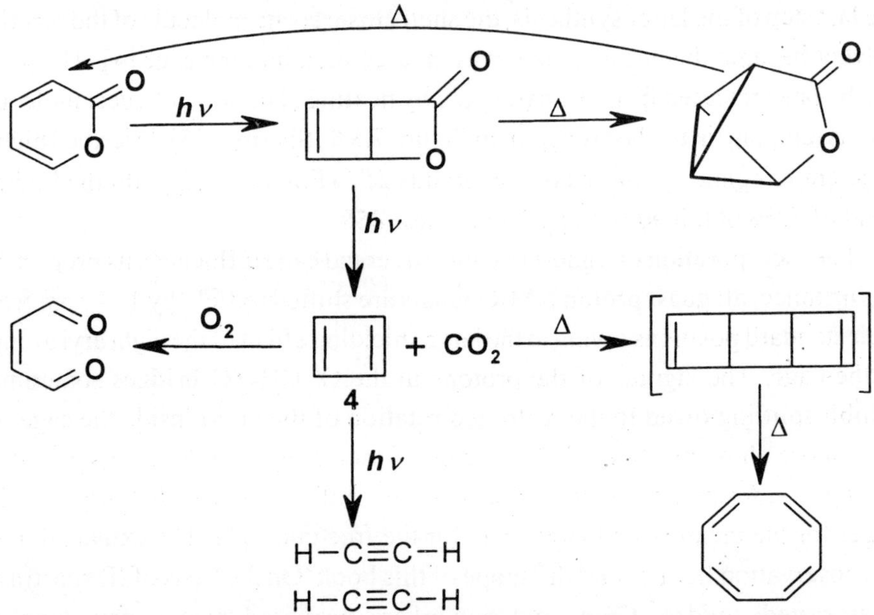

Figure 7.3.3. The scheme of formation of **4** and its chemistry.

allowed Cram and coworkers to carry out the reaction leading to highly unstable cyclobutadiene **4** that dramatically changed its stability upon incarceration [8]. This paved the way to, to-date a very rarely realized possibility, to obtain short-lived species stabilized in a molecular or supramolecular cage.

4 was dubbed by Cram "the Mona Lisa of organic chemistry in its ability to elicit wonder, stimulate the imagination, and challenge interpretive instincts" since "No other organic compound combines such a fleeting existence and so many different syntheses, with such a propensity for different chemical reactions, and with the variety of calculations of its structure." The molecule obtained according to Figure 7.3.3 was characterised in an argon matrix at 8 K as a short-lived species [9]. Carrying the reaction in hemicarcerand **5** allowed the Cram group to obtain the hemicarceplex **4@5** in which the size of the host cavity portals prevents the guest escape and dimerization.

Hemicarcerand **259** (X = CH_2CH_2Ph) differing from **5** in the number of CH_2 bridges, 4 vs. 3, and their lengths was applied by Warmuth to the photochemical

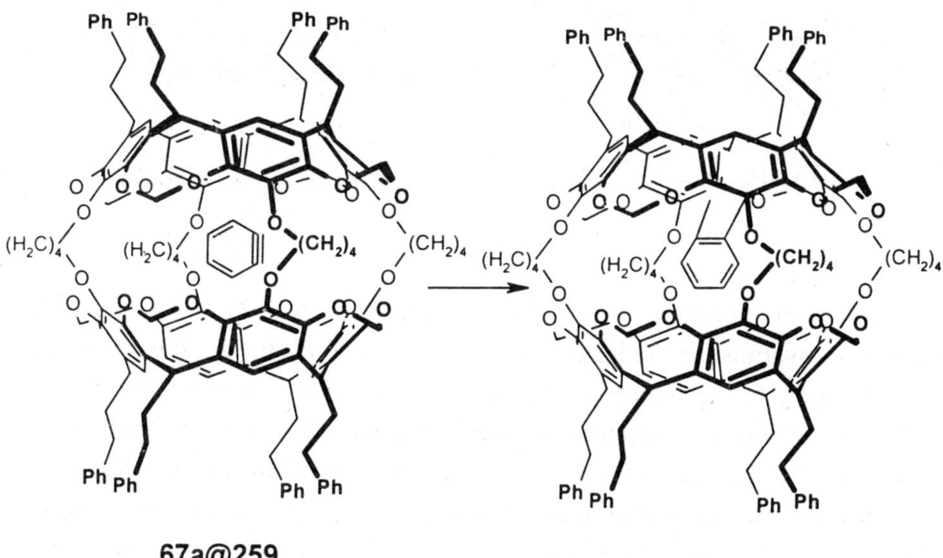

Figure 7.3.4. Diels-Alder reaction of encapsulated *o*-benzyne **67** with host cage **259**.

Figure 7.3.5. The scheme of formation of **260** and its chemistry.

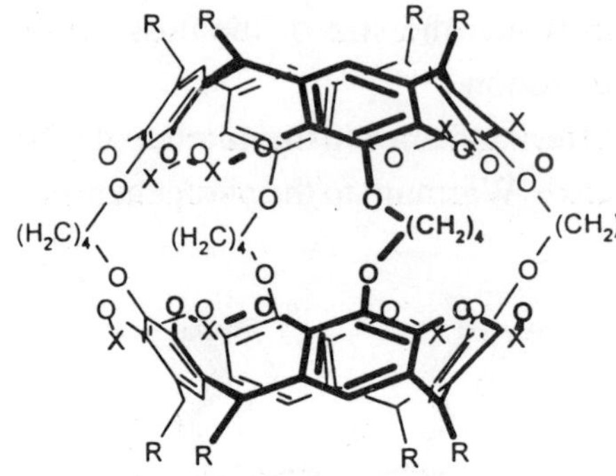

R = $(CH_2)_4CH_3$ X = CH_2, CD_2

261

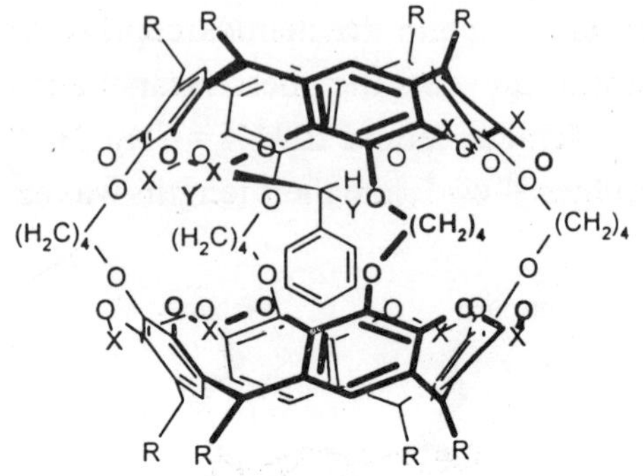

R = $(CH_2)_4CH_3$ X = CH_2, CD_2
Y = H,D

262

synthesis of another short-lived molecule *o*-benzyne **67a** at 77 K [10]. When the sample was warmed to 198 K, a product of the reaction of the guest with the host was detected [10, 11]. Interestingly, the first study (in which chemical shifts and carbon $^1J_{CC}$ coupling constants have been reported) did not allow Warmuth to decide whether the guest has the *o*-benzyne **67a** or cumulene **67b** structure. The equivocality was removed by quantum calculations [12] in favour of *o*-benzyne. The latter molecule **67a** was highly reactive and by warming from 77 K to room temperature it underwent the Diels-Alder addition to the host [13] (Figure 7.3.4).

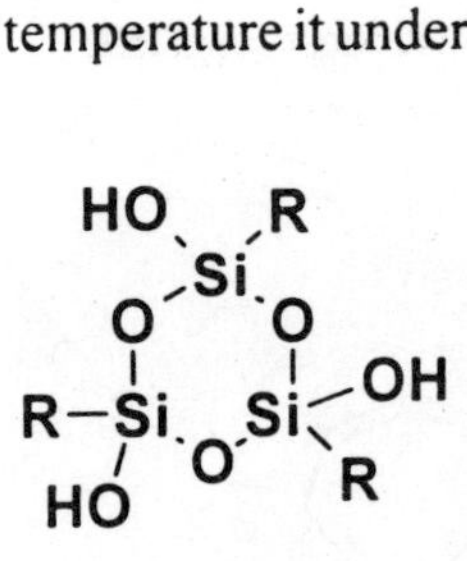

263

Another significant accomplishment in this domain was the synthesis of 1,2,4,5-cycloheptatetraene **260** in the inner phase of **261** also achieved by the Warmuth group [14] (Fig. 7.3.5). Similarly to cyclobutadiene, which was stable in hemicarceplex **4@5**, the otherwise unstable cumulene remained stable at room-temperature in **261**. Interestingly, the authors of Ref. 14 took advantage of the deuterium isotope effect that had slowed down the competing reaction of the guest with the host cage leading to **262**. The yield of cycloheptatetraene **260** in the last reaction was increased to ca. 60% not only by using the deuterated host but also by lowering the temperature of photolysis to 15.5 K and the addition of the triplet sensitizer $[D_8]$acetophenone to the bulk phase (15% v/v).

264

The Cram and Warmuth syntheses leading to the stabilization of hydrocarbons **4**, **67a** and **260** were carried out inside the covalently bonded hemicarcerands **5**, **259** and **261**, respectively, while two recent syntheses of other kinds of unstable species took place in self-assembled cages extending considerably the prospects of such studies. Kusukawa, Fujita and coworkers [15] succeeded in obtaining trimers of siloxanes **263** in the self-assembled **264** cage. The former molecules and the corresponding trimers were considered as intermediates in the polycondensation of trialkoxysilanes (so-called sol-gel condensation) resulting in the formation of siloxane networks or ladder polymers [16]. Thus the study of **263** is of importance not only for gaining understanding of the condensation process but also for modeling silica gel

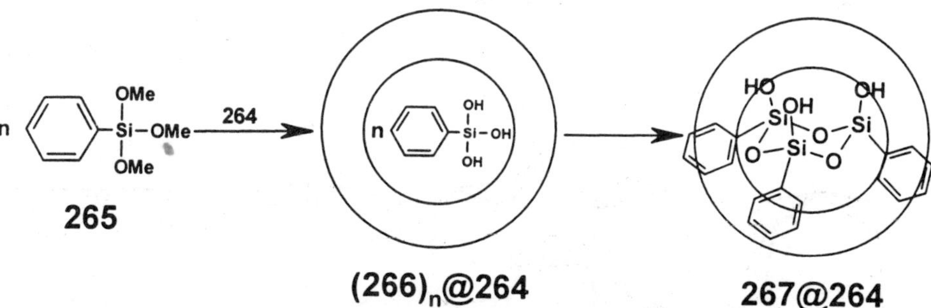

Figure 7.3.6. The condensation of **265** to **267** in the nanocage **264**.

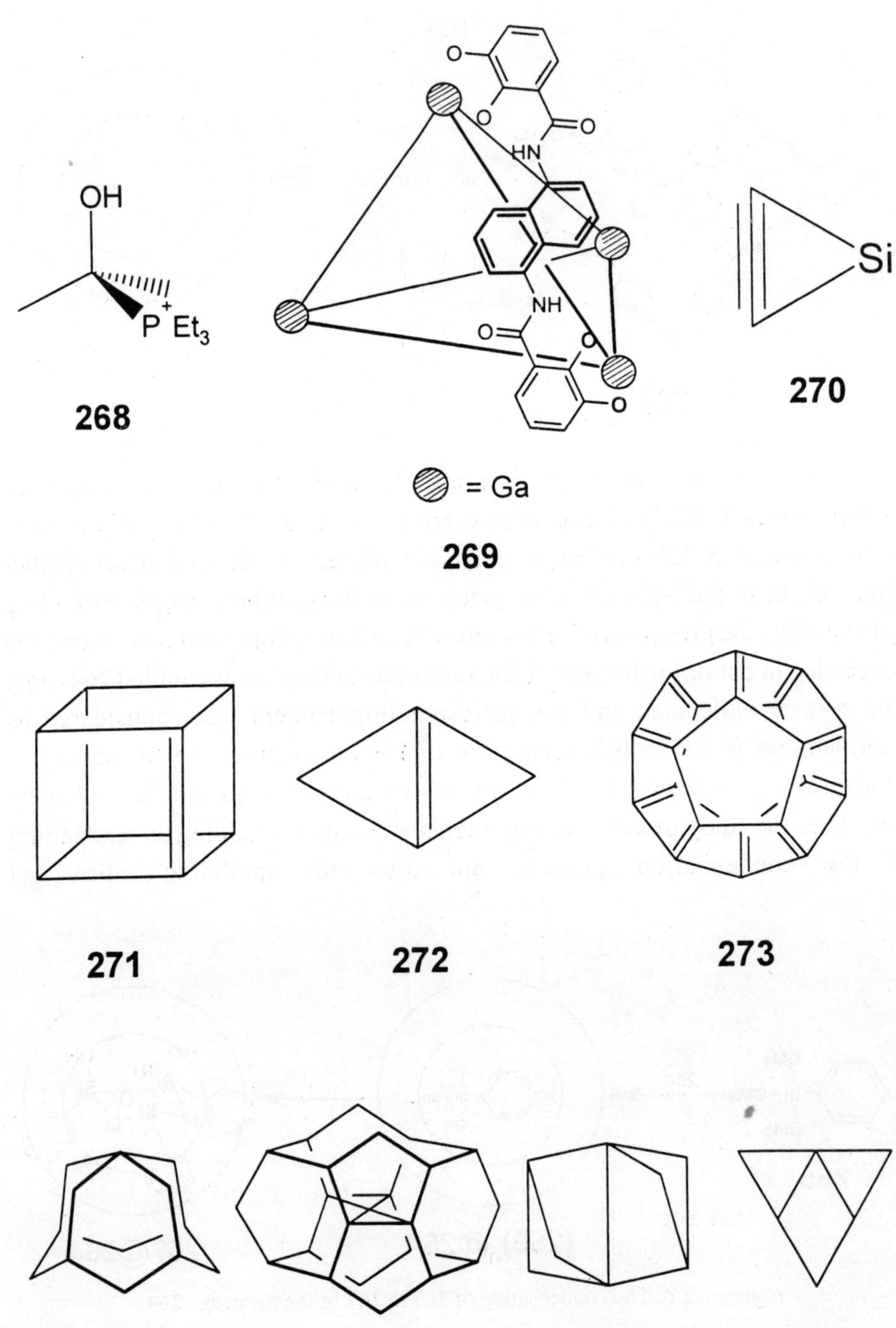
OH
P Et3
268
O
O
O
HN
NH
O
O
O
= Ga
269
Si
270
271
272
273
274
275
276
277

surface in a homogeneous system and the fabrication of silicon-based functional materials.

264 obtained by self-assembly of Pt(2,2'-bipyridyl)(NO_3)$_2$ and *tris*-(4-pyridyl)-1,3,5-triazine was shown to be stable under acidic and basic conditions. Phenyltrimethoxysilanes **265** which could enter the last cage were hydrolyzed there to oligomers of **266** (Figure 7.3.6) that yielded **267**. The products **267@264** could be isolated since the guest molecules were too big to escape from the host cage. When confined to the cage the otherwise unstable cyclic guest trimers remained intact for 1 month in water at room temperatures. They even survived the acidic conditions (pH < 1) necessary for the isolation of the encapsulated complex.

Another outstanding achievement in this domain was the synthesis of, of course unstable, $[(CH_3)_2C(OH)PC_2H_5)_3]^+$ cation **268** in the self-assembled cage **269** similar to **26** discussed in Section 2.1 [17].

Only a few reactions in 'molecular flasks' have been successfully carried out up to today, but they certainly mark the beginning of exciting development in the border area between organic, supramolecular and theoretical chemistry. The link between organic and supramolecular chemistry in this domain is obvious, while numerous calculations of cyclobutadiene **4** and the decisive role of theoretical studies in the assignment of benzyne **67a** structure obtained by Warmuth, as **67a@259** mentioned earlier in this Section, exemplify the role of theoretical studies in this area. Several nonstandard molecules which have been recently proposed as plausible synthetic targets on the basis of quantum calculations may be short-lived species that can be accessible only in the form of supramolecular complexes stabilized by the host cages. On the other hand, the attainability of short-lived species stabilized in the complexes which until now could be examined only at very low temperatures in noble gases' matrices extends tremendously the possibilities of their study. Silacyclopropyne **270** with the ultimate deformation of C≡C bond [18], cubene **271** [19], highly strained **272** [20] and fully unsaturated dodecahedrane (*i. e.* fullerene C_{20}) **273** [21] are only few examples of the latter species. On the other hand, bowlane **274** [22] which should have a pyramidal carbon atom, dimethylspiro[2.2]octaplane **275** [23] (with the planar configuration on one carbon) developed on its basis, [1.1.1]geminane **276** [24] and **277** which should have a linear C–C–C arrangement [25] as well as compounds of light noble gases (neon, helium) [26] are only few examples of hypothetical species still awaiting their syntheses, eventually in supramolecular cages. Thus 'molecular flasks' expand enormously the limits of what is possible in synthestic chemistry.

References

1. D. J. Cram, J. M. Cram, *Container Molecules and Their Guests*, The Royal Society of Chemistry, Cambridge, Great Britain, 1994.
2. D. J. Cram, Science, 1983, 219, 1177.
3. D. J. Cram, S. Karbach, Y. H. Kim, L. Baczynskij, G. W. Kalleymeyn, J. Am. Chem. Soc., 1985, 107, 2575; D. J. Cram, S. Karbach, Y. H. Kim, L. Baczynskij, K. Marti, R. M. Sampson, G. W. Kalleymeyn, J. Am. Chem. Soc., 1988, 110, 2554.
4. L. M. Tunstad, J. A. Tucker, E. Dalcanale, J. Weiser, J. A. Bryant, J. C. Sherman, R. C. Helgeson, C. B. Knobler, D. J. Cram, J. Org. Chem., 1989, 54, 1305.
5. R. G. Chapman, N. Chopra, E. D. Cochien, J. C. Sherman, J. Am. Chem. Soc., 1994, 116, 369; A. Jasat, J. C. Sherman, Chem. Rev., 1999, 99, 931.
6. J. A. Bryant, M. T. Blanda, M. Vincenti, D. J. Cram, J. Am. Chem. Soc., 1991, 113, 2167.
7. J. C. Sherman, C. B. Knobler, D. J. Cram, J. Am. Chem. Soc., 1991, 113, 2194.
8. D. J. Cram, M. E. Tanner, R. Thomas, Angew. Chem. Int. Ed. Engl., 1991, 30, 1024.
9. O. L. Chapman, C. L. MacIntosh, J. Pacansky, J. Am. Chem. Soc., 1973, 95, 614; C. Y. Lin, A. Krantz, J. Chem. Soc. Chem. Commun., 1972, 1111.
10. R. Warmuth, Angew. Chem. Int. Ed. Engl., 1997, 36, 1347.
11. R. Warmuth, J. Chem. Soc. Chem. Commun., 1998, 59; B. R. Beno, C. Sheu, K. N. Houk, R. Warmuth, D. J. Cram, J. Chem. Soc. Chem. Commun., 1998, 301.
12. H. Jiao, P. R. Schleyer, B. R. Beno, K. N. Houk, R. Warmuth, Angew. Chem. Int. Ed. Engl., 1997, 36, 2761.
13. R. Warmuth, J. Chem. Soc. Chem. Commun., 1998, 59.
14. R. Warmuth, M. A. Marvel, Angew. Chem. Int. Ed. Engl., 2000, 39, 1117.
15. M. Yoshizawa, T. Kusukawa, M. Fujita, K. Yamaguchi, J. Am. Chem. Soc., 2000, 122, 6311.
16. R. H. Baney, M. Itoh, A. Sakakibara, T. Suzuki, Chem. Rev., 1995, 95, 1409.
17. M. Ziegler, J. L. Brumaghim, K. N. Raymond, Angew. Chem. Int. Ed. Engl., 2000, 39, 4119.
18. G. Maier, H. Pacl, H. P. Reisenauer, M. Meudt, R. Janoschek, J. Am. Chem. Soc. 1995, 117, 12712.

19. P. E. Eaton, M. Maggini, J. Am. Chem. Soc. 1988, 110, 7230.

20. B. M. Branan, L. A. Paquette, D. A. Hrovat, W. T. Borden, J. Am. Chem. Soc. 1992, 114, 774.

21. H. Prinzbach, A. Weller, P. Landenberger, F. Wahl, J. Worth, L. T. Scott, M. Gelmont, D. Olevano, B. von Issendorff, Nature, 2000, 407, 60.

22. H. Dodziuk, J. Mol. Struct., 1990, 239, 167.

23. D. R. Rasmussen, L. Radom, Angew. Chem. Int. Ed. Engl., 1999, 38, 2876.

24. H. Dodziuk, J. Leszczyński, K. Jackowski, J. Org. Chem., 1999, 64, 6177.

25. H. Dodziuk, J. Leszczyński, K. S. Nowiński, J. Org. Chem., 1995, 60, 6860.

26. L. Khriachtchev, M. Petersson, N. Runeberg, J. Lundell, M. Rasanan, Nature, 2000, 406, 874.

7.4 Cyclodextrins, and Their Complexes [1]

7.4.1 Introduction

Cyclodextrins, CDs, **278** and their complexes clearly fall into the domain of host-guest chemistry and the importance of these macrocycles is enhanced by their numerous practical applications [1b, 1c]. However, they form an independent area and there is little mutual interaction between crown ethers, cryptands, calixarenes, hemispherands, and other types of manmade macrocycles discussed earlier in this chapter and CDs that are obtained by using a biotechnological procedure. As indicated by their formulae and the computer model presented in Fig. 7.4.1, CDs have a cavity in which other, mostly smaller, molecules (or ions or even radicals) may reside. The encapsulation of guest molecules by CDs is selective, forming the basis for their applications. As sometimes happens with novel ideas, the first notion of small molecules located inside larger CDs expressed by Cramer was met with fierce

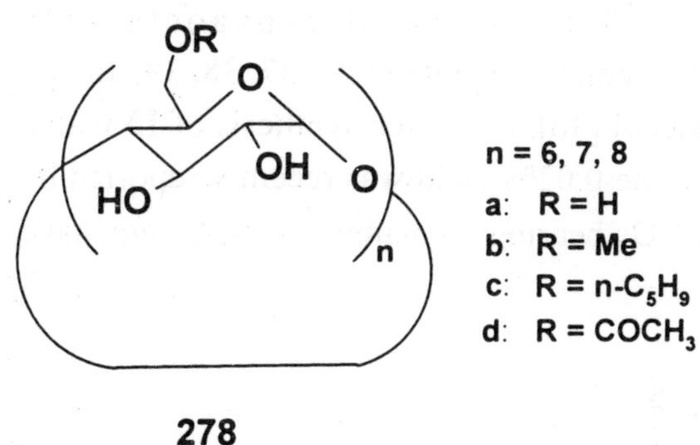

n = 6, 7, 8
a: R = H
b: R = Me
c: R = n-C_5H_9
d: R = $COCH_3$

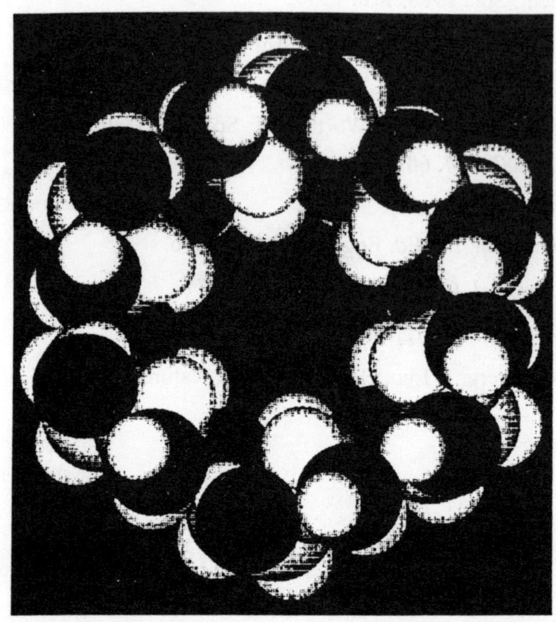

Figure 7.4.1. Computer model of α-CD **13** displaying the cavi-ty in which guest can enter.

criticism [2]. As he recalled many years later, one of his older authoritative colleagues even wanted to expel such a fallacious young man from the academic community for expressing unacceptably eccentric ideas.

CDs are macrocyclic oligosaccharides built of α-D-*gluco*pyranoside units interconnected by α-(1,4) bonds. They are obtained by enzymatic degradation of dextrins (Figure 7.4.2). The most important in this group of compounds **278a** with n = 6, 7 and **8** bear the name native CDs and are denoted as α-, **13**, β-, **11**, and γ-**68** CDs, respectively [3a]. Their full IUPAC designation is cumbersome. For instance, for **13** it is 5,10,15,20,25,30,35-heptakis(hydroxymethyl)-2,4,7,9,12,14,17,19,22,24,27,29,32,34-tetradecaoxaoctacyclo [31.2.2.2^{3,6}.2^{8,11}.2^{13,16}.2^{18,21}. 2^{23,26}.2^{28,31}]nonatetracontane-36, 37, 38, 39, 40, 41, 42, 43, 44, 45, 46, 47, 48, 49-tetradecol [3b]. A 21 step synthesis of **13** with a 0.3% total yield and that of **68** with the 0.02% yield was recently reported by Ogawa and coworkers [4]. Larger CDs bearing the names δ-, ε-, ζ-, *etc.* have

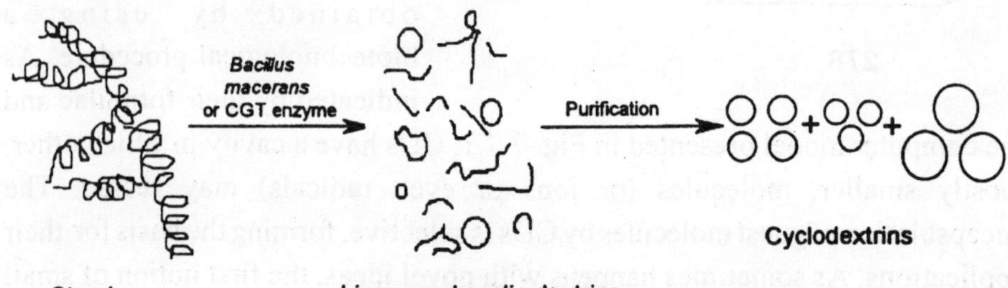

Figure 7.4.2. Scheme of biotechnological process of CD formation.

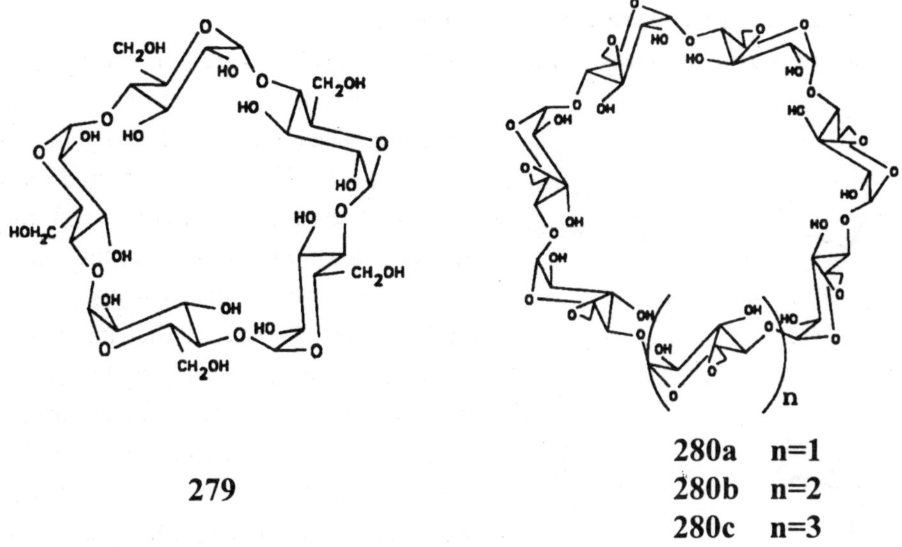

been obtained either by tedious chromatographic separations [5a] or by taking advantage of sophisticated biochemical process [5b]. They have been obtained in small quantities and are not well characterized. The largest CD for which X-ray structure has been determined is that with 26 glucoside units [5c] and gigantic analogues with more than 100 units have been reported [5d]. For many years, on the basis of molecular mechanics calculations [6a] CDs having less than 6 *gluco*pyranosidic rings were sentenced to nonexistence. However, the Nakagawa group successfully carried out the synthesis of **279** [6b]. Interestingly,

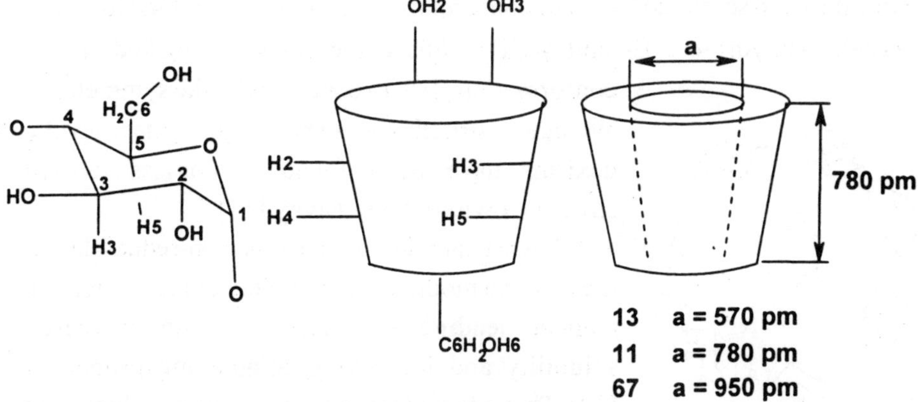

Figure 7.4.3. Atom numbering shape and size of α-, β-, and γ–CDs.

three CD derivatives were even found in Nature [7].

On the basis of X-ray studies α-, β- and γ-CDs were for a long time considered to have a rigid truncated cone structure. They are schematically shown in Fig. 7.4.3 with the primary hydroxyl groups situated at the narrower rim and the secondary ones located at the wider rim [1]. As a result, the hydroxy groups render polar character to the rims while glycosidic oxygen atoms lying in a plane and H3 and H5 hydrogen atoms pointing inside the cavity secure relatively apolar character of the cavity. Recent NMR results have shown that the molecules are flexible even in the solid state [8] in line with some other experimental data [9a], model calculations [9b] and the ease of selective complex formation that would be impossible for the rigid structure. Therefore the symmetrical structures **13, 11** and **68** with C_n symmetry axes (n = 6, 7 or 8, respectively) and the planar ring of glucoside oxygen atoms do not refer to the real but to the averaged structures. The interglucoside C1OC4 bridges are responsible for CDs flexibility but *gluco*pyranoside rings usually assume the rigid so-called $^4C_1$ chair conformation shown in Fig. 7.4.4. However, in *per*-3,6-anhydro-α-, β- and γ-CDs **280a-c** the rings are locked in $^1C_4$ conformation [10]. In contrast with the symmetrical averaged structures, very large macrocyclic dextrins have been found to possess twisted structures without any cavity [11].

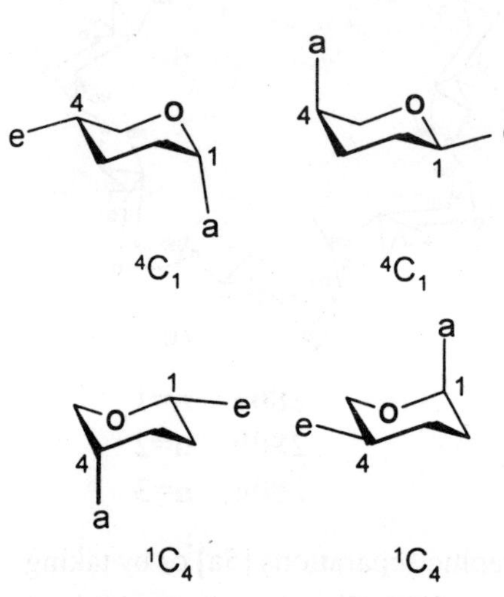

Figure 7.4.4. Conformations of the *gluco*pyranoside ring.

CDs are chemically stable and non-reducible and they can be modified in complete or regioselective manner leading to analogues with increased solubility and interesting complexing properties [12]. They can undergo various reactions which may involve cleavage of OH, CO, CH or CC bonds. It

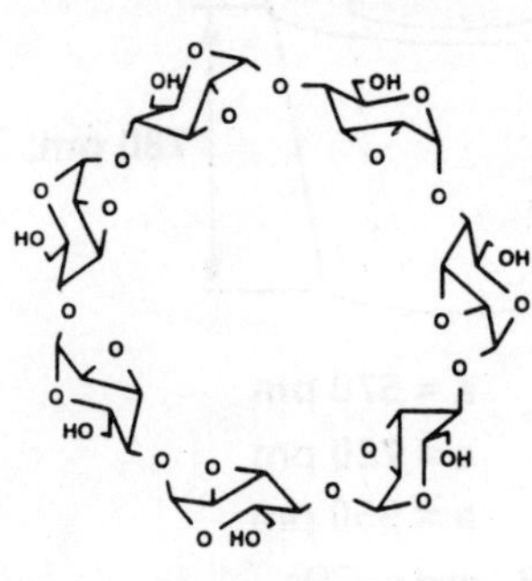

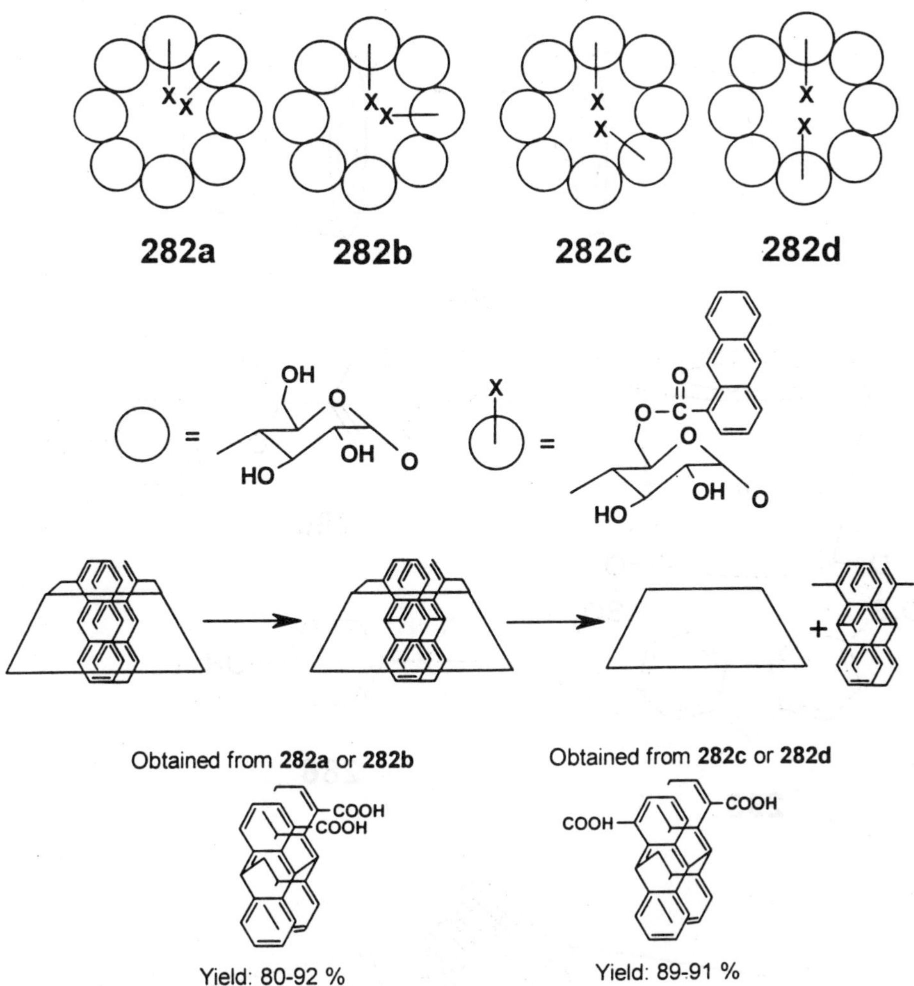

Figure 7.4.5. Regioselective formation of anthacene dimers in γ-CD cavity.

should be stressed, however, that the preparation of CDs bearing several substituents at distinct positions is often extremely difficult and requires a tedious chromatographic purification. On the other hand, completely *O*-methylated and *O*-acetylated CDs are obtained in good yields, thus they are commercially available. (**11** is known to be poorly soluble in water. Surprisingly, its heptakis-(2,6-*O*-dimethylated) derivative with less OH groups is better soluble than the parent compound.) **278a** with $n = 6-8$ are soluble in water but insoluble in common organic solvents, whereas permethylated CDs **278b**, with all OH groups

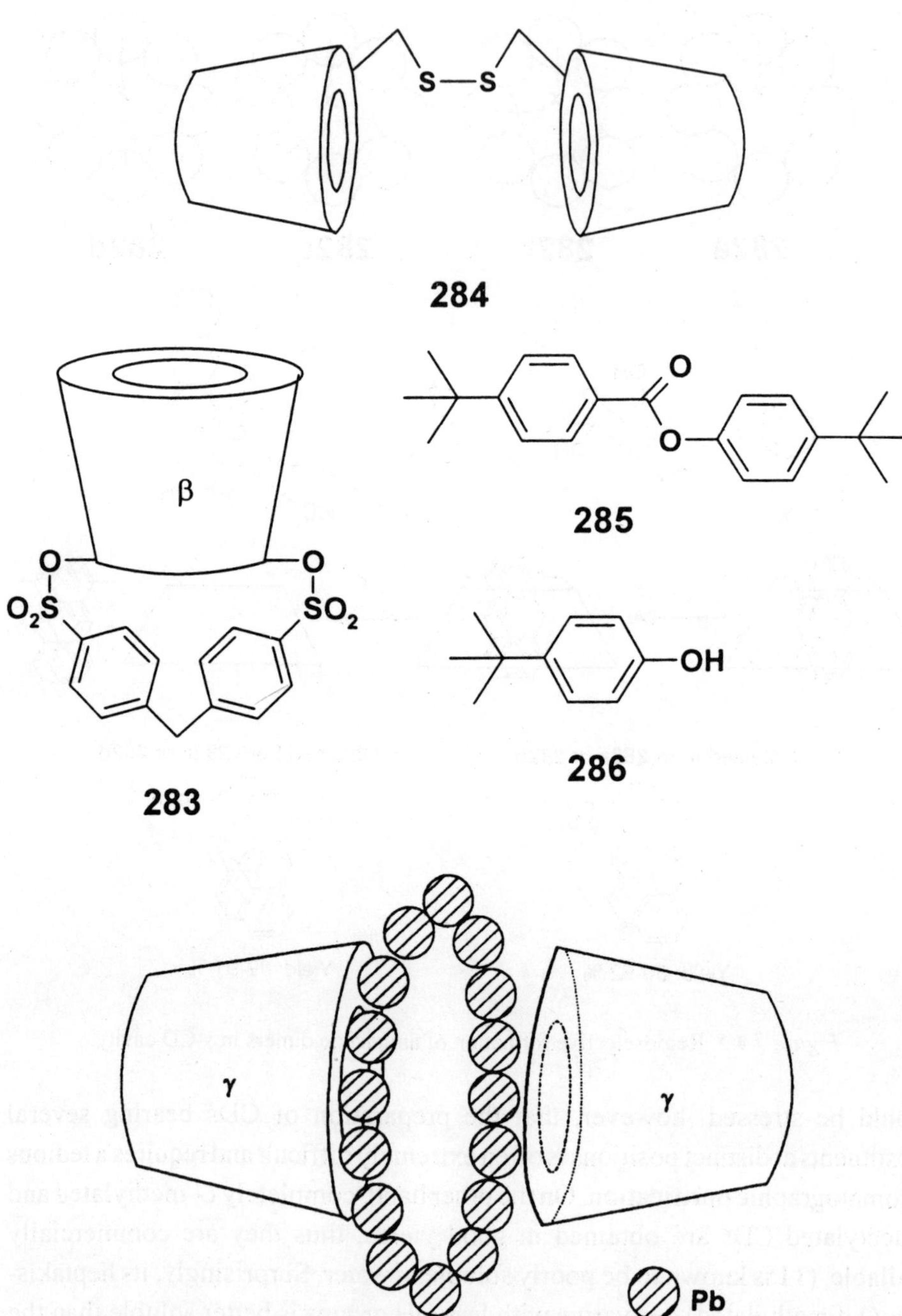

Figure 7.4.6. Schematic view of hexadecanuclear complex of sixteenfold deprotonated γ-CDs.

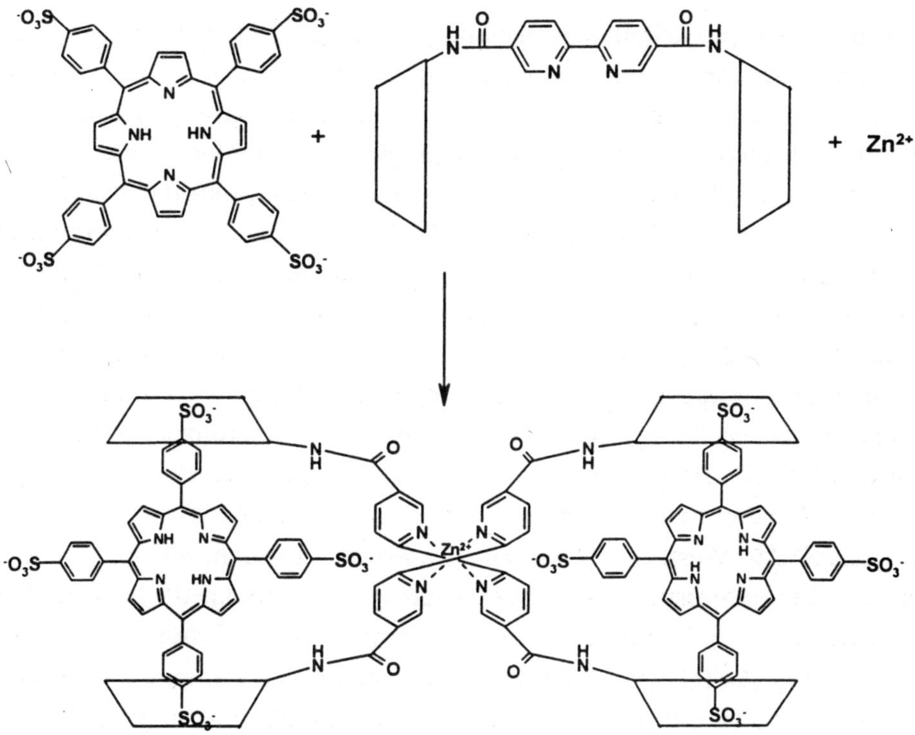

287

Figure 7.4.7. The formation of 2:2 bis-cyclodextrin complex with sulphonated porphyrin.

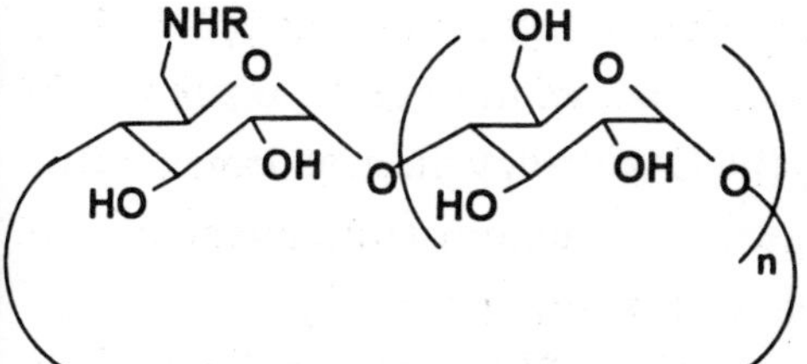

n = 6, 7 R = Glycyl-L-Phenylalanyl

288

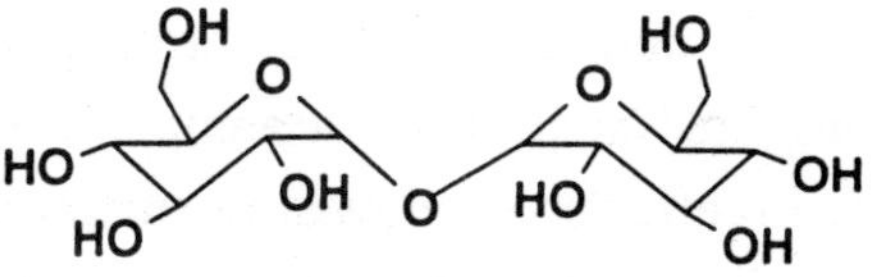

289

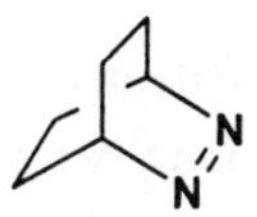

290

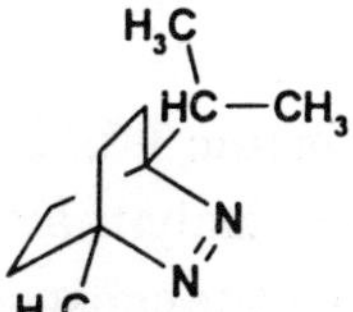

291

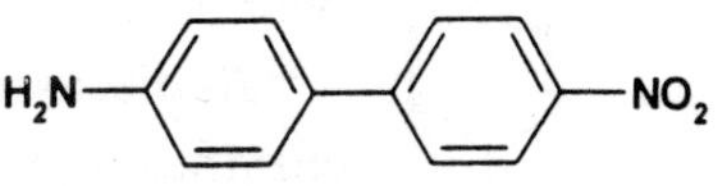

292

substituted, are soluble both in water and the latter solvents. This property can be of importance for potential applications in extraction, catalysis, and so forth. Similarly to **278b**, heptakis-(2,3,6-*O*-pentyl)-β-CD **278c** is an important additive to a mobile phase in chiral chromatography [13]. On the other hand, peracetylated CD **278d** is soluble only in organic solvents. Some interesting CD derivatives include per-2,3-epoxy- β-CD **281** [14], regioselectively disubstituted γ-CDs **282a-d** (Figure 7.4.5) which can undergo stereoselective photodimerization of self-included anthracene substituents [15] and so-called capped CDs like **283** [16] which are believed, actually without any sound proof, to exhibit better enantioselectivity than their parent compounds. CD dimers with a carefully chosen bridge are known to complex much stronger appropriate guests [17] than their monomers. For instance, **284** complexes *p-t*-butyl-phenyl groups of **285** more than 1000 times stronger than the parent **13** does with *p-t*-butylphenol **286**. Aesthetically pleasing sixteenfold deprotonated γ-CD dimer, formed with the corresponding number of lead cations, is presented schematically in Figure 7.4.6 [18]. A rare case of the CD complex with 2:2 stoichiometry is provided by the β-CD dimer **287** with a bipyridine derivative (Figure 7.4.7) [19]. CDs appended with peptides, like **288** [20a], or capped with a sugar unit, like α,α-trehalose **289** [20b], are expected to find applications as drug carriers. Exciting rotaxanes and catenanes involving CDs are presented in Chapter 1 and Section 8.1. Interestingly, CDs can form nanotubes either by covalent [21] or noncovalent self-assembly [22].

As chiral molecules, CD hosts are one of the best chiral selectors [23]. They can also induce circular dichroism signal in an achiral guest. As shown by Zhang and Nau [24] for the complexes of β-CD with bicyclic azaalkanes **290** or **291**, this effect may allow one to determine the orientation of the latter molecule in the CD cavity. An interesting example of the influence of **11** on the guest conformation was reported by Brett and coworkers [25]. They have shown that *p*-amino-*p'*-nitrobiphenyl **292**, which is planar in the solid state, becomes nonplanar in the solid state complex with β-CD.

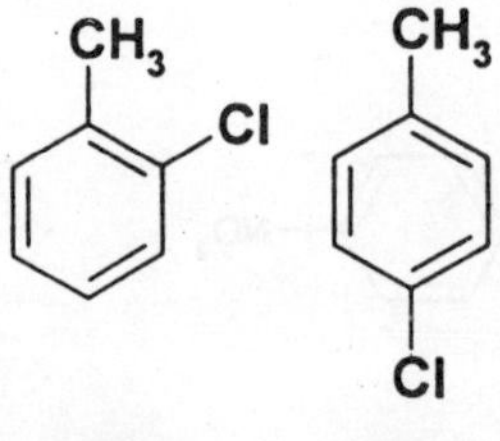

As discussed in Section 3.5, CDs have been used as enzyme mimics. Their catalytic activity usually results in only modest reaction rate enhancements, but in a very few specific cases these rates may reach values typical of enzymes. They are usually much more effective at influencing reaction stereoselectivity than in the enhancement of the reaction yields. The first possibility may be illustrated by the influence of β-CD on the chlorination of toluene [26]. Without the additive this reaction yields as a result a mixture of *ortho-* **293** and *para*-chlorotoluenes **294**. However, in the presence of **11** only the latter isomer is formed.

7.4.2 CD complexes as one of the few supramolecular systems that have found numerous applications

The industrial applications of CDs are possible owing to the easy manufacture of native CDs and their low cost (over 1500 tons per year and few dollars per kg for **11**). CDs are known to discriminate between different molecules, constitutional isomers and enantiomers [27]. However, to our knowledge they are not used for preparative chromatography on an industrial scale in spite of thousands of successfully separated mixtures. On the other hand, enantioselective chromatography using these macrocycles as chiral stationary or mobile phases is a powerful analytical technique allowing one to carry out stereochemical analysis of natural compounds, in order to determine the enantiomeric excess in asymmetric syntheses and to prove the enantiomeric purity of pharmaceuticals and chiral reagents. In addition, CDs are invaluable in the studies of metabolism of chiral compounds both in clinical tests and environmental research as well as in investigations of the mechanism of reactions involving conversion of chiral compounds (inversion or retention of configuration, racemization and chirality transfer in the rearrangement reactions).

Several CDs' applications have been discussed in Chapter 6. Their lack of toxicity forms the basis for their applications in pharmaceutical, agrochemical, and food industries as well as in toiletry and cosmetics [1b, 1c]. They are also used or have been proposed to be applied, as sensitizers and stabilizers of dyes in photography, for impregnating paper, as fluorescent and other sensors [28], as corrosion inhibitors and rust proofing materials, UV stabilizers and antioxidants. One of the most spectacular is the prospective use of CDs as biodegradable plastics [29]. The complex of nitroglycerine with β-CD marketed in Japan was mentioned in Chapter 1 while some other CDs applications have been presented in Chapter 6.

7.4.3 Predicting molecular and chiral recognition of CDs on the basis of model calculations

Molecular and chiral recognition by CDs is often modelled computationally [30] in view of their theoretical and practical importance. However, the mechanical complexity of CDs which have a very complicated energy hypersurface exhibiting several low-lying energy minima separated by low energy barriers [9, 23b] complicate such studies. Nevertheless, user-friendly programs producing delightful molecular models create a false impression of omnipotence of CD theoretical studies in unprepared researchers. Such an opinion is supported by the uncritical review by Lipkowitz [30] which, in addition to some vague comments concerning an incompetent use of the programs, presents unlimited prospects for such computations. In particular, the problem of the accuracy and reliability of the results obtained and their dependence on the parameterization used was not discussed there as also was not the way in which the comparison of calculated results with the corresponding experimental data should be carried out. (Our recent results indicate that the latter data strongly depend on the experimental technique used for their determination [31].) Experiments, for instance chromatographic studies, which are often performed in mixed solvents yield free energy differences, whereas steric energy differences provided by molecular mechanics calculations [32] can be compared only with enthalpy differences. In addition, no difference between the modelling of molecular and chiral recognition is made there, although the latter process is usually associated with much smaller energy differences which are impossible to be interpreted in terms of any physical model [33]. These problems were recently discussed by the Dodziuk group [34]. The complexes of **11** with the decalin isomers **295**, experimentally studied earlier in the same group [35], provide a unique opportunity for analysing molecular and chiral recognition in the same system since invertomers of *cis*-decalin **295a** and **295b** are enantiomers as well.

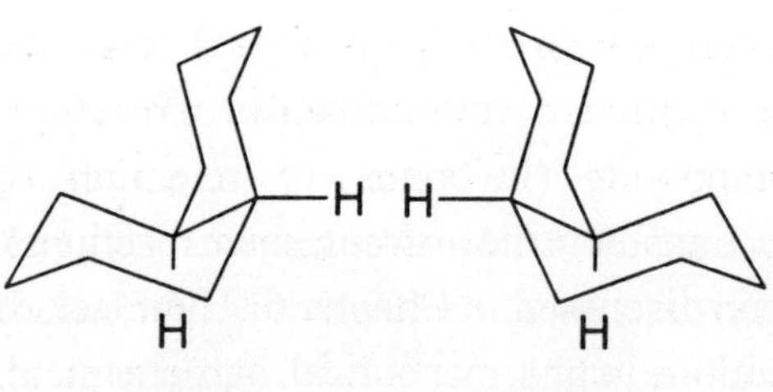

H H H H

295a **295b**

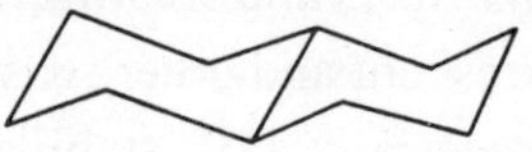

295c

Molecular mechanics calculations [34a] using four different force fields and five values of electric permittivity for the complexes of **295** with **11** revealed that the results obtained do depend on the assumed parameters values which cannot be reliably chosen. Moreover, in several cases the energy differences characterizing molecular recognition were smaller than the corresponding values describing chiral recognition. The molecular dynamics [36] calculations [34b] in the same system seemed to indicate that the average energies calculated by this method may qualitatively describe molecular recognition. However, long simulations for diastereomeric complexes of **13** and α-pinene **27** enantiomers [34c] have shown that not only the value of the energy difference but also its sign depends on the lengths of the simulation times. Therefore much longer simulation times are needed to carry out reliable modelling of molecular and chiral recognition by CDs.

The Free Energy Perturbation method provides the most accurate values of energy differences by simulations [37]. However, to our knowledge this method was only once applied to CD complexes [38], but too little information about the procedure applied, provided in this paper, precludes the assessment of its reliability.

References

1. (a) There is no up-to-date general monograph on CDs. A special issue of The Chemical Reviews, 1998, 98, (5) covers several aspects of CDs chemistry while the books listed below mainly present their applications; (b) *Comprehensive Supramolecular Chemistry*, v. 3, J. Szejtli, Ed., Kluwer Academic Publishers, Elsevier, Oxford, 1996; J. Szejtli, Cyclodextrin Technology, Kluwer, Dordrecht, 1988; (c) *Cyclodextrins and Their Industrial Uses*, D. Duchene, Ed., Edition du Sante, Paris, France, 1987; *New Trends in Cyclodextrins and Derivatives*, D. Duchene, Ed., Edition du Sante, Paris, France, 1991; (d) W. Saenger, Angew. Chem. Int. Ed. Engl., 1980, 19, 344; W. Saenger, in *Inclusion Compounds*, J. T. Atwood, J. E. D. Davies, D. D. MacNicol, Eds., v. 2 , Academic Press, London, 1984, p. 231; K. Harata, *ibid.*, v. 5, 1991, p. 311.

2. J. F. Stoddart, Carbohydr. Res., 1989, 192, X.

3. **278a** with $n = 6-8$ are also called cyclohexa-, cyclohepta- and cyclooctaamylose, respectively.

4. Y. Takahashi, T. Ogawa, Carbohydr. Res., 1987, 164, 277; 1987, 169, 127; P. Angibeau, J. P. Utille, Synthesis, 1991, 9, 737; G. Gatusso, S. A. Nepogodiev, J. F. Stoddart, Chem. Rev., 1998, 98, 1919.

5. (a) T. Endo, H. Nagase, H. Ueda, S. Kobayashi, T. Nagai, Chem. Pharm. Bull., 1997, 45, 532; T. Endo, H. Nagase, H. Ueda, A. Shigihara, S. Kobayashi, T. Nagai, *ibid.*, 1856; (b) Y.

Terada, M. Yanase, H. Takata, H. Takaha, J. Biol. Chem., 1997, 272, 15729; (c) K. Gessler, I. Uson, T. Takaha, N. Krauss, S. M. Smith, S. Okada, G. M. Sheldrick, W. Saenger, Proc. Natl. Acad. Sci. USA, 1999, 96, 4246; (d) T. Takaha, M. Yanase, S. Takata, S. Okada, S. M. Smith, J. Biol. Chem., 1996, 271, 2902.

6. (a) P. R. Sundarajan, V. S. R. Rao, Carbohydr. Res., 1970, 13, 351; (b) T. Nakagawa, U. Koi, M. Kashiwa, J. Watanabe, Tetrahedr. Lett., 1994, 35, 1921.

7. M. G. Usha, R. J. Wittebort, J. Am. Chem. Soc., 1992, 114, 1541: Y. Inoue, T. Okuda, R. Chujo, Carbohydr. Res., 1985, 141, 179; Y. Yamamoto, M. Onda, Y. Takahashi, Y. Inoue, R. Chujo, Carbohydr. Res., 1988, 182, 41.

8. M. Entzeroth, R. E. Moore, W. P. Niemczura, M. L. G. Patterson, J. N. Shoolery, J. Org. Chem., 1986, 51, 5307.

9. (a) F. V. Bright, G. C. Camena, J. Huang, J. Am. Chem. Soc., 1990, 112, 1343; A. F. Bell, L. Hecht, L. D. Barron, Chem. Eur. J., 1997, 3, 1292; T. Steiner, W. Saenger, J. Am. Chem. Soc., 1991, 113, 5676; (b) H. Dodziuk, K, Nowinski, J. Mol. Struct. (THEOCHEM), 1994, 304, 61.

10. P. R. Ashton, P. Ellwood, I. Staton, J. F. Stoddart, Angew. Chem. Int. Ed. Engl., 1991, 30, 80; J. Org. Chem., 1991, 56, 7274; A. Gadelle, J. Defaye, Angew. Chem. Int. Ed. Engl., 1991, 30, 78; H. Yamamura, K. Fujita, Chem. Pharm. Bull,, 1991, 636.

11. K. Gessler, I. Uson, T. Takaha, N. Krauss, S, M, Smith, S. Okada, G. M. Sheldrick, W. Saenger, Proc. Natl. Acad. Sci. USA, 1999, 96, 4246.

12. A. R. Khan, P. Forgo, K. J. Stine, V. T. D'Souse, Chem. Rev., 1998, 98, 1977; G. Gatusso, S. A. Nepogodiev, J. F. Stoddart, 1998, 98, 1919; L. Jicsinsky, E. Fenyvesi, H. Hashimoto, A. Ueno, *Comprehensive Supramolecular Chemistry*, 1996, v. 3, p. 58; G. Wenz, Angew. Chem. Int. Ed. Engl., 1994, 33, 803.

13. J. E. H. Köhler, M. Hohla, M. Richters, W. A. König, Angew. Chem. Int. Ed. Engl., 1992, 31, 319.

14. A. R. Khan, L. Barton, V. T. D'Souse, J. Org. Chem., 1996, 61, 8301.

15. A. Ueno, F. Moriwaki, Y. Iwama, I. Suzuki, T. Osa, T. Ohta, S. Nozoe, J. Am. Chem. Soc., 1991, 113, 7034.

16. I. Tabushi, K. Shimokawa, K. Fujita, Tetrahedron Lett., 1977, 18, 1527.

17. R. Breslow, S. Halfon, B. Zhang, Tetrahedron, 1995, 51, 377.

18. P. Klüfters, J. Schumacher, Angew. Chem. Int. Ed. Engl., 1994, 33, 1863.

19. F. Venema, A. E. Rowan, R. J. M. Nolte, J. Am. Chem. Soc. 1996, 118, 268.

20. (a) F. Djeddaini-Pilard, N. Azaroual-Bellanger, M. Gosnat, D. Vernet, B. Perly, J. Chem. Soc. Perkin Trans. 2, 1995, 723; (b) V. Cucinotta, G. Grasso, G. Maccarrone, A. Mazzaglia, G. Vecchio, Proceedings of the Ninth International Symposium on Cyclodextrins, J. J. Torres-Labandeira, J. L. Vila-Jate, Eds. Kluwer Academic Publishers, Dordrecht, 1999, p. 501.

21. G. Pistolis, A. Malliaris, J. Phys. Chem. 1996, 100, 15562.

22. A. Harada, J. Li, M. Kamachi, Nature, 1993, 364, 516.

23. (a) H. Dodziuk, *Modern Conformational Analysis. Elucidating Novel Exciting Molecular Structure*, VCH Publishers, NY, 1995, p. 122; (b) p. 219.

24. X. Zhang, W. M. Nau, Angew. Chem. Int. Ed. Engl., 2000, 39, 544.

25. T. J. Brett . S. Liu, P. Coppens, J. J. Stezowski, J. Chem. Soc. Chem. Commun., 1999, 551.

26. R. Breslow, P. Campbell, J. Am. Chem. Soc., 1969, 91, 3084.

27. S. Li, W. C. Purdy, Chem. Rev., 1992, 92, 1457; J. Szejtli, B. Zsadon, T. Csehati, in *Ordered Media in Chemical Separations*, W. L. Hinze, D. W. Armstrong, Eds., American Chemical Society, Washington, DC, 1987, p. 201; W. A. König, *Gas Chromatographic Enantiomer Separation with Modified Cyclodextrins*, Hütig, Heidelberg, Germany, 1992, S. M. Han, D. W. Armstrong, in *Chiral Separations by HPLC*, A. M. Krstulovic, Ed., Wiley, New York, 1989, Chapter 10, p. 208; S. G. Allenmark, *Chromatographic Enantioseparations. Methods and Applications*, Ellis Horwood, Chichester, 1988; *Chiral Separations by Liquid Chromatography*, G. Subramanian, Ed., VCH, Weinheim, 1994.

28. M. Fukushima, T. Osa, A. Ueno, J. Chem. Soc. Chem. Commun., 1991, 15; Chem. Lett., 1991, 709.

29.M. Yoshinaga, Chem. Abstr., 1992, 117, 214839b.

30. K. B. Lipkowitz, Chem. Rev., 1998, 98, 1829.

31. H. Dodziuk, F. P. Schmidtchen, W. Kozminski, G. Dolgonos, K. S. Nowinski, poster at 13th International Symposium Chirality-2001, 15-18th July, Orlando, USA.

32. E. Osawa, H. Musso, Angew. Chem., 1983, 22, 1.

33. W. H. Pirkle, T. C. Pochapsky, Chem. Rev., 1989, 89, 347.

34. (a) H. Dodziuk, O. Lukin, K, S. Nowinski, J. Mol. Struct. (THEOCHEM), 2000, 503, 221; (b) H. Dodziuk, O. Lukin, Pol. J. Chem., 2000, 74, 997; (c) H. Dodziuk, O. Lukin, Chem. Phys. Lett., 2000, 327, 18.

35. H. Dodziuk, J. Sitkowski, L. Stefaniak, J. Jurczak, D. Sybilska, J. Chem. Soc. Chem. Commun., 1992, 207.

36. W. F. van Gunsteren, H. J. Berendsen, Angew. Chem., 1990, 29, 992.

37. J. Marelius, T. Hansson, J. Aqvist, Int. J. Quant. Chem., 1998, 69, 77; M. L. Lamb, W. L. Jorgensen, Curr. Opinion Chem. Biol., 1997, 1, 449.

38. D. Salvatierra, X. Sanchez-Ruiz, R. Garduno, E. Cervello, C Jaime, A. Virgili, F. Sanchez-Ferrando, Tetrahedron, 2000, 56, 3035.

7.5 Endohedral Fullerene Complexes, Nanotubes and Other Fullerene-based Supramolecular Systems

Fullerene chemistry is a very new but rapidly developing research field [1]. Its beginning was marked by theoretical predictions in the early 1970s [2] which were confirmed only in the mid 1980s [3] gaining the Nobel Prize for the discoverers of C_{60} **40** (also denoted [60]fullerene) Kroto, Curl and Smalley. Interestingly, this molecule is the fifth (besides diamond, graphite, carbene and carbyn) allotrope form of carbon [4]. Until now fullerenes and their complexes are prepared not by standard chemical syntheses but by the methods such as arc discharges or laser evaporation of graphite (mixed with metal oxides or carbides [5a], if necessary). The Krätschmer group succeeded in developing an efficient preparation procedure allowing one to obtain **40** in gram quantities [5b]. As mentioned in Section 2.4, generation of C_{60} is a very efficient all-or-nothing covalent self-assembling process.

Fullerenes are fascinating molecules capable of inclusion complex formation since, as will be shown below, they can not only host ions or molecules inside their cages but also play the guest role of being buried inside a capsule formed by two γ-cyclodextrin **68** or calix[5]arene **221** (n = 5) molecules. In spite of considerable efforts no total organic synthesis of **40** has yet been reported. Interestingly, attempts to patent C_{60} in the USA have failed since it has been found in Nature. They occur not only in some ores, coals, and meteorites [6] but 2 to 40 different fullerenes are also generated by burning dinner candles [1a].

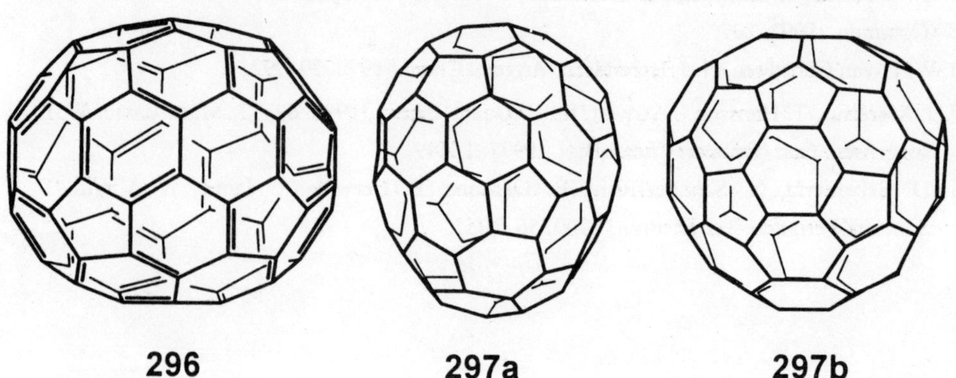

296 **297a** **297b**

Most stable fullerenes usually obey IPR (the isolated pentagon rule) [7a] which states that fullerenes having adjacent five-membered rings are less stable than their counterparts with isolated rings. However, not only C_{36} with a few adjacent five-membered rings was reported [7b] but also C_{20} **273**, the smallest known fullerene type of compound, built solely of pentagons was synthesized [7c]. C_{60} of icosahedral symmetry with all carbon atoms equivalent is the parent compound for all fullerene family because of its remarkable stability. The next member of the family, C_{70} **296**, was detected early since it is formed in considerable amount together with C_{60}. For higher fullerenes several structural isomers of different symmetries are possible for a given number of carbon atoms.

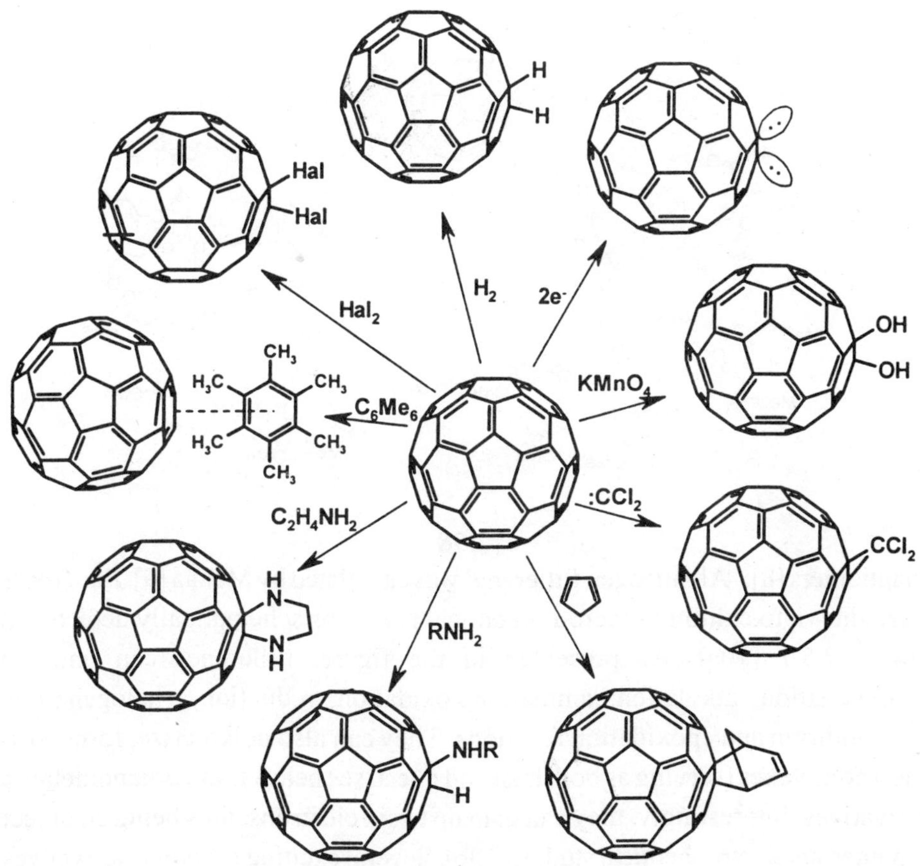

Figure 7.5.1. Various chemical reactions of C_{60}.

A few of them have been unequivocally assigned until now but the pace of research in this field allows one to expect that many others will be detected and characterized soon. C_{76} **297a** of D_2 symmetry (another possible isomer has T_d symmetry **297b**) is remarkable since it is chiral and was resolved into

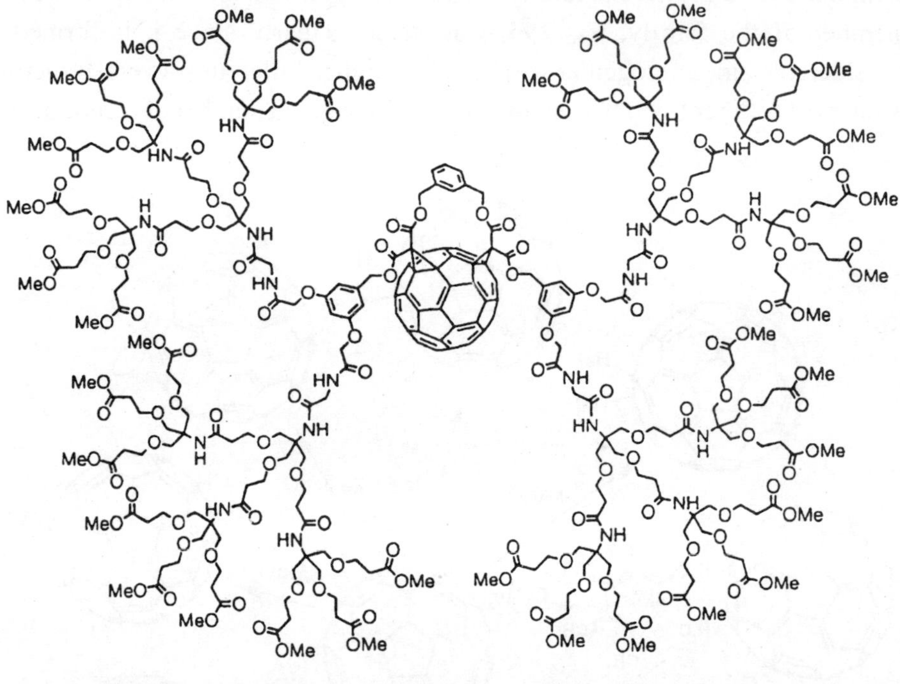

298

enantiomers [8]. 'All nitrogen fullerene' was calculated by Manaa [9]. Fullerenes were shown to exhibit numerous chemical reactions schematically depicted in Figure 7.5.1 [10a]. As presented in the figure, fullerenes can undergo hydrogenation, alkylation, amination, oxidation, reduction, halogenation, cycloaddition and epoxidation reactions. They can also polymerize, form host-guest complexes (serving as both host and guest, see below), and organometallic derivatives. Interestingly, they can gain up to six electrons, thus being an object of numerous electrochemical studies [10b]. Several exciting fullerene derivatives are shown in **298-303** [10c]. Crown ether-C_{60} conjugate forms a self-assembled monolayer on a gold surface (Figure 7.5.2) taking advantage of the recognition of ammonium ion by the ether [10d].

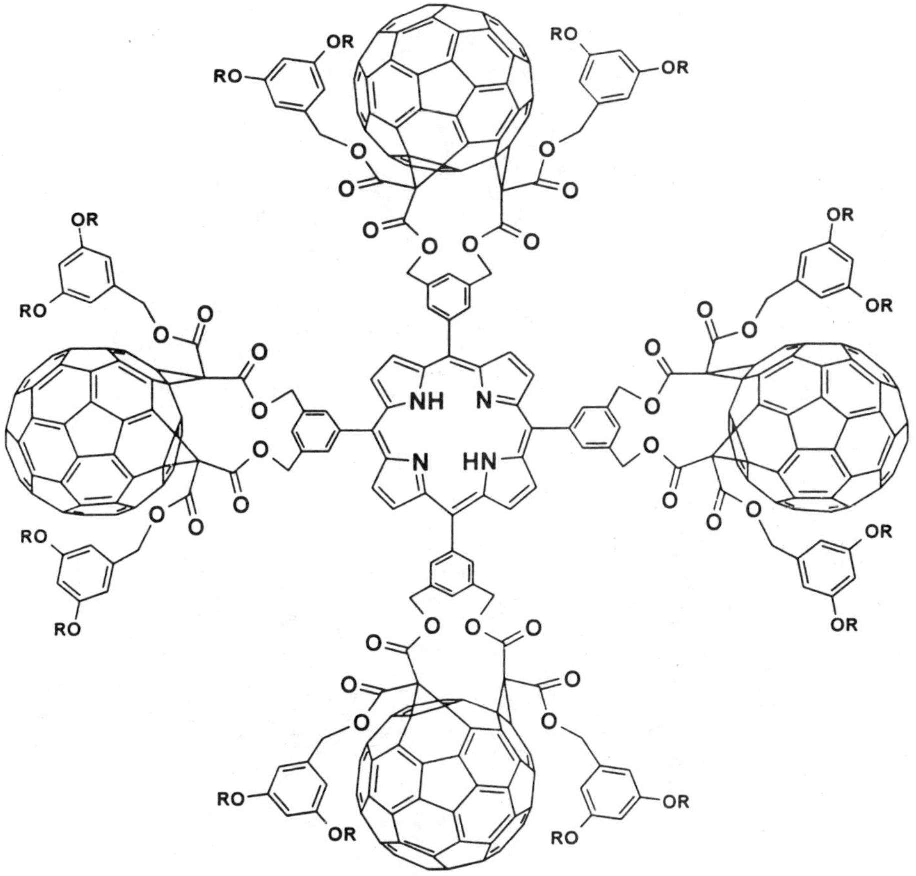

RO
OR
RO
OR
OR
RO
OR
OR
O
O
O
O
O
O
NH
N
N
HN
RO
OR
RO
OR
$R = C_{12}H_{25}$

299

O
O
O
O
X
X
X = $C(COOEt)_2$

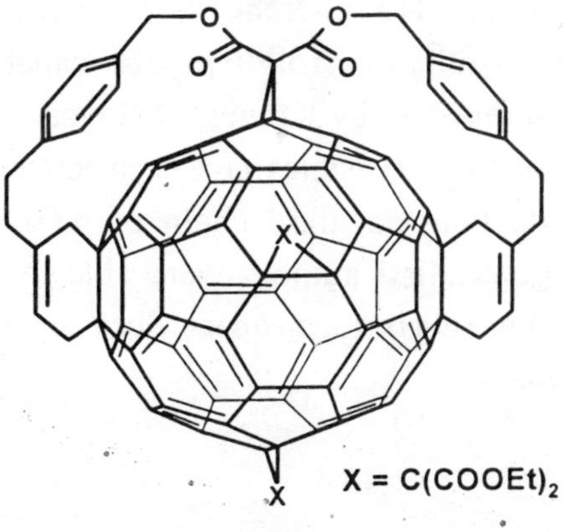

300

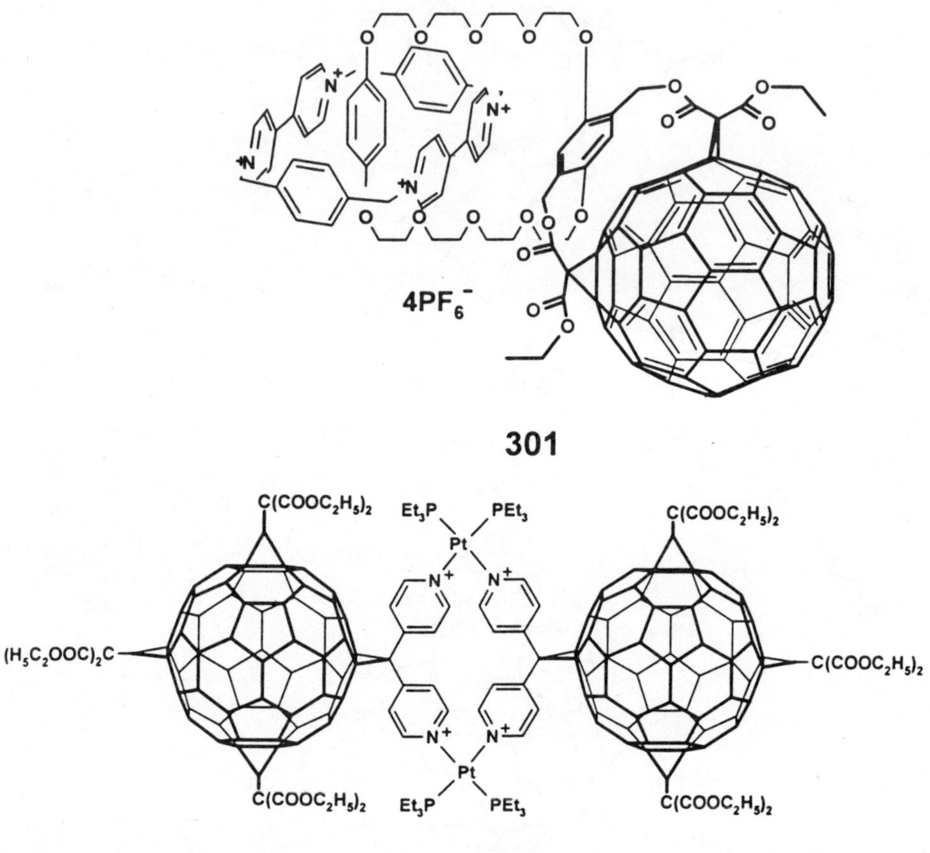

Noteworthy, chlorin coupled to fullerene **304** has been obtained by one-pot synthesis [10e]. 2-(*n*-alkyl)fulleropyrrolidines **305** capable of formation of Langmuir films has been reported by Kutner, D'Souza and coworkers [10f], while self-assembled nanorods and vesicles have been obtained by the Tour group [10g] using **306**. By attaching an alkyl chain of different length through pyrrolidine linker to C_{60}, the first authors were able to study the effect of protonation of the pyrrolidine ring nitrogen on the electrochemical behaviour and Langmuir film properties.

Owing to its almost spherical shape, at room temperature **40** executes random rotations even in the solid state making X-ray determination of its bond lengths impossible. Thus their first, quite inaccurate, measurement was carried out by using NMR [11] showing that the length of the CC bonds connecting six-member rings was different from that of the bonds connecting five-membered and six-membered rings. The size of C_{60} cavity, ca. 700 pm in diameter, allows for the encapsulation of practically any element of the periodic table inside it [12]. However, mainly lantanides (and to a lesser extent alkaline metals, noble gases, nitrogen, and a few other elements) have been found to form endohedral complexes denoted as $M_m@C_{2n}$. The fact that the guest is situated inside the host cage was first unequivocally proved by X-ray synchrotron powder diffraction of the solid $Y@C_{82}$ [13a]. The study has not only manifested the encapsulation of the lantanide inside the cage but has also shown that it is not situated at the cage's center. Hence, a large permanent dipole moment of $Y@C_{82}$ which has

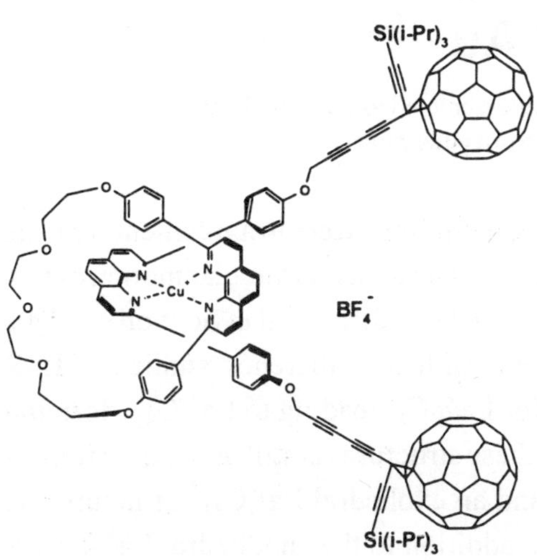

Si(i-Pr)$_3$ BF$_4^-$ Si(i-Pr)$_3$

303

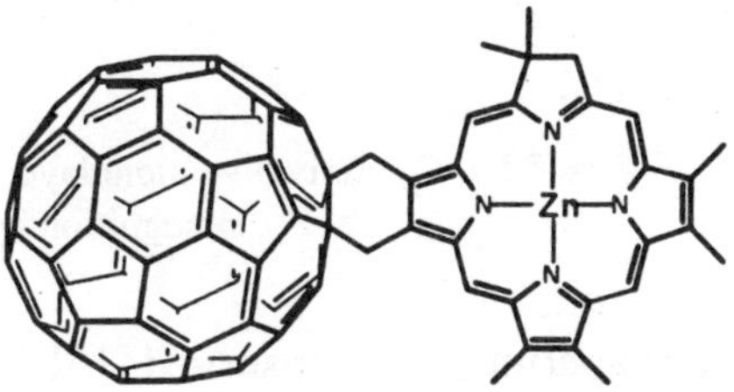

304

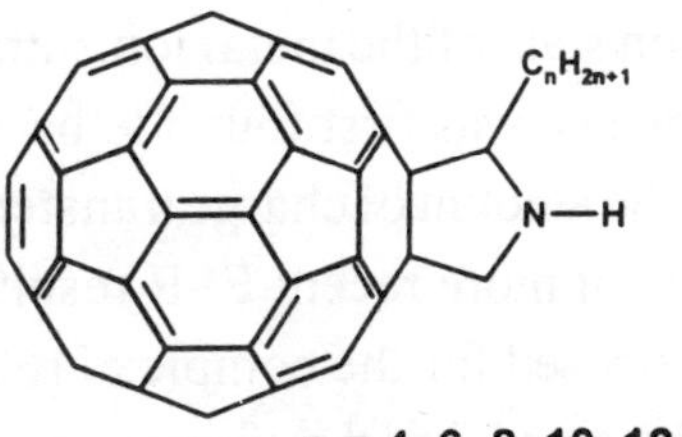

C_nH_{2n+1} N—H

n = 4, 6, 8, 10, 12

305

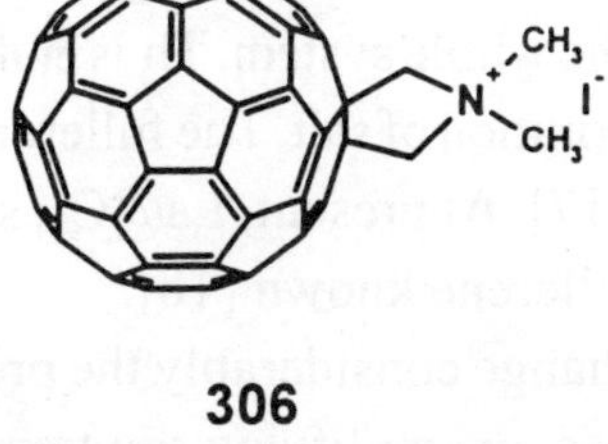

CH_3 N$^+$ CH_3 I$^-$

306

C(COOEt)$_2$ N

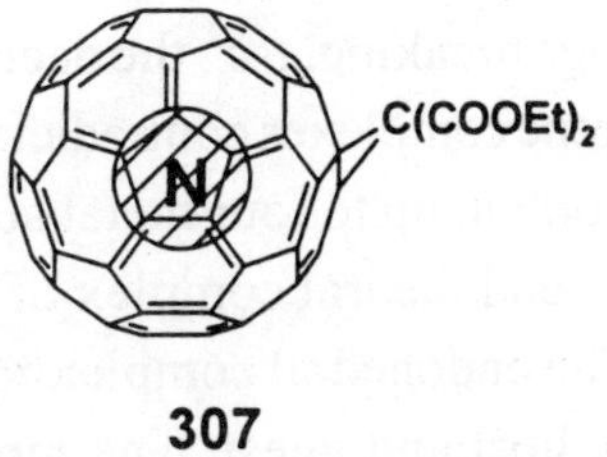

307

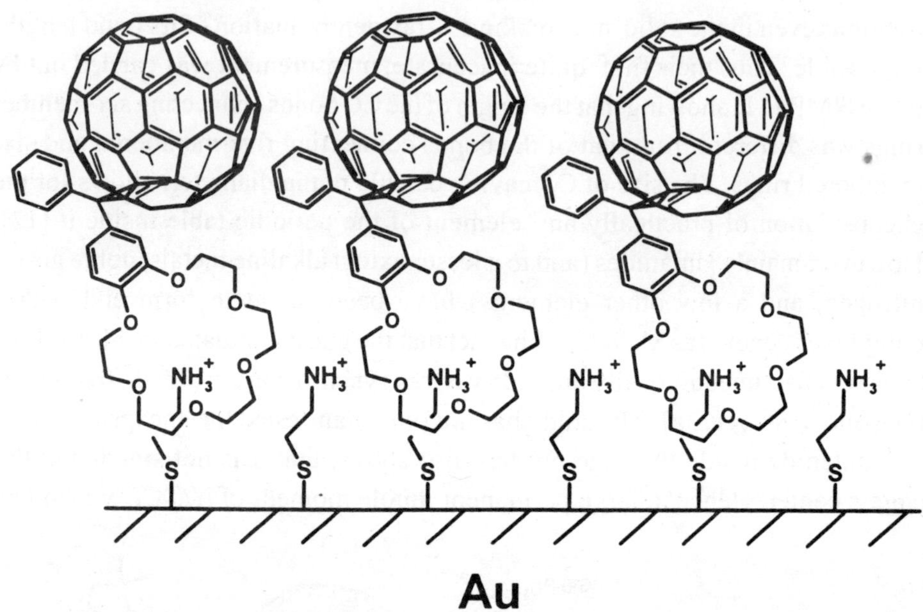

Figure 7.5.2. C_{60} derivative monolayer self-assembled as a result of ammonium cation recognition by crown ether moiety.

been experimentally observed [13b]. These findings were in agreement with the calculations preceeding the experiment [14]. The dynamic circular motion of two guest La cations inside C_{80} **80** cage [15] have been discussed in Section 3.5. The oxidation state of the metal ion entrapped inside the fullerenes, studied by ESR experiments, was first found to be +3 for La@C_{82} leading to $La^{3+}C_{82}^{3-}$ formula [16a]. The amount of charge transfer in these structures is still under question on the basis of more recent ESR results, and an exohedral $La^{2+}C_{82}^{2-}$ structure has been proposed for the complex [16b] in addition to the endohedral $La^{3+}@C_{82}^{3-}$ one advocated by the former research groups. It should be stressed that endohedral fullerene complexes with ions represent an interesting example of ionic compounds that cannot be dissociated into their constituting ions without the cage breaking, *i. e.*, the decomposition of the whole system. Thus endohedral fullerene complexes contradict the classical definition of salt. The fullerene cages can contain up to four metal atoms (or ions) [17]. At present, La@C_{44} seems to be the endohedral complex of the smallest fullerene known [18].

The endohedral complex formation can change considerably the properties of the host and guest. The most striking cases, exemplifying the tremendous influence the host cage can play, are probably the trapping of diatomic molecules of noble gases He_2 or Ne_2 inside C_{70} **296** [19a], on the one hand, and that of atomic nitrogen inside the same host [19b], on the other. Interestingly, in the latter case of N@C_{70}, nitrogen was found by ESR to be in the quartet ground state $^4S_{3/2}$ [20] confirming extraordinary inertness of the inner fullerene surface. Contrary to this inertness, the outer surface exhibits considerable reactivity. An observation of an elementary particle muon inside **296** should also be mentioned [19c]. Amazingly, a permanent distortion of the C_{60} cage can influence the wavefunction of the guest. The effect was documented by EPR investigation in solution [20] for the endohedral monoadduct N@$C_{60}(COOEt)_2$ **307** [21], obtained by nucleophilic cyclopropanation of N@C_{60} with diethyl bromomalonate.

Beside of those with the noble gases molecules and Sc_2@C_{84} [22], no other endohedral fullerene complexes involving molecular guests have been proven up-to-date. Sc_3N@C_{80} is an equivocal case since some experimental data on this system have been interpreted in terms of the structure containing bonds between scandium atoms and the host cage [23a]. (The most recent data on Sc_3N@C_{66} and Sc_3N@C_{68} did not specify the structure [23b, c]). On the basis of molecular mechanics calculations [24] only H_2 and H_2O molecules were found to be capable of forming the endohedral complexes with the parent C_{60}.

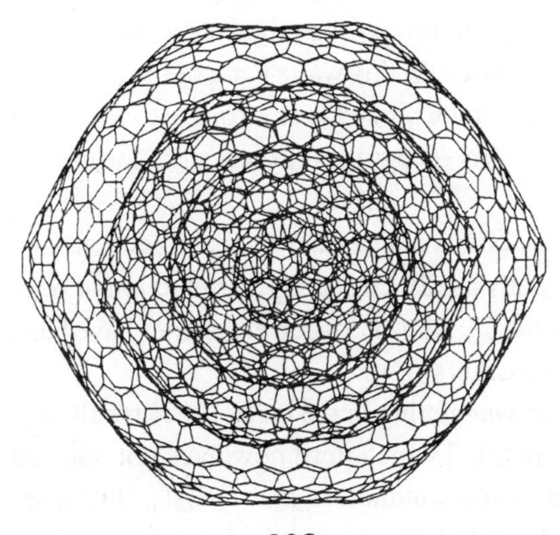

308

As mentioned earlier, in spite of efforts by several research groups, C_{60} has not yet been prepared by a conventional chemical synthesis. Such an approach would eventually enable the incorporation of a guest inside the cage during the synthesis. However, the synthesis of C_{60} with a hole by the Rubin group [25] paves the way to endohedral fullerene complexes with organic molecules.

Nested (or onion-like) fullerenes built of several cages buried inside one another such as C_{60}@C_{240}@C_{540}@C_{960} **308**, were found by using a Tunneling

Electron Microscopy, TEM, technique [26a]. Molecular mechanics calculations have shown that very weak, but numerous, nonbonded interactions are responsible for their formation [26b]. The communication on the formation of $C_{80}@C_{240}@C_{560}$ with all fullerenes of icosahedral symmetry [26c] seems ill-founded since: (a) only C_n cages with $n = 60k^2$ have I_h symmetry [26d]; (b) according to our theoretical study, $C_{80}@C_{240}@C_{560}$ should be less stable than $C_{60}@C_{240}@C_{540}$ [26b].

Other onion-like structures reported recently include those built of tungsten sulphide and other inorganic materials [26e].

Understandably, guest inclusion can stabilize the fullerene cage. For instance, genuine C_{66} and C_{68} are not very stable, and they have not been isolated in considerable amounts. However, endohedral complexes of these fullerenes have been reported [23]. Efficient 1:2 complexation of C_{60} with γ-cyclodextrin [27a] or calix[5]arenes [27b] may open the way for very efficient purification of fullerenes. Another method of the purification making use of antibodies has been proposed in Ref. 28: in the efforts to create smart nanotools mimicking an amazing effciency of Mother Nature, Chen and coworkers used animal produced antibodies to separate C_{60}. They first tethered a derivative of C_{60} to a large protein and injected the product into a rabbit or mouse. Out of a wide variety of antibodies extracted from bloodstream, a single type of antibody recognizing very selectively C_{60} was separated and multiplied. In this way 50 mg of the pure anti-C_{60} monoclonal antibody was obtained which binded very strongly to SWNT or C_{60}. The authors suggested the use of this recognition process for sorting out different types of nanotubes.

Fullerene itself exhibits catalytical activity and, as mentioned in Section 6.4.2 it can increase the activity of other catalysts. Electrocatalytic reduction of α,ω-diiodoalkanes $I(CH_2)_mI$ (m = 1 - 8) at the C_{60} film-modified electrodes have been reported by D'Souza, Kutner and coworkers [29].

As mentioned in Section 6.3, band waggoning in the early 1990s resulted in several proposals for uses of fullerene [30a] which until now have not reached fruition. High-temperature fullerene superconductors and a $C_{60}F_{60}$ lubricant illustrate the point. The latter, ball shaped perfluorinated hydrocarbon, should have revolutionized industry by minimizing friction losses of energy. The compound was synthesized. However, it proved unstable. Moreover, it decomposed in air with HF formation, precluding any industrial application [30b]. Concerning the use of doped C_{60} as a prospective room-temperature superconductor which should minimize energy losses by electricity transmission, it seems also a failure today, since the effect was observed at much higher temperature for inorganic materials [31a]. However, a new material consisting of hole-doped C_{60} raises hopes of the achievability of fullerene superconductivity at temperatures in excess of 100 K [31b]. In addition, fullerenes with 'a hole' and 'door' have been thought of as a designed drug carrier capable of controlled discharge of the drug from its cavity only when the door is open [29]. This proposal has not been accomplished but a C_{60} derivative promises to be an effective photodynamic drug destroying tumor tissue owing to its ability efficiently to generate singlet oxygen [32]. The possibility of applying lanthanide metallofullerenes as new diagnostic tools or therapeutic radiopharmaceuticals is also vividly explored [33]. One of the most promising C_{60} uses is its application as a scanning tunneling microscope tip [34]. Several other applications of fullerenes and their complexes have been patented. Amongst others, the use of fullerenes in diamond production that should be much easier, since it would not require high pressure [35].

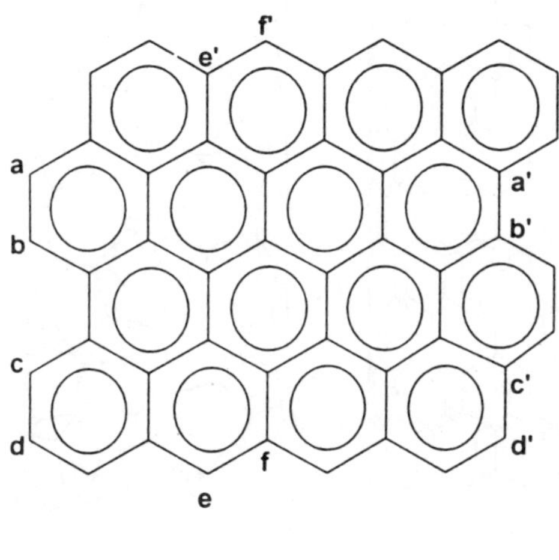

309

The twelve pentagons of the fullerene structure ensure its curvy shape. A net built only of carbon atom hexagons can be planar like graphite **309** or be rolled to form various types of single-wall nanotubes (Figure 7.5.3): armchair **310** (by connecting *e* with *e'*, *f* with *f'*, etc.), zigzag **311** (by connecting *a* with *a'*, *b* with *b'*, etc.), and chiral **312** ones (by connecting, for instance, *a* with *c'*, *b* with *d'*, etc.) [36]. In achiral nanotubes some Csp^2Csp^2 bonds have parallel or perpendicular arrangement with respect to the tube axis while there are no such bond arrangements in the helical nanotubes. The prediction that nanotubes should exhibit metallic conductivity or operate as semiconductors depending on the cylinder diameter and its wrapping angle [37] has been recently confirmed by two

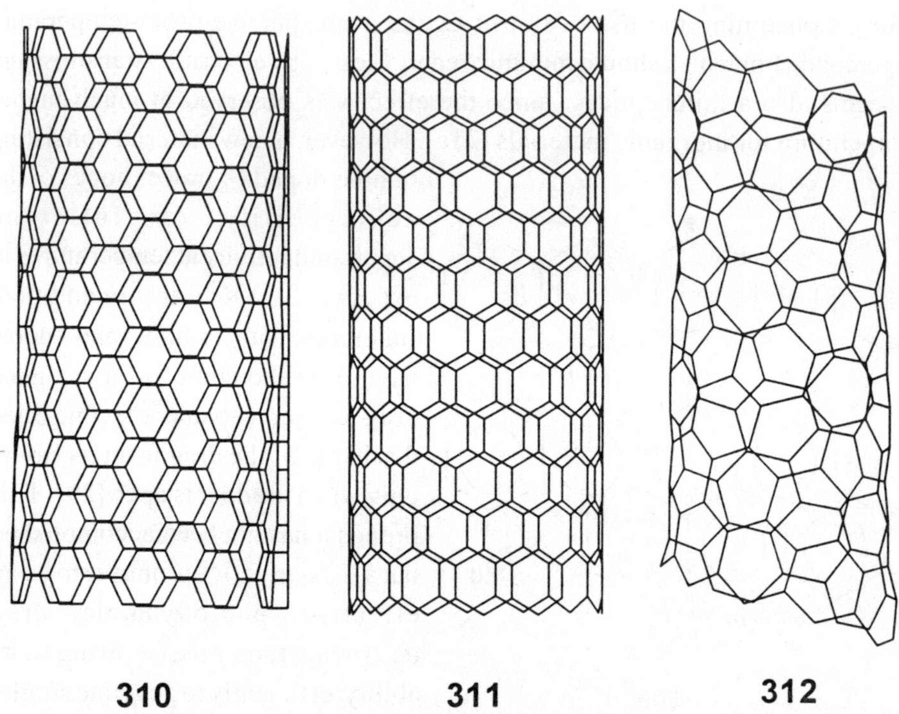

Figure 7.5.3. Armchair **310**(left), zigzag **311**(middle) and chiral **312** (right) nanotubes.

research groups [38]. Thus conducting nanotubes could be used as wires. The properties of nanotubes are changed and, perhaps, could be tuned by insertion of atoms inside them.

One should differentiate between single-wall (SWNT) and multi-wall (MWNT) nanotubes. By analogy with nested fullerenes, *i.e.* giant fullerenes containing smaller ones inside one another like the Russian dolls, MWNT consist of several coaxial layers of tubes. Understandably, it is much more difficult to obtain them in the uniform form. Considerable effort has been put into reproducible fabrication of nanotubes of a given size and structure since they typically exhibit dispersions of these parameters which influence their electrical conductivity and other properties. Present methods used to synthesize SWNTs result in tubes with different diameters and helicity. Scanning tunneling microscopy, STM, is a convenient tool for studying these critical parameters.

Nanotubes may form various higher interesting structures such as ropes [39a], bundles [39b], circles, and catenanes [39c] and even crops [39d].

Nanotubes are obtained by the methods analogous to those used for the preparation of fullerenes [5]. Most of these methods are based on the sublimation of carbon in an inert atmosphere, such as an electric arc discharge, laser ablation, or a solar technique. Interestingly, nanotubes with an extra large diameter of 80–120 nm were obtained by explosive synthesis [40a]. The most effective synthetic route to more typical nanotubes (of several nm in diameter) is chemical vapour deposition [40b] allowing one to obtain not only high quality nanotubes with high yields but also to align those deposited on a planar support. Their subsequent isolation and purification by high performance liquid chromatography, HPLC, is a tedious process. Until recently the yields were up to 1% of the produced soot corresponding to microgram up to milligram quantities. The small production yields, and, sometimes, air sensitivity limit studies of endohedral fullerene complexes and their applications. This restriction seems to be overcome by the Smalley group [41a] who make SWNTs from carbon monoxide according to the reaction

$$2\,CO \rightarrow C\,(SWNT) + CO_2$$

typically carried out at ca. 900° C in 10 to 40 atm of CO in the presence of $Fe(CO)_5$ or another metal carbonyl. The nanotubes obtained can consist of up to 99% of SWNT with no amorphous carbon or graphite and only about 1% of the catalytic metal. However, the reactor produces 250 mg per hour, that is, 6 gram per day, and there are hopes that the process can be sped up to allow for production of 100 g of SWNT per day. The production of large arrays of well aligned carbon nanotubes was also a formidable task of technological importance which only recently seems to be solved [41b].

Capped with halves of fullerene pseudospheres the nanotubes can be regarded as elongated fullerenes, although today one rather calls fullerenes spheroidal nanotubes. In the first experiments on nanotube formation various multi-wall nanotubes were obtained. The mechanism of their build-up is still little understood but today one can control the process to produce and manipulate single-wall nanotubes. Moreover, as discussed in Section 6.3.2.2 an electronic device consisting of only one semiconducting, single-wall nanotube about 1 nm in diameter has been recently built [42]. In this device the nanotube is draped over two platinum electrodes situated 400 nm apart on a SiO_2 layer covering a silicon substrate, which plays the role of a gate electrode. It should be stressed that the switch in this case is molecular, however, other elements of the device fabricated by using current microtechnology render the whole device's dimensions comparable to those of conventional silicon transistors. Thus the device, of hybrid molecular and solid state character, is a great achievement, but enormous technical difficulties are still to be overcome on the way to industrial applications of monomolecular devices. Nevertheless, it brings Lehn's idea of chemionics, that is molecular electronics [43], closer to realization.

Similarly to fullerenes, nanotubes may include atoms or small molecules either into their walls or inside them leading to new materials. H_2, small metal crystals and nanoclusters embedded in carbon nanotubes which promise numerous applications have been briefly discussed in Sect. 4.2.4 and 6.3. For instance, lithium- or potassium-doped carbon nanotubes are potent hydrogen sorbents capable of storing 14 weight percent of H_2 at room temperature [44]. Its reversible release at higher temperatures could provide a basis for H_2 storage, which presents the main technical obstacle preventing the use of this highly effcient energy carrier. Nanotubes can form even more complicated complexes such as those containing several fullerenes, which were dubbed 'pea pots' [45a]. A linear hydrocarbon chain in multi-wall nanotube has also been found [46]. Moreover, similarly to $La_2@C_{80}$ **80** (discussed in Section 3.5) the fullerenes can be filled with rapidly moving two lanthane ions [45b] observed by Transmission Electron Microscopy, TEM, technique.

As briefly discussed in Chapter 6, miniaturization of electronic devices is an imperative [47]. This requires the ability to manipulate molecular clusters or even single molecules. Nygard and coworkers' measurements of electrical transport on single-wall carbon nanotubes [48] are one of the first examples of such manipulation. De Heer and coworkers' prototype of the nanobalance, based on multi-wall carbon nanotubes, is another example of a nanodevice. It operates on the basis of observation that the fundamental frequency and higher harmonics of MWNT depended on the weight of the nanocluster attached to the tube [49]. Nanotweezers composed of two nanotubes is another tool for construction of molecular sized devices [50]. Nanotube based planar displays which should replace liquid crystalline displays are close to realization [51], while molecular pressure sensors [52] exemplify numerous nanotube applications in sensing. High-temperature superconductors, redox-active systems, ferromagnetic compounds, drug carriers are representative examples of many other proposed applications of nanotubes marking a definite trend of shifting the research focus from fullerenes to nanotubes.

Different types of nanotubes built of self-assembled organic molecules have been discussed in Section 4.2.4. Inorganic nanotubes, for instance, built of vanadium oxide [53] are also known [54].

References

1. M. S. Dresselhaus, G. Dresslhaus, P. C. Ecklund, *Science of Fullerenes and Carbon Nanotubes*. 1996, Academic Press, San Diego, p. 11.
2. E. Osawa, Kagaku (Kyoto), 1970, 25, 854; D. A. Botchvar and E. G. Galpern, Dokl. Akad. Nauk. SSSR, 1973, 209, 610.
3. H. W. Kroto, J. R. Heath, S. C. O'Brien, R. F. Curl, R. E. Smalley, Nature, 1985, 162.
4. Ref. 1, Chapter 2.
5. (a) Ref. 4, Chapter 5; (b) W. Krätschmer, K. Fostiropoulos, D. Huffmann, Chem. Phys. Lett., 1990, 170, 167.
6. T. K. Daly, P. R. Buseck, P. Williams, C. F. Lewis, Science, 1993, 259, 1599; P. R. Buskeck, S. J. Tsipursky, R. Hettich, Science, 1993, 259, 1599; R. Becker, R. J. Poreda, T. E. Bunch, Proc. Natl. Acad. Sci. USA, 2000, 97, 2979.
7. (a) T. G. Schmalz, W. A. Seitz, D. J. Klein, G. E. Hite, Chem. Phys. Lett., 1986, 130, 203; H. W. Kroto, Nature, 1987, 329, 529; (b) C. Piskoti, J. Yarger, A. Zettl, Nature, 1998, 393, 771. C_{36} probably the smallest fullerene obtained in bulk; (c) H. Prinzbach, A. Weiler, L. Landenberger, F. Wahl, J. Wörth, L. T. Scott, M. Gelmont, D. Olevano, B. V. Issendorff, Nature, 2000, 407, 6800.
8. A. Herrmann, F. Diederich, Helv. Chim. Acta, 1996, 79, 1741.
9. N. R. Manaa, Chem. Phys. Lett., 2000, 331, 262.
10. (a) R. Taylor, D. R. M. Walton, Nature, 1993, 363, 685; (b) F. Zhou, C. Jehoulet, A. J. Bard, J. Am. Chem. Soc., 1992, 114, 11004; J. Chlistunoff, D. Cliffel, A. J. Bard, Chapter 7 in *Handbook of Conductive Molecules and Polymers*, v. 1, H. S. Nalva, Ed., Wiley, New York, 1997; (c) J.-F. Nierengarten, T. Habicher, R. Kessinger, F. Cardullo, F. Diederich, Helv. Chim. Acta, 1997, 80, 2238; J.-F. Nierengarten, C. Schall, J.-F. Nicoud, Angew. Chem. Int. Ed. Engl., 1998, 37, 1934; L. Isaaks, F. Diederich, R. F. Haldimann, Helv. Chim. Acta, 1997, 80, 317; P. R. Ashton, F. Diederich, M. Gomez-Lopez, J.-F. Nierengarten, J. A. Preece, F. M. Raymo, J. F. Stoddart, Angew. Chem. Int. Ed. Engl., 1997, 36, 1448; T. Habicher, J.-F. Nierengarten, V. Grämlich, F. Diederich, Angew. Chem. Int. Ed. Engl., 1998, 37, 1916; F. Diederich, C. Dietrich-Buchecker, J.-F. Nierengarten, J.-P. Sauvage, J. Chem. Soc. Chem. Commun., 1995, 781; (d) F. Arias, L. A. Godinez, S. R. Wilson, A. E. Kaifer, L. Echegoyen, J. Am. Chem. Soc., 1996, 118, 6086; (e) F.-P.

Montforts, O. Kutzki, Angew. Chem. Int. Ed. Engl., 2000, 39, 599; (f) W. Kutner, K. Noworyta, G. R. Deviprasad, F. D'Souza, Mol. Materials, 2000, 13, 295; (g) A. M. Cassell, C. L. Asplund, J. M. Tour, Angew. Chem. Int. Ed. Engl., 1999, 38, 2403.

11. (a) W. Krätschmer, L. D. Lamb, K. Fostiropoulos, D. R. Huffman, Nature, 1990, 347, 354; P. W. Stephens, L. Mihaly, P. L. Lee, R. L. Whetten, S. M. Huang, R. Kaner, F. Diederich, K. Holczer, Nature, 1991, 351, 632; (b) C. S. Yannoni, P. P. Bernier, D. S. Bethune, G. Meier, J. R. Salem, J. Am. Chem. Soc., 1991, 113, 3205.

12. Y. Rubin, Top. Curr. Chem., 199, 1999, 67.

13. (a) H. Shinohara, H. Yamaguchi, N. Hayashi, H. Sato, M. Ohkohchi, Y. Ando, Y. Saito, J. Phys. Chem., 1993, 97, 4259; M. Takata, B. Umeda, E. Nishibori, M. Sakata, Y. Saito, M. Ohno, H. Shinohara, Nature, 1995, 377, 46; (b) H. Shinohara, M. Inakuma, M. Kishida, S. Yamazaki, T. Hashizume, T. Sakurai, J. Phys. Chem., 1995, 99, 13769.

14. W. Andreoni, A. Curioni, Phys. Rev. Lett., 1996, 77, 834; K. Laasonen, W. Andreoni, M. Parrinello, Science, 1992, 258, 1916.

15. T. Akasaka, S. Nagase, K. Kobayashi, M. Waelchli, K. Yamammoto, H. Funasaka, M. Kako, T. Hoshino, T. Erata, Angew. Chem. Int. Ed. Engl., 1997, 36, 1643.

16. (a) S. Suzuki, S. Kawata, H. Shiromaru, K. Yamauchi, K. Kikuchi, T. Kato, Y. Achiba, J. Phys. Chem., 1992, 96, 7159; (b) S. Nagase, K. Kobayashi, T. Akasaka, Bull. Chem. Soc. Japan, 1996, 69, 2131.

17. R. E. Smalley, Materials Science and Engineering, 1993, 1, 319.

18. F. D. Weiss, J. L. Elkind, S. C. O'Brien, R. F. Curl, R. E. Smalley, J. Am. Chem. Soc., 1988, 110, 4464.

19. (a) J. Laskin, T. Peres, C. Lifshitz, M. Saunders, R. J. Cross, A. Khong, Chem. Phys. Lett., 1998, 285, 7; A. Khong, H. A. Jimenez-Vazquez, M. Saunders, R. J. Cross, J. Laskin, T. Peres, C. Lifshitz, B. Strongin, A. B. Smith III, J. Am. Chem. Soc., 1998, 120, 6380; (b) A. Weidinger, M. Waiblinger, B. Pietzak, T. A. Murphy, Appl. Phys. A, 1998, 66, 287; (c) K. Prassides, T. J. S. Dennis, C. Christides, E. Roduner, H. W. Kroto, R. Taylor, D. R. M. Walton, J. Phys. Chem., 1992, 96, 10600.

20. M. Waiblinger, B. Pietzak, A. Weidinger, Phys. Stat. Sol. A, 2000, 177, 81.

21. A. Hirsch, Synthesis, 1995, 895.

22. C. R. Wang, T. Kai, T. Tomiyama, T. Yoshida, Y. Kobayashi, E. Nishibori, M. Takata, M. Sakata, H. Shinohara, Angew. Chem. Int. Ed. Engl., 2001, 40, 397.

23. (a) R. Dagani, Chem. Eng. News, 1999, 77, 54; S. Stevenson, G. Rice, T. Glass, K. Harich, F. Cromer, M. R. Jordan, J. Craft, E. Hadju, R. Bible, M. M. Olmstead, K. Maitra, A. J. Fischer, A. L. Balch, H. C. Dorn, Nature, 1999, 401, 55; (b) C -R. Wang, T. Kai, T. Tomiyama, T. Yoshida, Y. Kobayashi, E. Nishibori, M. Takata, M. Sakata, H. Shinohara, Nature, 2000, 408, 426; (c) R. R. Reisz, H.-D. Sues, Nature, 2000, 408, 427.

24. H. Dodziuk, G. Dolgonos, O. Lukin, Carbon, 2001, 39, 1911.

25. G. Shick, T. Jarrosson, Y. Rubin, Angew. Chem. Int. Ed. Engl., 1999, 38, 2360.

26. (a) Q. Ru, M. Okamoto, Y. Kondo, K. Takayanagi, Chem. Phys. Lett., 1996, 259, 425; (b) H. Dodziuk, G. Dolgonos, O. Lukin, Chem. Phys. Lett., 2000, 329, 351; (c) V. Z. Mordkovich, Chem. Mater., 2000, 12, 2813; (d) M. Y. Kornilov, T. V. Ljubchuk, V. I. Zamkovyi, M. O. Shigorin, V. V. Plakhotnyk, S. D. Isaev, *Fullerenes and Atomic Clusters, 4th Biannial International Workshop in Russia*, IWFAC'99, October 4-8, 1999, St. Petersburg, Russia, poster 140; (e) A. Rothschild, J. Sloan, R. Tenne, J. Am. Chem. Soc., 2000, 122, 5169; M. Remskar, Z. Skraba, P. Stadelmann, F. Levy, Adv. Mater., 2000, 12, 814.

27. (a) T. Andersson, K. Nilsson, M. Sundahl, O. Westman, O. Wennerstrom, J. Chem. Soc. Chem. Commun., 1992, 604; (b) J. L. Atwood, G. A. Koutsantonis, C. L. Raston, Nature, 1994, 368, 229.

28. B.-X. Chen, S. R. Wilson, M. Das, D. J. Coughlin, B. F. Erlanger, Proc. Natl. Acad. Sci. USA, 1998, 95, 10809; B. C. Braden, F. A. Goldbaum, B.-X. Chen, A. N. Kirschner, S. R. Wilson, B. F. Erlanger, Proc. Natl. Acad. Sci. USA, 2000, 97, 12193.

29. F. D'Souza, J.-P. Choi, Y.-Y. Hsieh, K. Shriver, W. Kutner, J. Phys. Chem. B, 1998, 102, 212.

30. (a) J. F. Stoddart, Angew. Chem. Int. Ed. Engl., 1991, 30, 70; (b) R. Taylor, A. G. Avent, T. J. Dennis, J. P. Hare, H. W. Kroto, D. M. R. Walton, J. H. Holloway, E. G. Hope, G. J. Langley, Nature, 1992, 355, 27.

31. (a) J. G. Bednorz, K. A. Müller, Z. Phys. B, 1986, 64, 189; (b) J. H. Schön, C. H. Kloc, B. Batlogg, Nature, 2000, 408, 549.

32. Y. Tabata, Y. Ikada, Pure Appl. Chem., 1999, 71, 2047.

33. L. J. Wilson, D. W. Cagle, T, P. Thrash, S. J. Kennel, S. Mirzadeh, J. M. Alford, G. J. Ehrhardt, Coordin. Chem. Rev., 1999, 192, 199.

34. K. F. Kelly, D.Sarkar, S. J. Oldenburg, G. D. Hale, N. J. Halas, Synthetic Metals, 1997, 86, 2407; C. L. Cheung, J. H. Hafner, C. M. Lieber, Proc. Natl. Acad. Sci. USA, 2000, 97, 3809.

35. V. I. Sokolov, I. V. Stankevich, Uspekhi Khim., 1993, 62, 455 (in Russian).

36. P. M. Ajayan, Chem. Rev., 1999, 99, 1787; M. Terrones, W. K. Hsu, H. W. Kroto, D. R. M. Walton, Top. Curr. Chem., 1999, 199, 189.

37. N. Hamada, S.-I. Sawada, A. Oshiyama, Phys. Rev. Lett., 1992, 68, 1579; R. Saito, M. Fujita, M. S. Dresselhaus, Phys. Rev. Lett., 1992, 60, 2204; J. W. Mintmire, B. I. Dunlop, C. T. White, Phys. Rev. Lett., 1992, 68, 631.

38. J. W. G. Wildöer, L. C. Venema, A. G. Rinzler, R. E. Smalley, C. Dekker, Nature, 1998, 391, 59; T. W. Odom, J.-L. Huang, P. Kim, C. M. Lieber, Nature, 1998, 391, 62.

39. (a) P. Delaney, H. J. Choi, J. Ihm, S. G. Louie, M. L. Cohen, Phys. Rev. B, 1999, 60, 7899; (b) C. Laurent, E. Flahaut, A. Peigney, A. Rousset, New J. Chem., 1998, 1229; (c) M. Jacoby, Chem. Eng. News, 1999, Oct. 4, 31; R. Martel, H. R. Shea, P. Avouris, J. Chem. Phys. B, 1999, 103, 7551; (d) J. Liu, H. I. Dai, J. H. Hafner, D. T. Colbert, R. E. Smalley, S. J. Tans, C. Dekker, Nature, 1997, 385, 780.

40. (a) C. Journet, P. Bernier, Appl. Phys. A, 1998, 67, 1; (b) R. Dagani, Chem. Eng. News, 1999, 77, 31.
41. (a) R. Dagani, Chem. Eng. News, 2000, 78, 36; (b) Z. F. Ren, Z. P. Huang, J. W. Xu, D. Z. Wang, J. H. Wang, *Electronic Properties of Novel Materials - Science and Technology of Molecular Nanostructures*, H. Kuzmany, J. Fink, M. Mehring, S. Roth, Eds., American Institute of Physics, New York, 1999, p. 263.
42. S. J. Tans, A. R. M. Verschueren, C. Dekker, Nature, 1998, 393, 49.
43. J.-M. Lehn, Angew. Chem. Int. Ed. Engl., 1988, 27, 89.
44. P. Chen, X. Wu, J. Lin, K. L. Tan, Science, 1999, 285, 91.
45. (a) J. Sloan, R. E. Dunin-Borkowski, J. L. Hutchison, K. S. Coleman, V. C. Williams, J. B. Claridge, A. P. E. York, C. Xu, S. R. Bailey, G. Brown, S. Friedrichs, M. L. H. Green, Chem. Phys. Lett., 2000, 316, 195; (b) B. W. Smith, D. E. Luzzi, Y. Achiba, Chem. Phys. Lett., 2000, 331, 137.
46. Z. Wang, X. Ke, Z. Zhu, F. Zhang, M. Ruan, J. Yang, Phys. Rev. B, 2000, 61, 2472.
47. C. Joachim, J. K. Gimzewski, A. Aviram, Nature, 2000, 408, 541.
48. J. Nygard, D. H. Cobden, M. Bockrath, P. L. McEuen, P. E. Lindelof, Appl. Phys. A, 1999, 69, 297.
49. P. Poncharal, Z. L. Wang, D. Ugarte, W. A. De Heer, Science, 1999, 283, 1513.
50. Y. Saito, S. Uemura, Carbon, 2000, 38, 169.
51. R. Baum, Chem. Eng. News, 1999, Dec 13, 7.
52. J. R. Wood, H. D. Wagner, Appl. Phys. Lett., 2000, 76, 2883.
53. M. E. Spahr, P. Bitterli, R. Nesper, M. Müller, F. Krumeich, H.-U. Nissen, Angew. Chem. Int. Ed. Engl., 1998, 37, 1339.
54. W. Tremel, Angew. Chem. Int. Ed. Engl., 1999, 38, 2175.

7.6 Dendrimers [1]

The name *dendrimers* was coined from Greek *dendron* meaning tree and *meros* denoting part, since the oligomers like poly(amidoamine) PAMAM dendrimers **313** (Figure 7.6.1) exhibit a repetitive branching responsible for their unique character. *Arborols* or *cascade molecules* are also used to denote these oligomers. One can distinguish three characteristic parts of them: a core; repetitive units the number of which defines the dendrimer generation; and end groups (peripheral units). Numerous molecules can be used as dendrimer cores

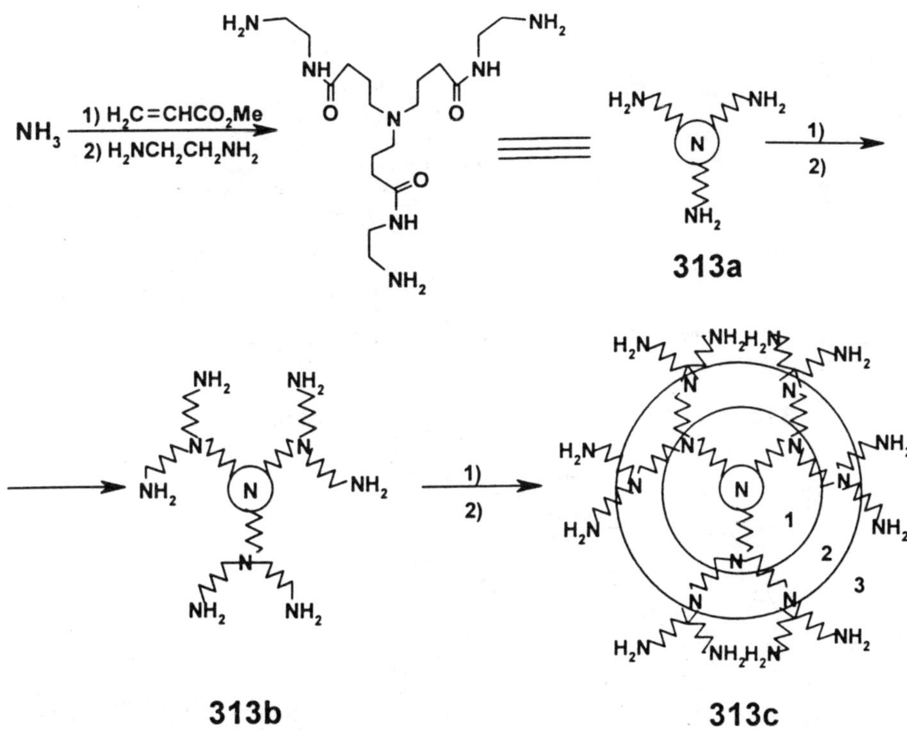

Figure 7.6.1. Reaction scheme of the syntheses of the first **313a**. second **313b** and third **313c** generations of the PAMAM dendrimer.

of different multiplicity *m* like ammonia, pentaerithritol, benzene, adamantane, porphyrin, or even fullerene. Two strategies to obtain dendritic architectures have been proposed:

1. *Divergent method* [2] consisting in the successive attachment of one set of branching units, characterized by their multiplicity *n*, after another leading to the multiplication of the number of peripheral groups. Thus, in case of the PAMAM dendrimer **313** there are three amine groups in the so called first generation G1 multiplication of the number of peripheral groups. Thus in the case of the PAMAM dendrimer **313** there are three amino groups in the so-called first-generation G1 dendrimer **313a**, 6 in the second-generation dendrimer **313b**, 12 in the third-generation dendrimer **313c**, and so forth. (Note that a different system of denoting generations was originally used by Tomalia and coworkers [1a].)

2. Opposite to the divergent method, the *convergent method* [3] consists in the stepwise synthesis of dendrimer branches first. They are then attached to the core yielding the dendrimer. Today the name dendrimer has been extended to

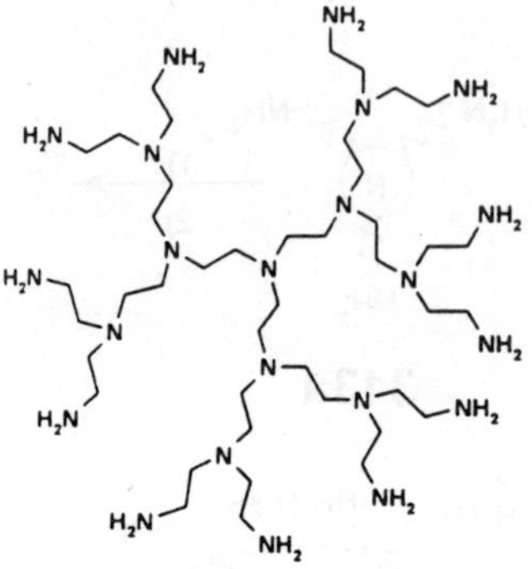

314

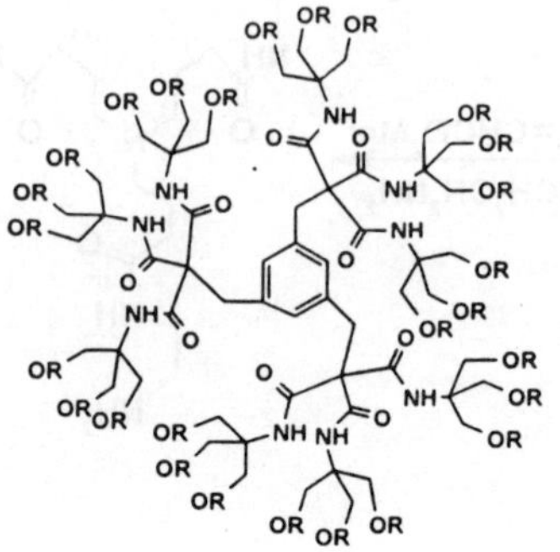

316

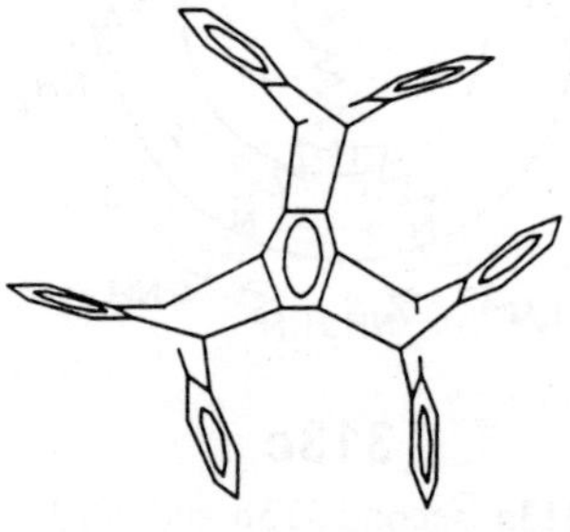

315

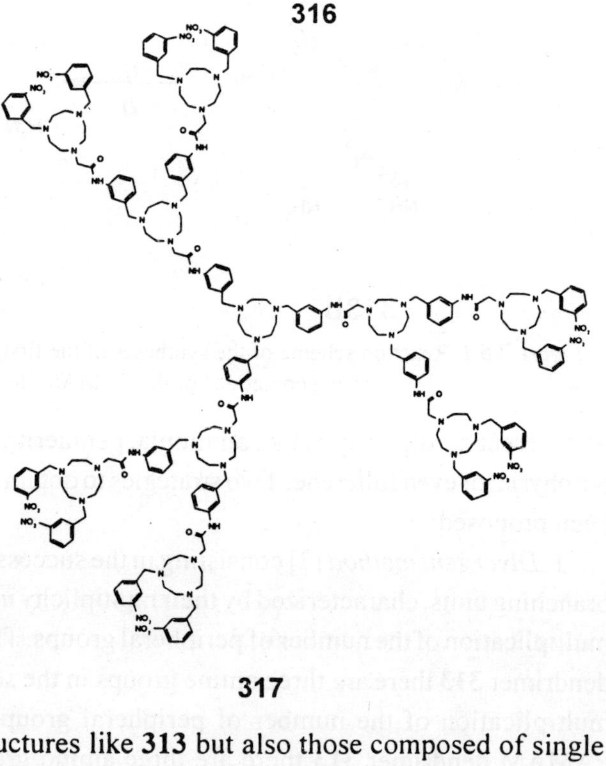

317

cover not only globular structures like **313** but also those composed of single branches called dendrons, which may more easily find applications.

The number of attached groups increases considerably with the increase of generation number. Thus even by a planar core the substituents go out of plane and bigger dendrimers have a globular structure. For instance **313** dendrimers start from a disc-like shape for generations 1 to 3 and reach a nearly symmetrical

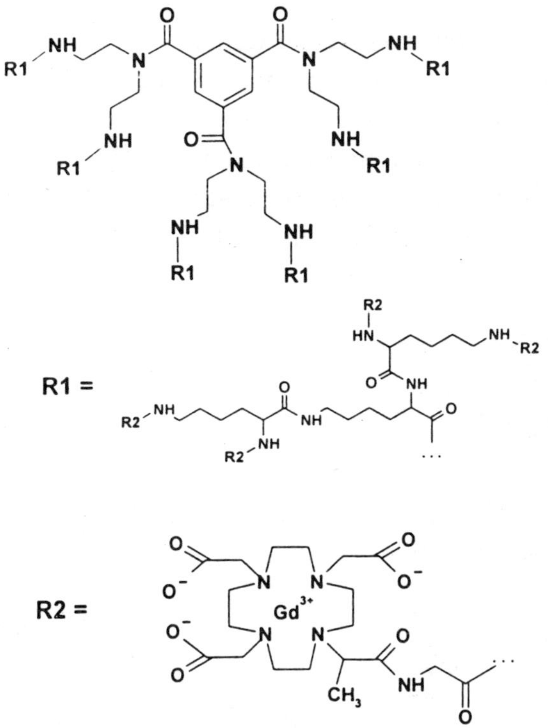

R1 =

R2 =

318

R =

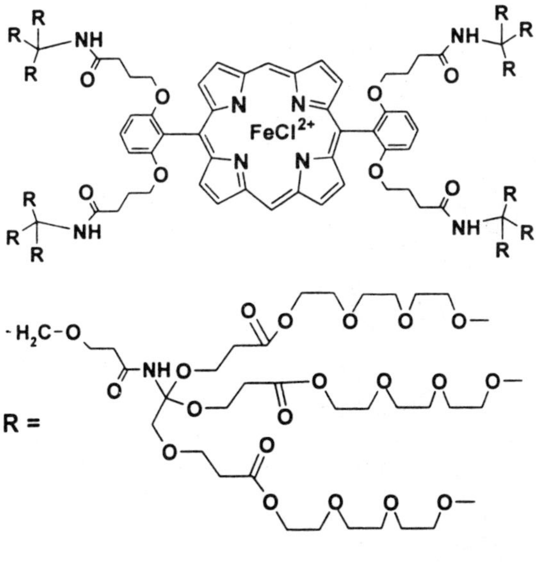

319

spheroids at generations 6 and higher. The surface end groups are so numerous that, after reaching about the 10th generation, the steric overcrowding prevents the formation of further generations without defects. This so-called *starburst effect,* predicted on the basis of mathematical analysis [4], leads to the accumulation of defects by the build-up of further generations. On the other hand, the convergent synthesis allows one to remove the unwanted by-products after each step of the branch growth. However, the steric problem will reappear when the reaction of the branch segments with the core is carried out. In their seminal review [1a] Tomalia and coworkers stated that there is an analogy between fractal structures (Cantor dust, Koch snowflakes, etc.) and dendrimers. However, the steric overcrowding in hypothetical bigger dendrimers of high generation preventing their existence renders this analogy unfounded. It seems also inaccurate to use the term polymers for dendritic structures as is done in the review, since dendrimers are oligomers with well-defined molecular weight.

Due to their unique globular shape, viscosity and the possibility to accumulate numerous functional end groups at the surface, dendrimers occupy an intermediate position between simple organic molecules and (hyperbranched) polymers. Several interesting dendritic molecules are presented in **314-316**: polyethylenimine **314** [5], iptycenes **315** [6], polyamidoalcohol **316** [2b,7].

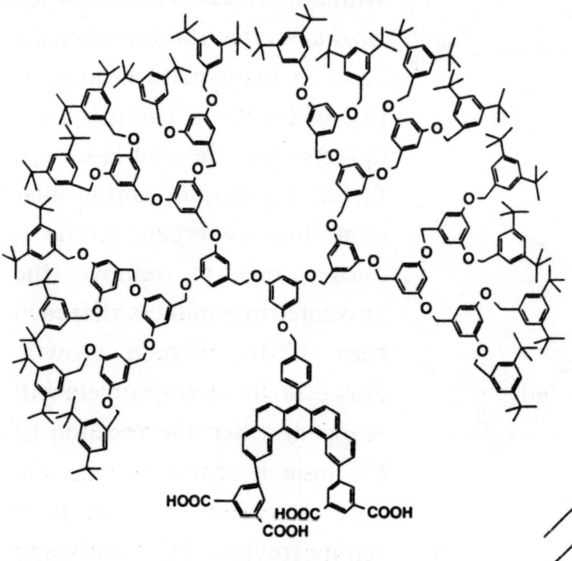

HOOC HOOC COOH COOH

320

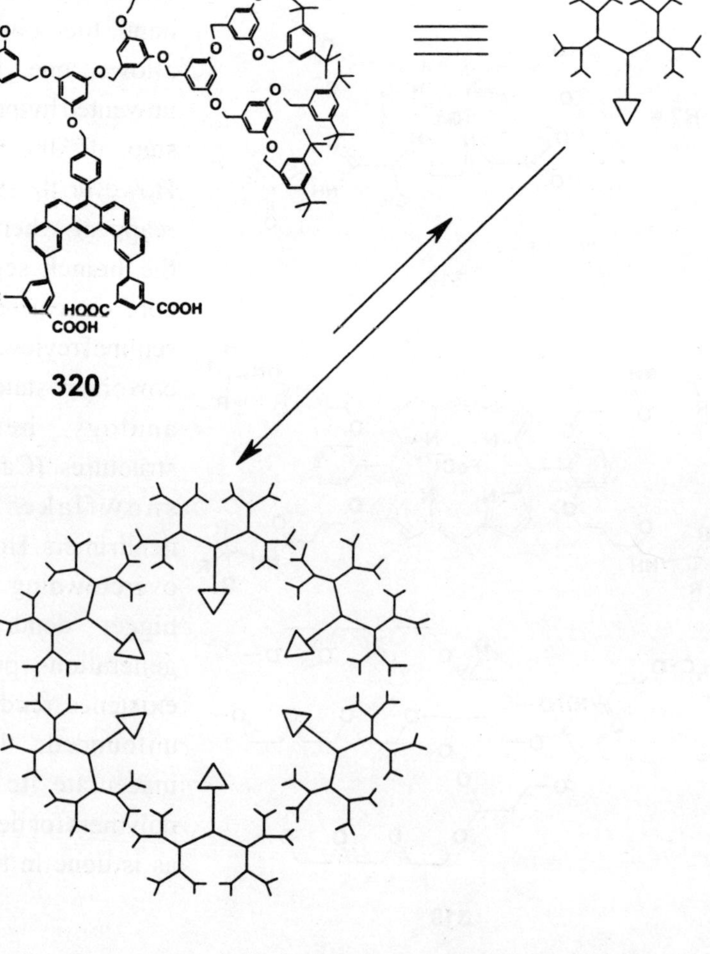

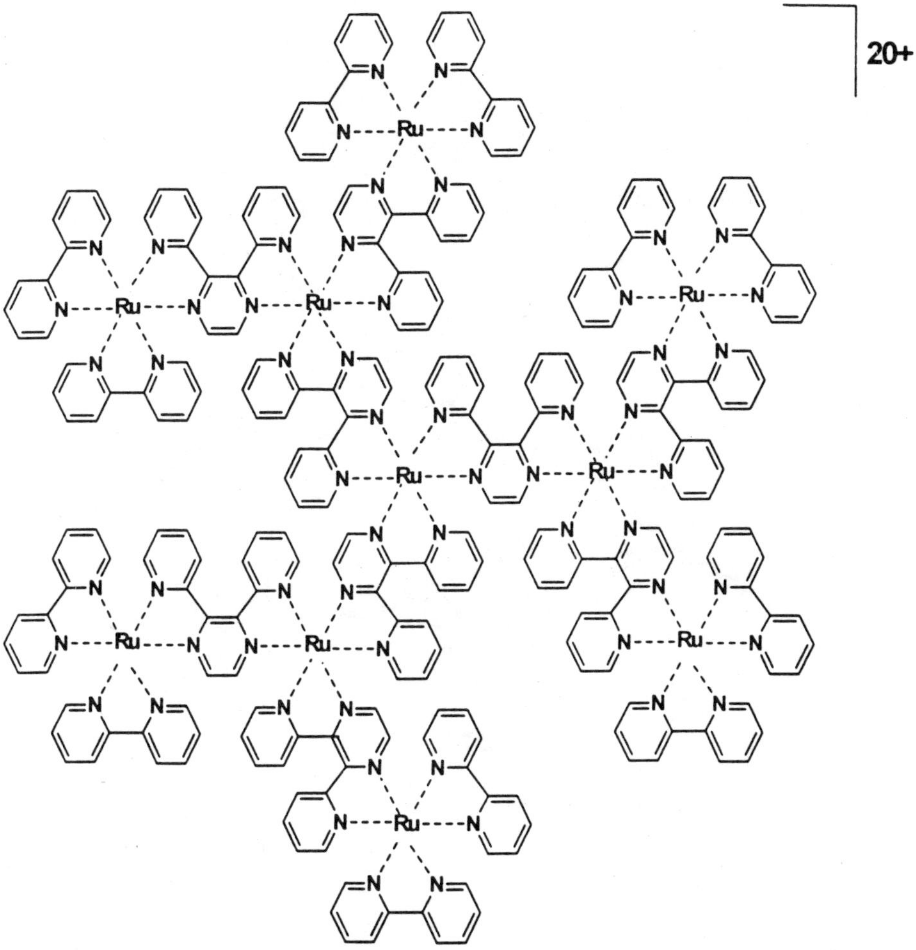

20+

321

Dendrimers based on 1,4,7-triazacyclononane macrocycle (capable of multiple coordination of Cu(II) and Ni(II) metal cations) like **317** have been synthesized by Beer and Gao [8].

Synthetic methods in dendrimer chemistry have developed in recent year to such an extent that one can obtain dendrimers with hydrophilic surface and hydrophobic inside. In such a way dendrimers can exhibit solubilizing effect analogous to that shown by micelles. Similarly to cyclodextrins **278** in the pharmaceutical industry discussed in Sect. 6.3.3, dendrimers can be used as drug

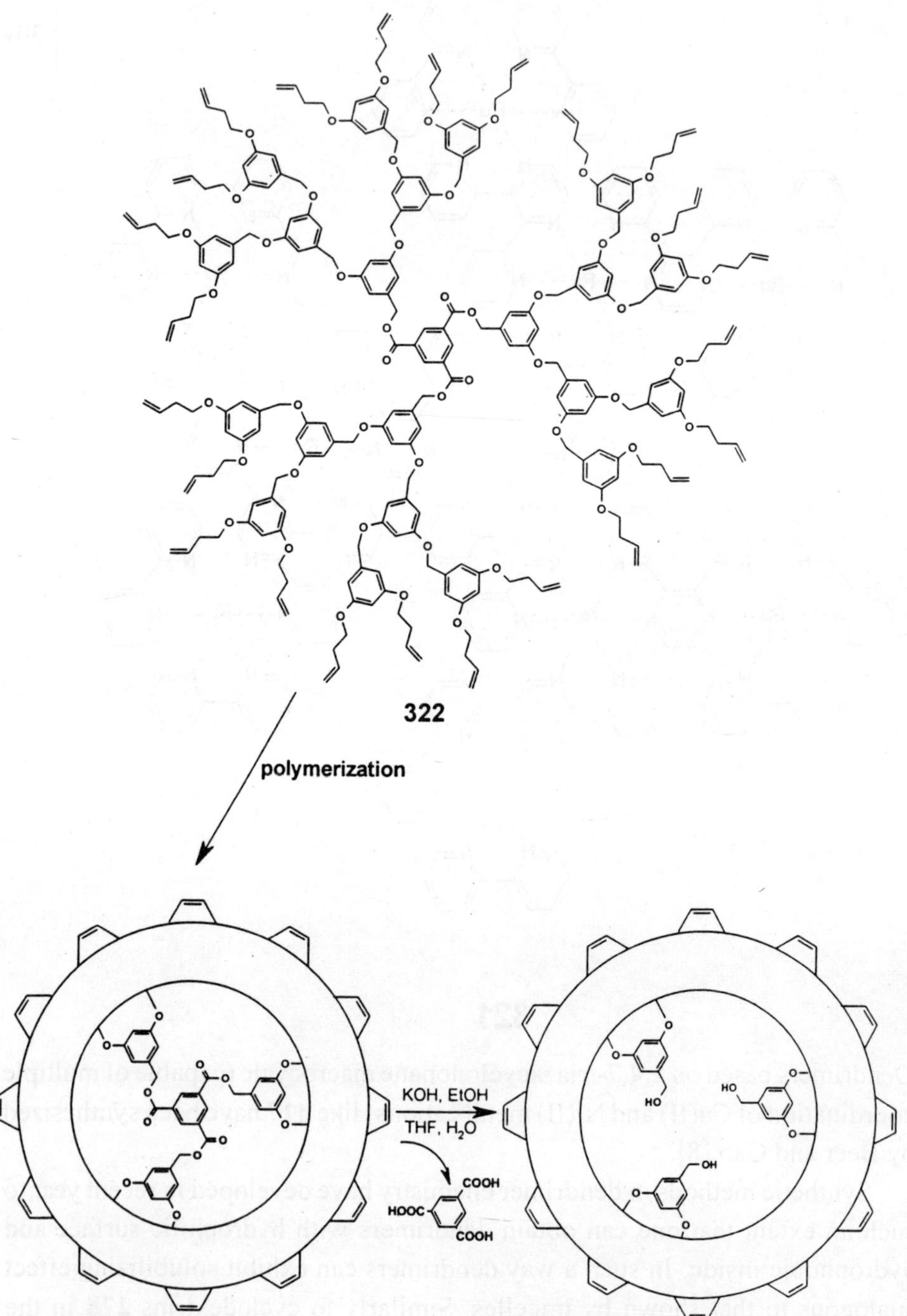

Figure 7.6.2. The scheme of the synthesis of the dendrimer lacking the core.

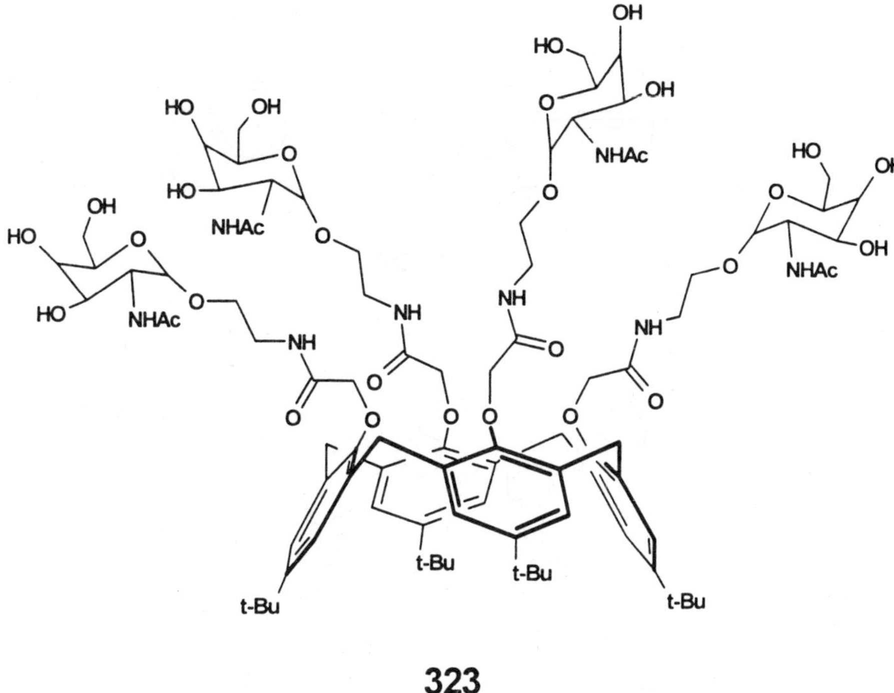

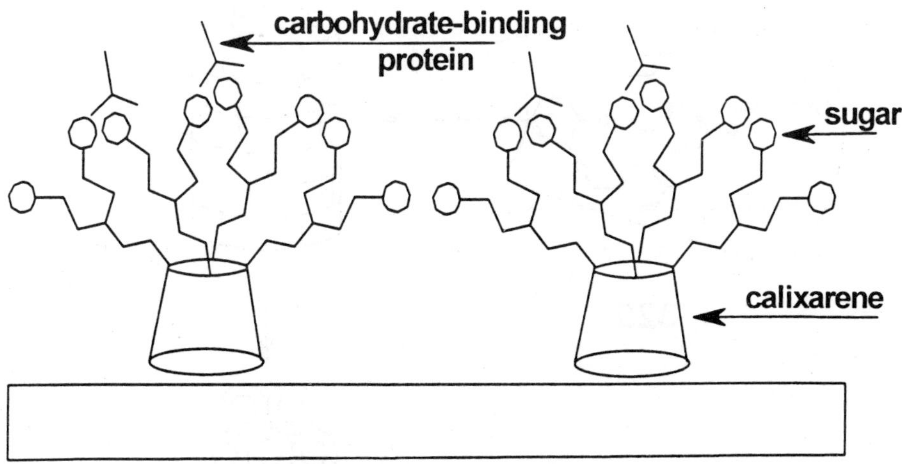

Figure 7.6.3. The scheme of the calixarene-based monolayer mimicking the recognition at a cell surface.

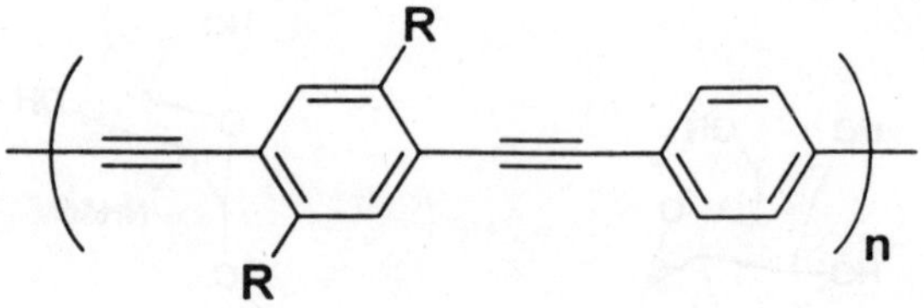

324

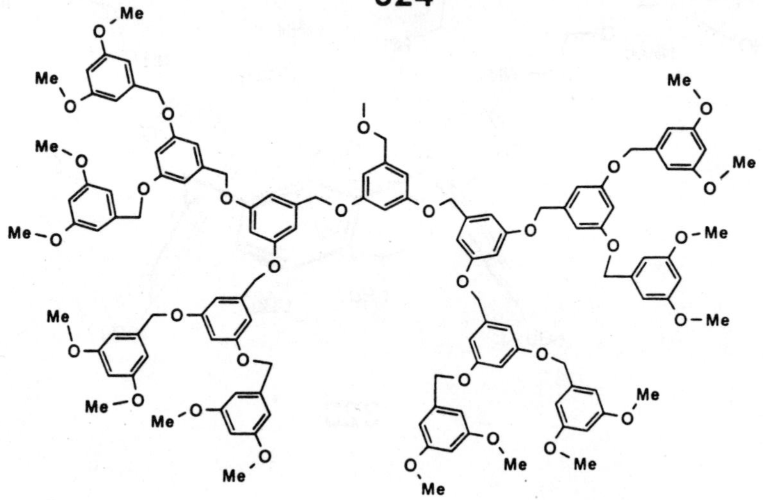

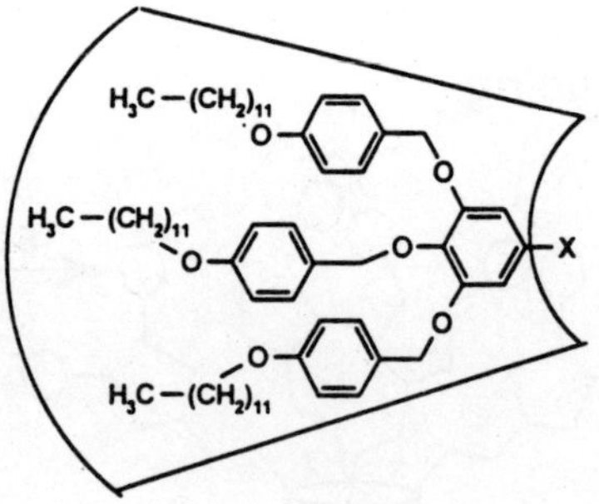

325

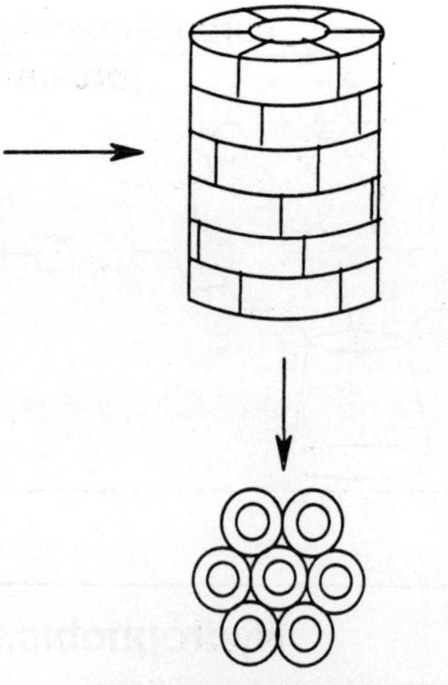

Hexagonal packing

Figure 7.6.4. A dendrimer branch self-assembling to hexagonal packing.

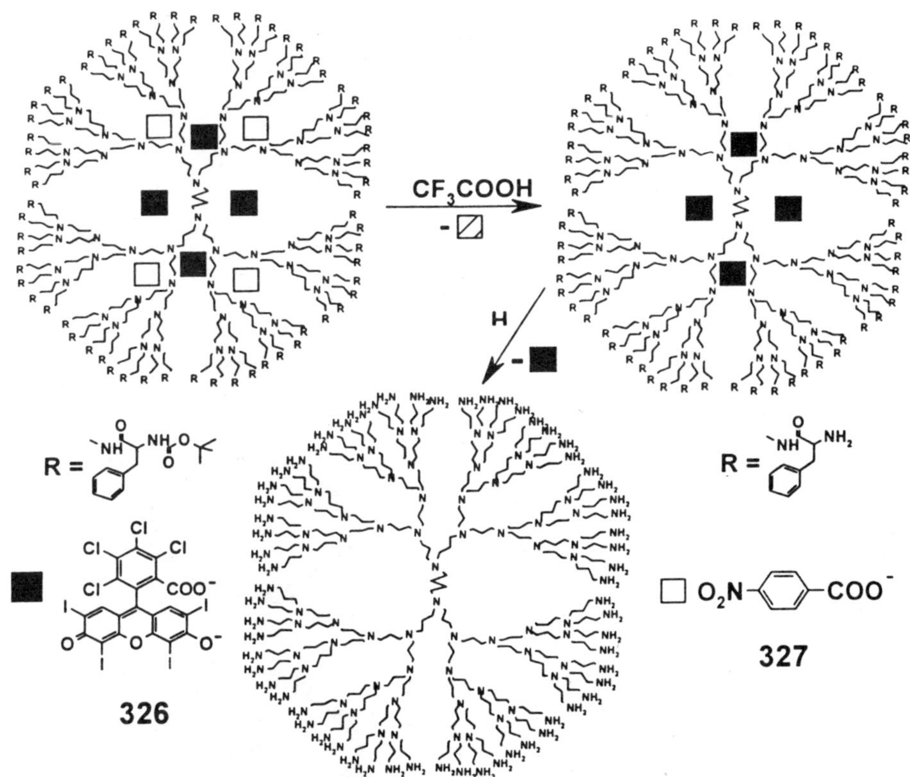

Figure 7.6.5. A dendrimer hosting two different kinds of the guest molecules that selectively releases them.

delivery systems [9]. One of the most vivid fields of dendrimer research is associated with their application in medical diagnostics [10], in particular, as *in vivo* and *in vitro* [11] contrast agent in X-ray, Magnetic Resonance Imaging MRI and Ultrasound analyses. The resonance imaging contrasting agent Gadomer-17 **318** [10] seems to be one of the most prospective application. The latter compound with a molecular weight of 17 kilodaltons, which was very difficult to obtain in monodispersed form, allows for a complete elimination of the heavy metal from the body. The second generation iron porphyrin **319** [12] synthesized in the Diederich group mimics globular heme and cytochrome proteins.

Dendrimers must not necessarily be obtained by covalent synthesis. Self-assembled hexameric dendrimer **320** with a large cavity in which a guest molecule could be included was reported by Zimmerman and coworkers [13]. Interesting luminescent and redox-active self-assembled dendrimers **321** have

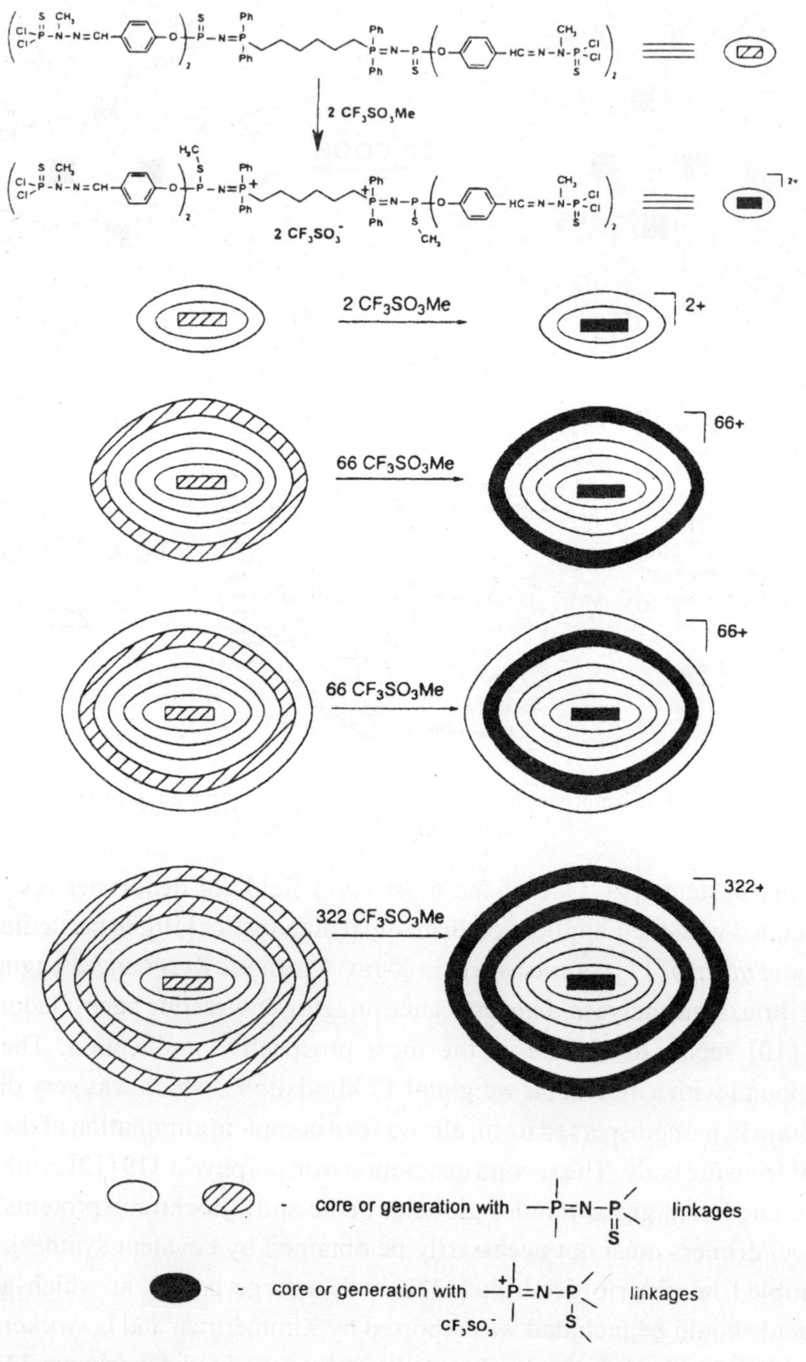

Figure 7.6.6. Phosphorus containing dendrimers forming multiply charged cations.

been obtained in Balzani's group [14]. There can even be a dendrimer without a core! Zimmerman and Wendland succeeded in removing the core of **322** by cross-linking the dendrimer's peripheral groups (Figure 7.6.2) [15].

As with other systems discussed in this book, only a few out of the numerous dendrimers synthesized up-to-date can be mentioned. A thin film formed by amphiphilic dendrimers bearing peripheral fullerene units **299** [16], a glycocalix mimic composed of amphiphilic *p-tert*-butylcalix[4]arene units bearing carbohydrate dendrons **323** attached to hydrophobic polystyrene surface (Figure 7.6.3) [17], and a dendritic rod with high molecular weight of 280,000 **324** that is soluble in organic solvents and capable to emit very efficiently blue light [18] as well as dendritic branch **325** which can self-assemble to hexagonal packing (Figure 7.6.4) [19] should be mentioned here. Another branch involving calixarene moiety self-assembles to the cubic packing [19].This brings us to the exciting topic: dendrimers in the chemistry of inclusion complexes, where they can play either the role of host or that of guest. The first is exemplified by a dendrimer (Figure 7.6.5) which can host both Rose Bengal **326** and *p*-nitrobenzoic acid **327** which can be selectively freed [20a] and by regioselective gold complexation by phosphorous-containing dendrimers [20b]. A complex of dendrimer **328** with multiple β-cyclodextrins **11** hosts reported by the Kaifer group [20c] exemplifies the second possibility.

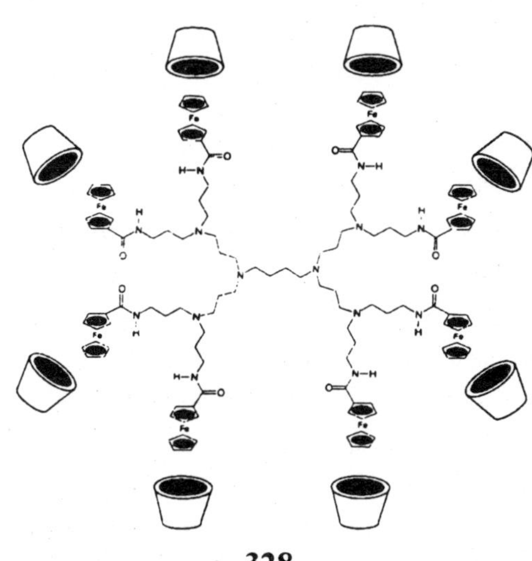

328

Dendrimers could be also applied as surface coatings [1c], light-harvesting antennae [21], and catalysts [22]. Exciting, but still poorly investigated, properties exhibit phosphor-containing dendrimers (Figure 7.6.6) investigated by Majoral [23]. Built of repetitive sequences of two types of branching units, they were found to provide microcompartments for chemical reactions carried out inside the dendrimers in analogy with the chemical reactions carried out in hemicarcerands discussed in Section 7.3. Moreover, different reactions could be carried out in different compartments.

References

1. (a) D. A. Tomalia, A. M. Naylor, W. A. Goddard, III, Angew. Chem. Int. Ed. Engl., 1990, 29, 138; (b) G. R. Newkome, C. N. Moorefield, F. Vögtle, *Dendritic Molecules: Concepts, Syntheses, Perspectives*, VCH, Weinheim, 1996; (c) *Dendrimers I*, Top. Curr. Chem., 2000, 197, F. Vögtle, Ed.; *Dendrimers II. Architecture, Nanostructure and Supramolecular Chemistry*, F. Vögtle, Ed., Top. Curr. Chem., 2000, 210; (d) M. Fischer, F. Vögtle, Angew. Chem. Int. Ed. Engl., 1999, 38, 885.
2. (a) E. Buhleier, W. Wehner, F. Vögtle, Synthesis, 1978, 155; (b) G. R. Newkome, Z.-Q. Yao, G. R. Baker, V. K. Gupta, J. Org. Chem., 1985, 50, 2003.
3. C. J. Hawker, J. M. J. Frechet, J. Chem. Soc. Chem. Commun., 1990, 1010; M. Sayaraman, J. M. J. Frechet, J. Am. Chem. Soc., 1998, 120, 12996.
4. P. G. de Gennes, H. J. Hervet, Phys. Lett., (Paris), 1983, 44, 351.
5. D. M. Hedstrand, P. Meister, D. A. Tomalia, unpublished results cited in Ref. 1a.
6. A. Bashir-Hashemi, H. Hart, D. L. Wart, J. Am. Chem. Soc., 1986, 108, 6675; H. Hart, A. Bashir-Hashemi, J. Juo, M. Meador, Tetrahedron, 1986, 42, 1641.
7. G. R. Newkome, Z.-Q. Yao, G. R. Baker, V. K. Gupta, P. S. Russo, M. J. Sanders, J. Am. Chem. Soc., 1986, 108, 849; G. R. Newkome, G. R. Baker, M. J. Sanders, P. S. Russo, V. K. Gupta, Z.-Q. Yao, J. E. Miller, J. Chem. Soc. Chem. Commun., 1986, 752.
8. P. D. Beer, D. Gao, J. Chem. Soc. Chem. Commun., 2000, 443.
9. M. J. Liu, K. Kono, J. M. J. Frechet, J. Control. Release, 2000, 65, 121.
10. W. Krause, N. Hackmann-Schlichter, F. K. Maier, R. Müller, Top. Curr. Chem., 2000, 210, 261.
11. *In vivo* means studies inside an organism, like visualization of bones by X-ray, while *in vitro*, that is, in a tube, refers to the study of, for instance, blood or urine.
12. P. J. Dandliker, F. Diederich, J.-P. Gisselbrecht, A. Louati, M. Gross, Angew. Chem. Int. Ed. Engl., 1995, 34, 2725.
13. S. C. Zimmerman, F. Zeng, D. E. C. Reichert, S. V. Kolotuchin, Science, 1996, 271, 1095.
14. S. Campagna, G. Denti, S. Serroni, A. Juris, M. Venturi, V. Ricevuto, V. Balzani, Chem. Eur. J., 1995, 1, 211.
15. S. C. Zimmerman, M. S. Wendland, J. Am. Chem. Soc., 1999, 121, 1389.
16. D. Felder, J.-L. Gallani, D. Guillon, B. Heinrich, J.-F. Nicoud, J. F. Nierengarten, Angew. Chem. Int. Ed. Engl., 2000, 39, 201.
17. R. Roy, J. M. Kim, Angew. Chem. Int. Ed. Engl., 1999, 38, 369.
18. T. Sato, D.-L. Jiang, T. Aida, J. Am. Chem. Soc., 1999, 121, 10658.

19. V. Percec, W.-D. Cho, P. E. Mosier, G. Ungar, P. J. P. Yeardley, J. Am. Chem. Soc., 1998, 120, 11061.

20. (a) J. F. G. A. Jansen, E. M. M. de Brabander-van den Berg, E. W. Meijer, J. Am. Chem. Soc., 1995, 117, 4417; (b) C. Larré, A.-M. Caminade, J.-P. Majoral, Chem. Eur. J., 1998, 4, 2031; (c) R. Castro, J. Cuadrado, B. Alonso, C. M. Casado, M. Moran, A. E. Kaifer, J. Am. Chem. Soc., 1997, 119, 5760.

21. S. L. Gilat, A. Adronov, J. M. J. Frechet, Angew. Chem. Int. Ed. Engl., 1999, 38, 1422.

22. M. Q. Zhao, R. M. Crooks, Angew. Chem. Int. Ed. Engl., 1999, 38, 364.

23. J.-P. Marjoral, A. M. Caminade, Chem. Rev., 1999, 99, 845.

7.7 Cyclophanes and Steroids That May Form Inclusion Complexes

7.7.1 Cyclophanes [1]

There is no uniformly accepted definition of cyclophanes. Usually one understands under this name a macrocyclic or macrobicyclic compound having built-in aromatic rings. However, IUPAC rules define (cyclo)phanes much more broadly; they include linear molecules containing rings that are not necessarily aromatic [2]. Here the presentation will be limited to cyclophanes in the first, narrower, meaning. The charge distribution in a benzene ring is characterized by a permanent dipole moment equal to zero and a large quadrupole effect leading to the regions of negative charge above and below the ring plane and to a ring of positive charge roughly coinciding with the hydrogen atom positions. Thus cyclophanes are capable of binding not only neutral molecules but also cations. Somewhat surprisingly, but understandably in view of the above reasoning, in the gas phase the binding of K^+ to benzene is stronger than to the water molecule in spite of the big dipole moment of the latter molecule. The binding by cyclophanes is strongly influenced by the hydrophobic effect [3], thus solvent effects play an important role in such binding. The Koga group [4] synthesized the cyclophane **329** water-soluble at pH < 2 due to protonation of the nitrogen atoms.

COO- COO- COO- COO- COO- COO- COO- COO- -OOC -OOC -OOC -OOC

329

330

331

332

333

334

It should be stressed that calixarenes, hemispherands and spherands, carcerands and hemicarcerands, and some other molecules discussed elsewhere in this book, belong to the cyclophane group of compounds.

Formulae **330-333** represent various interesting cyclophane structures [5].

Cyclotriveratrilene **334** (also a cyclophane) [6] is an interesting building block for the synthesis of complex systems. Its obtaining was reported as early as in 1915 [7a] but its structure was established only 50 years later [7b]. The molecule can exist in one of the crown conformations separated by the barrier of ca. 27 kcal/mol [6].

7.7.2 Steroids [8]

Steroids like cholesterol **335**, cholic acid **336**, or the sex hormone testosterone **337** play an important role in almost all living organisms. Therefore they are the subject of numerous studies, analysing, in particular, their transport through membranes, binding by specific enzymes and protein, and so forth. In supramolecular chemistry biomimetic studies of steroid complexes with cyclodextrins **278** are of special importance as well as the prospective application of such complexes in pharmaceutical industry as orally, sublingually, transdermally, intravenously and intracerebrally administered drugs. As discussed in Section 6.3.3, the complexation by cyclodextrins stabilizes the drug and enhances its solubility.

335

Steroids form complexes not only of 1 : 1 stoichiometry; cholecalciferol **338**, that is, vitamin D_3, forms a 1 : 2 complex with β-cyclodextrin **11** (Fig. 7.7.1). Synthetic receptors for specific steroids [9] or cholesterol imprinted polymers [10] have been reported. A few steroid based synthetic receptors like **339** complexing alkylglycosides [11] and steroid capped porphyrins (with OH groups pointing inside the cavity) for complexation of amines and monosaccharides [12] have been reported. Another macrocyclic system consisting of porphyrin and steroid moieties **340** is especially good receptor for (-)morphine **341** [13]. Steroid based receptors can form not only inclusion complexes but also higher aggregates. For instance, those bearing crown ether moiety **342** are amphiphillic

336

337

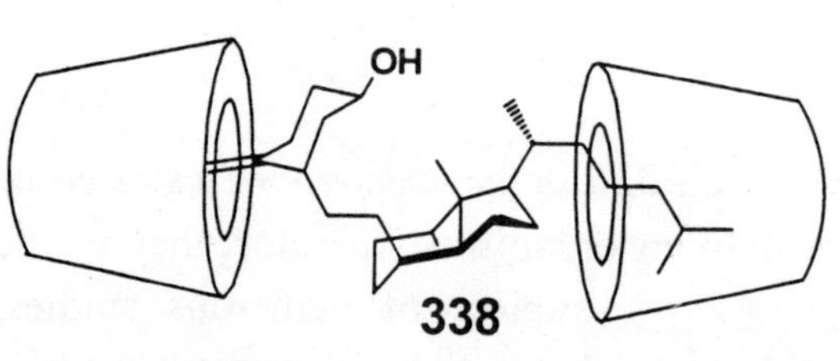

Figure 7.7.1. The complex of cholecalciferol **338** with two β-cyclodextrins.

339

340

341

342

and self-assemble forming vesicles [14]. In the solid state steroids can form clathrates of different structure [15].

References

1. (a) F. Diederich, *Cyclophanes*, The Royal Society of Chemistry, Cambridge, 1991; (b) D. A. Dougherty, *Comprehensive Supramolecular Chemistry*, Elsevier, 1996, v. 2, p. 195.
2. W. H. Powell, Pure Appl. Chem., 1998, 70, 1513.
3. W. Blokzijl, J. B. F. Engberts, Angew. Chem. Int. Ed. Engl., 1993, 32, 1545.
4. K. Odashima, A. Itai, Y. Iitaka, K. Koga, J. Am. Chem. Soc., 1980, 102, 2504.
5. A. R. Bernardo, J. F. Stoddart, A. E. Kaifer, J. Am. Chem. Soc., 1992, 114, 10624; C. O. Dietrich-Bucheker, J.-P. Sauvage, J.-M. Kern, J. Am. Chem. Soc., 1984, 106, 3043; C. O. Dietrich-Bucheker, J.-P. Sauvage, Tetrahedron, 1990, 46, 503; M. W. Hosseini, J.-M. Lehn, M. P. Mertes, Helv. Chim. Acta, 1983, 66, 2454; K. Odashima, K. Koga, *Comprehensive Supramolecular Chemistry*, Elsevier, 1996, v. 2, p. 143.
6. A. Collet, Tetrahedron, 1987, 43, 5725.
7. (a) G. M. Robinson, J. Chem. Soc., 1915, 102, 267; (b) A. S. Lindsey, J. Chem. Soc., 1965, 1685.
8. P. Wallimann, T. Marti, A. Fürer, F. Diederich, Chem. Rev., 1997, 97, 1567.
9. K. Kobayashi, Y. Asakawa, Y. Kikuchi, H. Toi, Y. Aoyama, J. Am. Chem. Soc., 1993, 115, 2648.
10. M. J. Whitcombe, M. E. Rodriguez, P. Villar, E. Vulfson, J. Am. Chem. Soc., 1995, 117, 7105.
11. R. P. Bonar-Law, A. P. Davis, Tetrahedron, 1993, 49, 9855.
12. R. P. Bonar-Law, J. K. M. Sanders, J. Chem. Soc. Chem. Commun., 1991, 574.
13. L. G. Mackay, R. P. Bonar-Law, J. K. M. Sanders, J. Chem. Soc. Perkin Trans. I, 1993, 1397.
14. G. W. Gokel, Chem. Soc. Rev., 1992, 21, 39.
15. E. Giglio, in *Inclusion Compounds*, J. L. Atwood, J. E. D. Davies, D. D. McNicol, Eds, Academic Press, London, 1984, v. 2, p. 207.

7.8 Anion Binding Receptors and Receptors with Multiple Binding Sites [1]

7.8.1 Cationic receptors for anions

The first unambiguous observation of selective noncovalent complexation of halide anions by a molecular cage **343** was published as early as in 1968 [2]. However, until the late 1980s the studies of inclusion complexes were focused mainly on cations and neutral molecules as guests. This situation has drastically changed in recent years due to the appreciation of the importance of anionic guests recognized by neutral receptors in Nature [3] on one hand, and the prospects of using complexes involving such guests as sensors [4], on the other.

Almost all elements can form stable singly charged anions in the gas phase [5]. Many elements are more stable at higher oxidation states in the condensed phases. The stability is enhanced in the presence of water or oxygen due to delocalization of the net charge over a few atoms in oxoanions formed. Owing to electrostatic stabilization by their environment, many multiply charged anions stable in the solid phases are unstable in the gas phase [6]. In polar protic solvents a considerable anion stabilization associated with huge solvation energies due to hydrogen bonding formation takes place. An analogous interactions of anions in water are much stronger and not fully understood [7].

There is a significant difference between cationic and anionic guests. The latter exhibit smaller electrostatic interactions with the environment but, owing to larger polarisabilities, their dispersion interactions are considerably stronger than those of cations. As a result, anions are more easily transferred from water to typical organic solvents of higher polarizability, stabilizing the softer anionic species. On the other hand, the common anions are considerably larger than cations, thus, bigger host structures are needed to accommodate them. However, for the ions of comparable sizes like F^- and K^+, the former are more strongly hydrated and considerably stabilized.

In most cases the receptors for anions are positively charged ions. However, they can also be neutral molecules which bind ions exclusively by hydrogen bonding, ions-dipole interaction or coordinate anions as Lewis acid centers of an organometallic ligand.

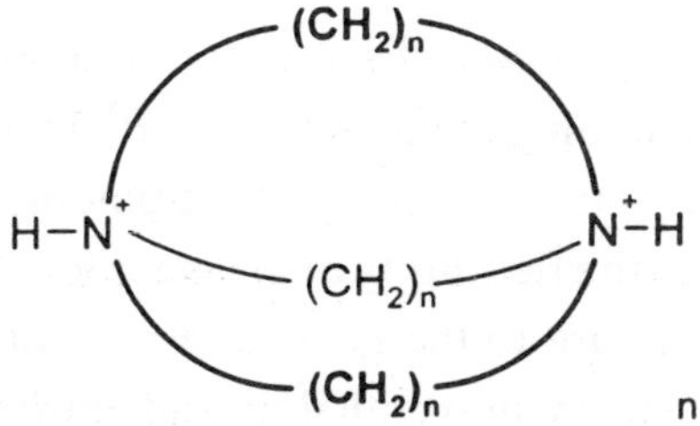

n = 9, 10

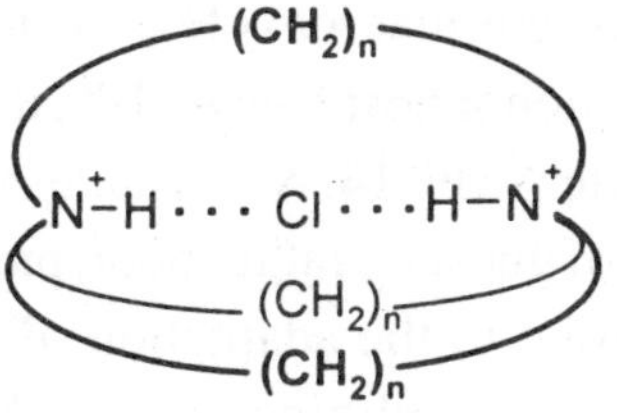

343a **343b**

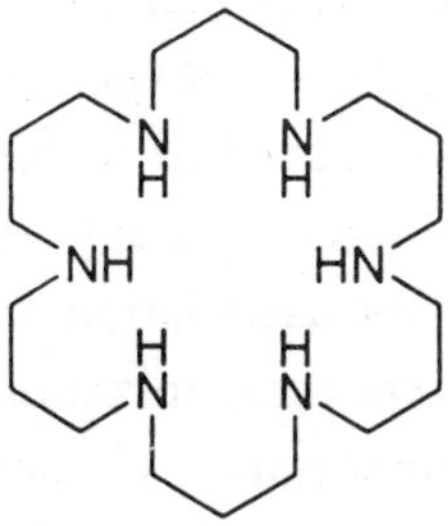

344a 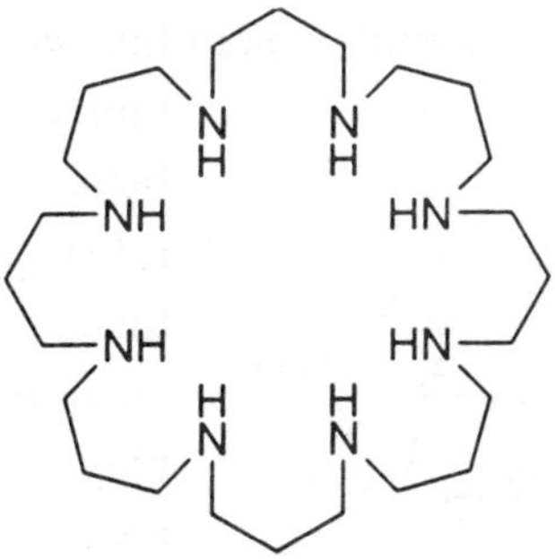 **344b** 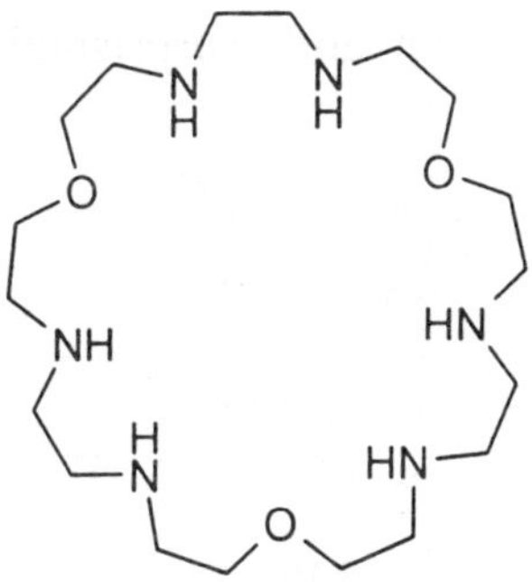 **344c**

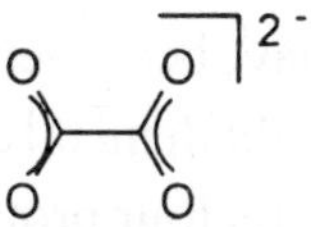

345a 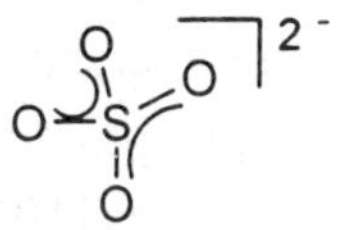 **345b** 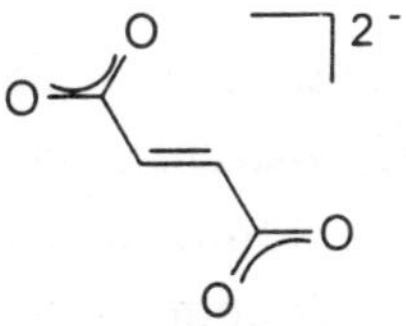 **345c** 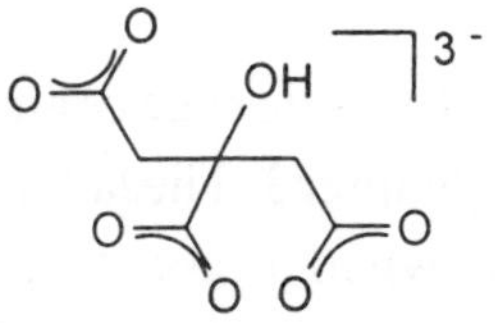 **345d**

Positively charged cyclic hosts like **343a** [2], polyprotonated azacrown ethers **344a-c** [8] and ATP^{4-} **14** complexing azacrown **16** form complexes with ions mainly due to ion pairing and hydrogen bonds. The complexes of **344** with common anions like oxalate **345a**, sulfate **345b** and fumarate **345c** are quite strong ($\log K_{as} = 2-4$) but usually not very selective. However, a few cases of a remarkable selectivity have been observed. For instance, the values of K_{as} for citrate **345d** in complexes with **344a**·6H and **344b**·8H differ by almost a factor of 1000 ($\log K_{as}$ equal to 4.7 and 7.6, respectively). Similarly, for a given polyammonium macrocycle **344** there is a significant discrimination among adenosine monophosphate AMP^{2-}, the corresponding diphosphate ADP^{3-} **15** and triphosphate ATP^{4-} **14**. Nevertheless, in general the selectivity of the binding in this group of guests is rather poor, partly due to the flexibility of the macrocycles that are prone to the adaptation of their structure to the guest at a very little energy cost. The splitting of the Gibbs free energy into enthalpic and entropic components can yield a deeper insight into the complexation mechanism. Such measurements by isothermal titration calorimetry carried out by Gelb *et al.* [9] for the complexes of chloride, bromide and some oxoanions with hexacyclene **346** in water revealed that the host-guest association in these complexes is entropy-driven and, surprisingly, the complex stability increases with increasing temperature. This unexpected behaviour was ascribed to the release of a water molecule in the course of the association. In general, solvent reorganization plays an important, if not always decisive, role in the complexes' stabilization, in addition to the direct mutual host-guest interactions.

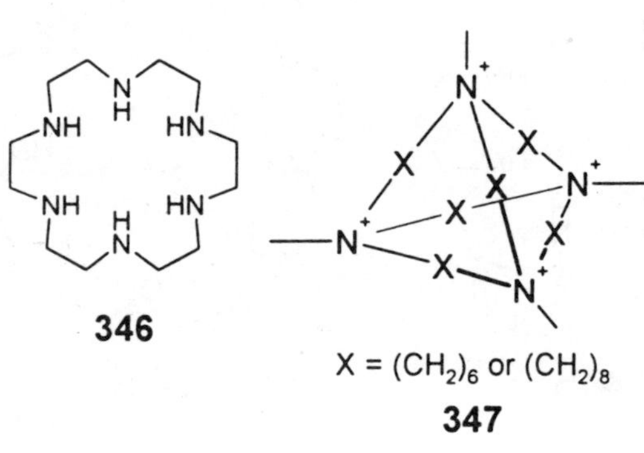

Bicyclic cage complexing N_3^- anion **55** and another, **53**, have been mentioned in Chapter 3. The latter cage was designed as the receptor for alkali metal cations but, when protonated, it was found to complex anions [10]. Its four protonated ammonium sites bind chloride counteranion by an array of four hydrogen bonds leading to a very high association constant and remarkable selectivity of the **53** host (chloride anion complexation is favoured over 1000 times than that of the bromide). X-ray studies confirmed inclusion of the chloride anion into the cage and its localization at the cage center [11]. The inclusion complex formation with 1:1 stoichiometry was also found for a variety of anions complexed with polyaza hosts **347** [12].

Guanidinium group, ubiquitous in anion-binding enzymes, stabilizes oxoanions not only through electrostatic attraction but also by two parallel hydrogen bonds documented in numerous X-ray structures of the corresponding salts like **348** [13]. Simple guanidinium compounds like **349** form complexes of

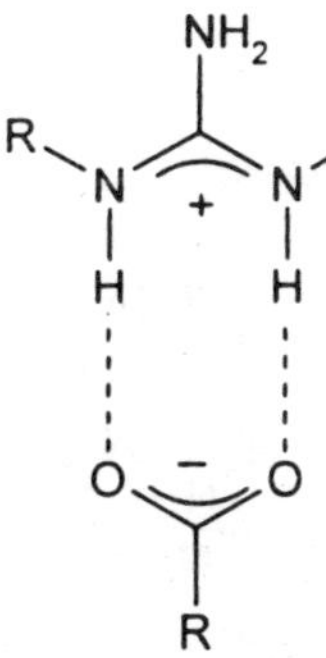

348a 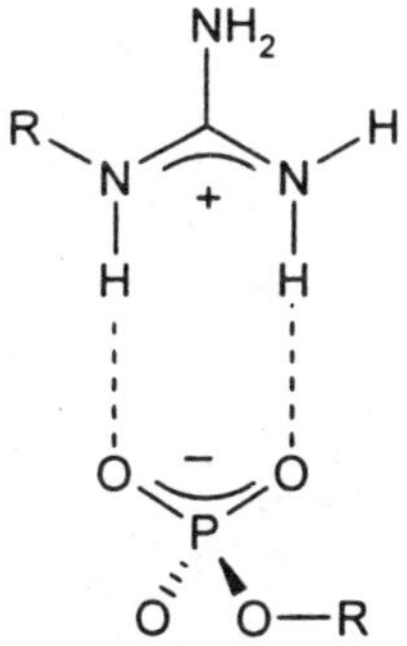 **348b** 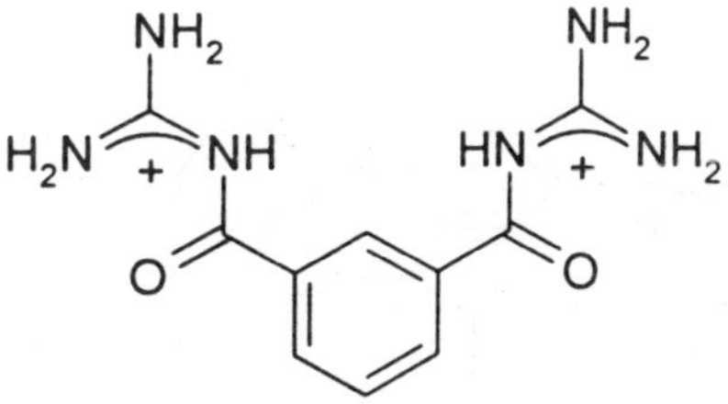 **349**

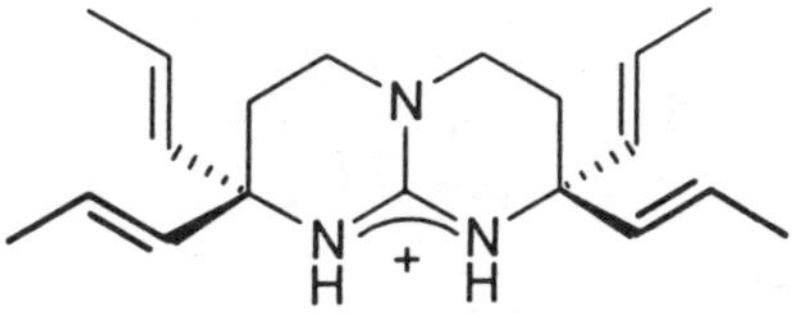

350

considerable stability (K_{as} = 5·10^4 M^{-1} in acetonitrile) with phosphodiesters and exhibited rate enhancement for transesterification reactions by a factor of 300 [14] while more preorganized host **350** complexed *p*-nitrobenzoate even more strongly (K_{as} = 1.4·10^5 M^{-1} in chloroform) [13b]. High enantioselectivity (of ca. 80% ee, enantiomeric excess) was found for **350** host in two-phase liquid extractions of *L*-tryptophane [15].

Nitrogen-bearing cyclophanes like **351** [16] and **352** [17] bind larger organic anions in water due to superposition of the hydrophobic effect and electrostatic attraction. The phenanthridinium hosts like **351** have been found to form the most stable nucleotide complexes known so far. On the other hand, free tetrapyrrolic porphyrins do not bind anions since their cavity is too small to take advantage of the convergent N-H dipoles for the complex stabilization [18]. However, expanded diprotonated porphyrins like sapphyrin **353** were shown to form stable complexes with phosphate [19a] and halide [19b] anions.

Native cyclodextrins **278a** form weak complexes (K_{as} is equal to 10–50 M^{-1}) with many inorganic ions [1a, 1e]. Considerably stronger complexes were found for protonated aminocyclodextrins. Stability constant of the complex of fully protonated heptamethylamino-β-CD **354** with ATP^{4-} **14** reaches the value of 3·10^6 M^{-1} [20]. As discussed in Section 3.5, modified CDs, *e. g.*, selectively substituted with amidozole groups, serve as enzyme mimics not because of the big

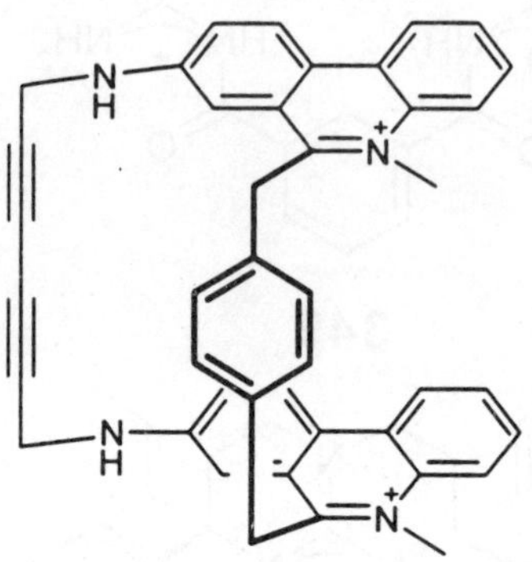

351

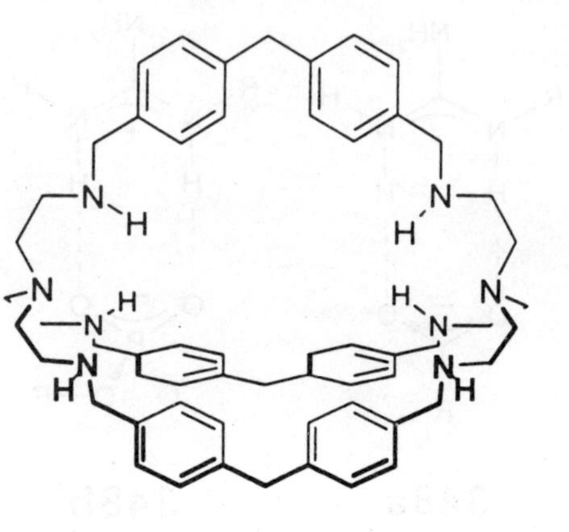

352

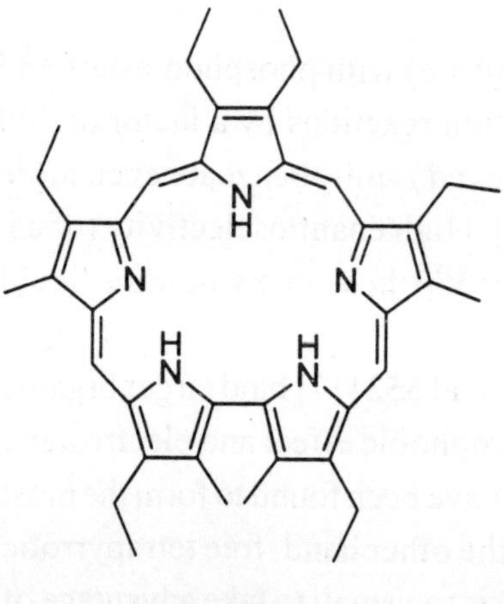

353

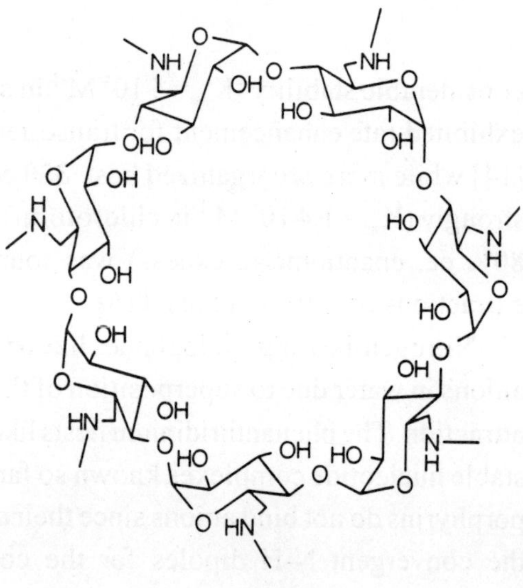

354

acceleration factors of the reactions involving anions but due to the achieved regioselectivity.

7.8.2 Neutral receptors for anions

There are two types of neutral anion-binding receptors: those that bind anions solely by hydrogen bonding or ion-dipole interaction and those that coordinate

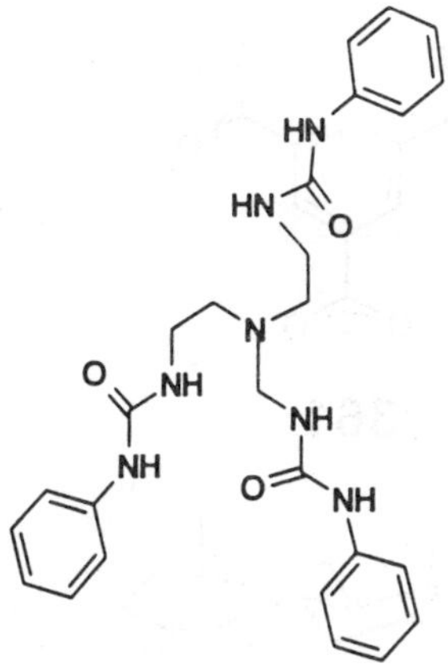

355 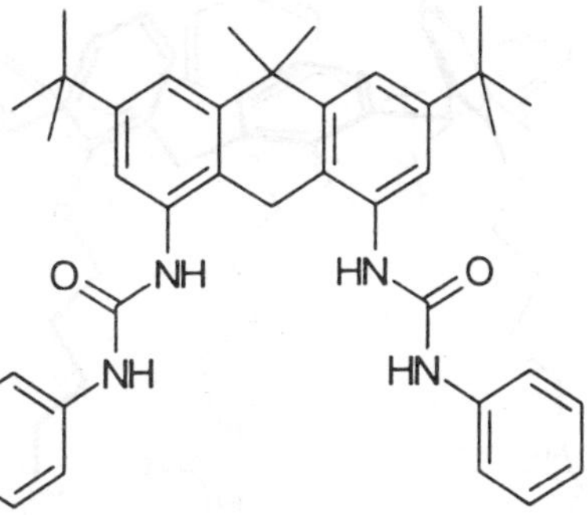 **356**

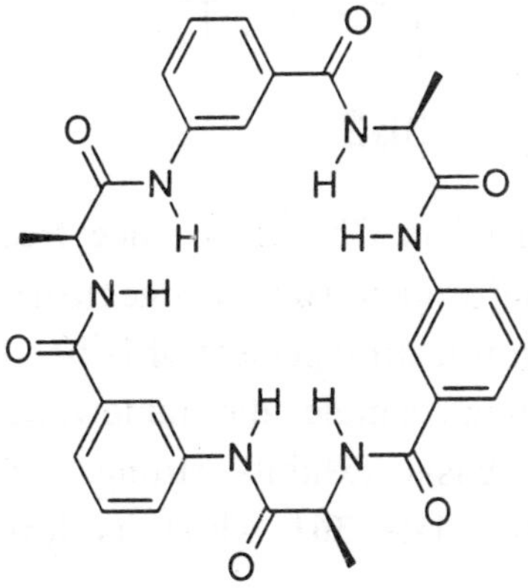

357 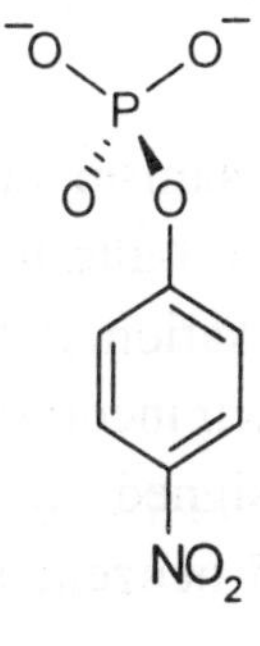 **358** 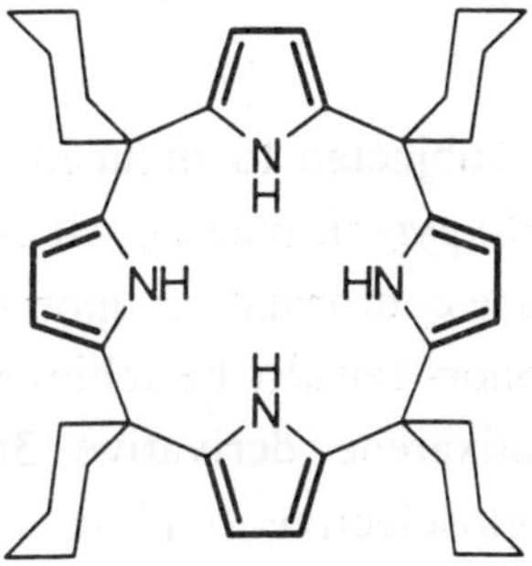 **359**

anions as Lewis acid centers of an uncharged organometallic ligand. Hydrogen bonding is responsible for the strong complexation of $H_2PO_4^-$ by **355** [21] and **356** [22]. Electrostatic ion-dipole interactions play an important role in the complex of cyclic peptide **357** with *p*-nitrophenylphosphate **358** which shows remarkably strong association $K_{as} = 1.2 \cdot 10^6\ M^{-1}$ [23]. Calixpyrrole **359** [24] preferentially binding fluoride anion over chloride or $H_2PO_4^-$ forms the complexes through four cooperative hydrogen bonds. The formation of the latter complex

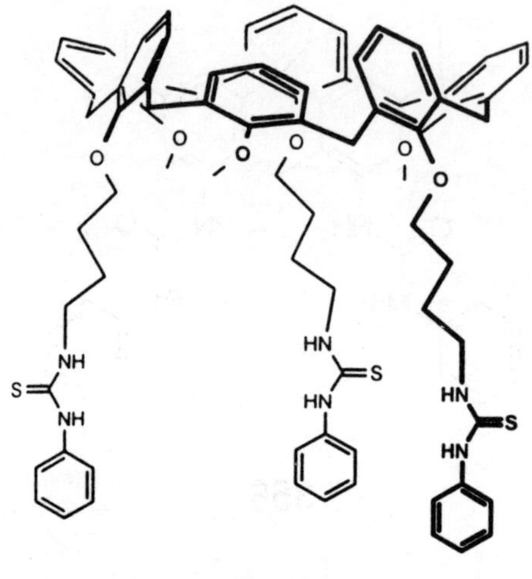

360

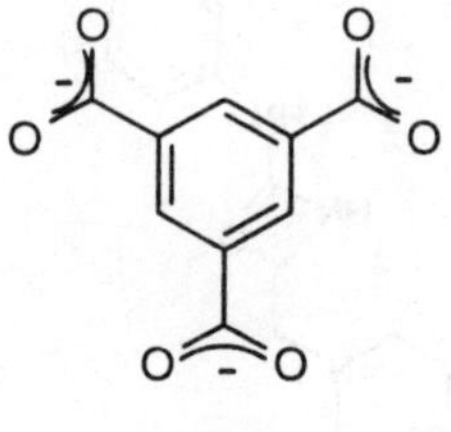

361

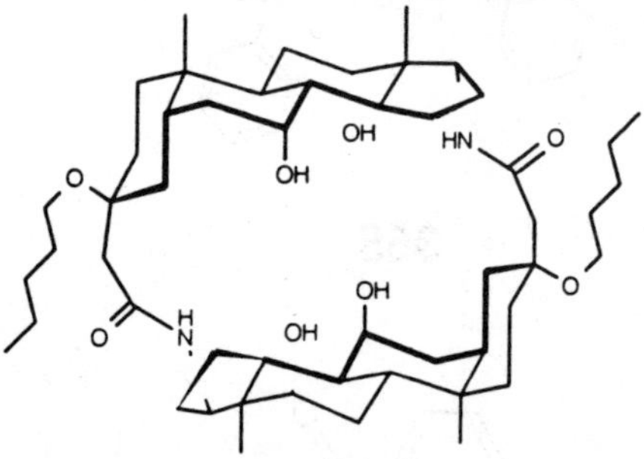

362

is subjected to induced fit mechanism introduced in Chapter 1 since free calixpyrrole macrocycle assumes 1,3-alternate conformation that changes to the cone conformation upon complexation. A favourable arrangement of H-bond donors can also be achieved by their incorporation into a molecular framework. Calixarene derivative **360** designed on this basis exhibits strong and regioselective binding of 1,3,5-benzenetricarboxylate **361** while its less

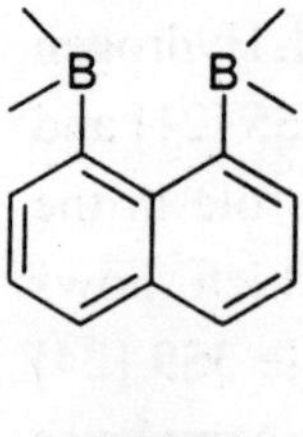

363

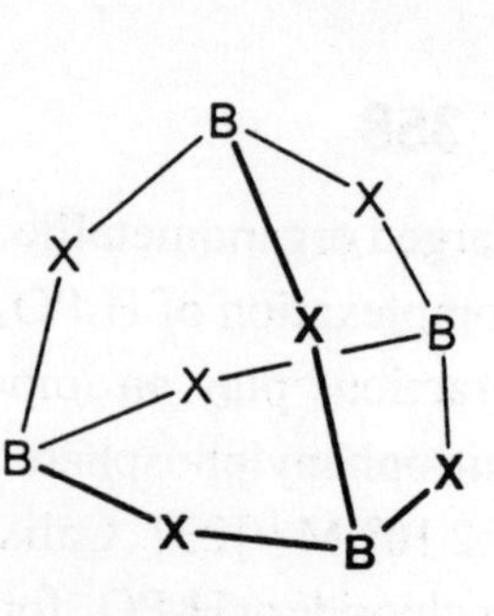

X = $(CH_2)_n$; n = 2-4

364

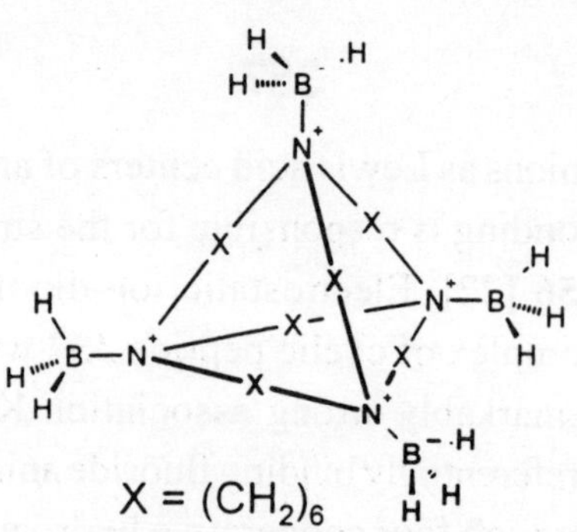

X = $(CH_2)_6$

365

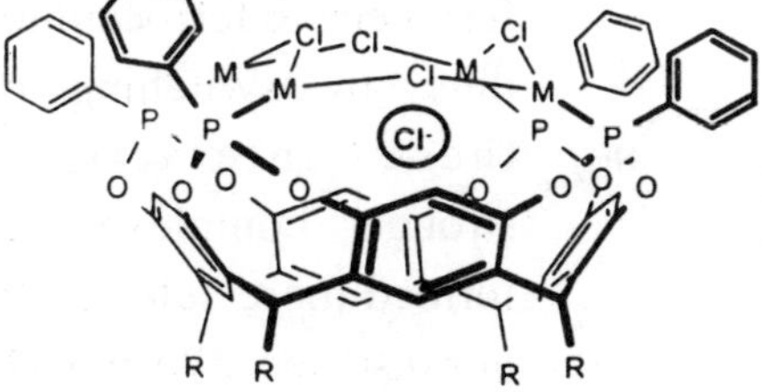

$R = CH_2CH_2C_6H_5$

366

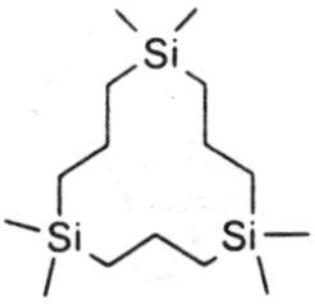

367

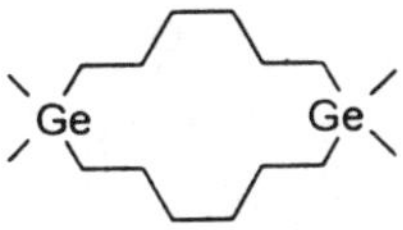

368

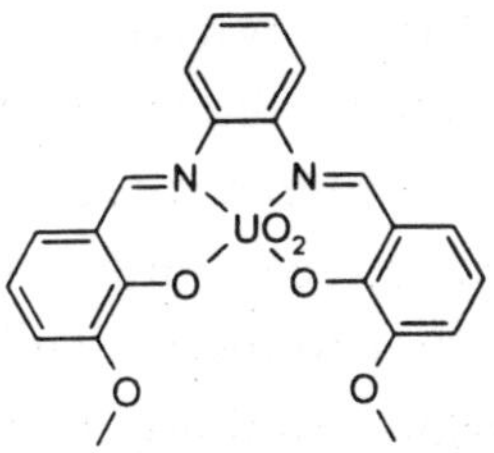

369

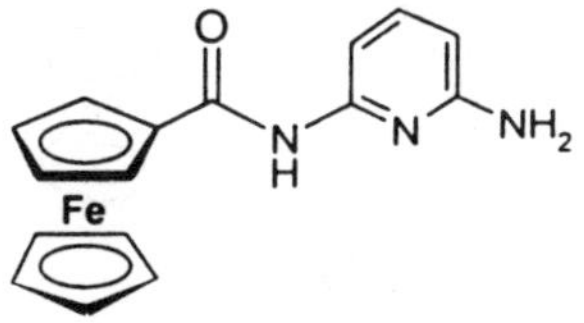

370

symmetric 1,2,3- and 1,2,4-isomers were bound 10 to 100 times weaker [25]. Analogous steroid-based macrocycle **362** [26] is also the receptor for anions making use of hydrogen bonds between the host and guests.

By analogy with cation complexing crown ethers like **47–50**, attachment of a defined number and type of Lewis acids to a rigidified molecular scaffold in such a way that their electron-efficient sites are exposed for the interaction with the lone-electron pairs of anions should lead to the receptors for anions. This design principle was called 'anticrown chemistry' [28]. Pure Coulombic interactions are sensitive to size, density and distance between the charges. Thus electroneutrality of anion receptors synthesized according to this design is of great advantage since much more subtle factors influencing the Lewis acid-Lewis base interaction (stereoelectronic factors, symmetry of molecular orbitals, softness, back-bonding ability, etc.) can be utilized in poly Lewis acid hosts. A proton sponge consisting of naphthalene having two boron Lewis acids in 1,8 positions **363** [27], macrotricyclic borane **364** [29] and borane tetraadduct **365** [30] are typical examples of such receptors for anions. Calixarenes functionalized at the upper rim like **366** also act as poly Lewis acid hosts for anions since they exhibit size-selective binding of halides [31]. More exotic anion receptors acting on the same principle are Si- or Ge-containing macrocycles, **367** [32] and **368** [33], respectively. Hydrogen bonds between $H_2PO_4^-$ guest and substituted salophen unit **369** with uranyl cation coordinating the anion make the binding stronger and more selective [34]. Due to the lack of electrostatic attraction with anions, neutral ferrocene derivatives **370–372** form only weak complexes with anions [35]. However, by oxidation

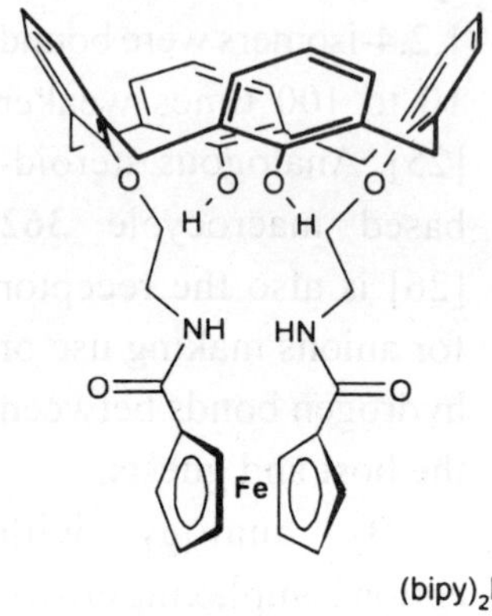

371

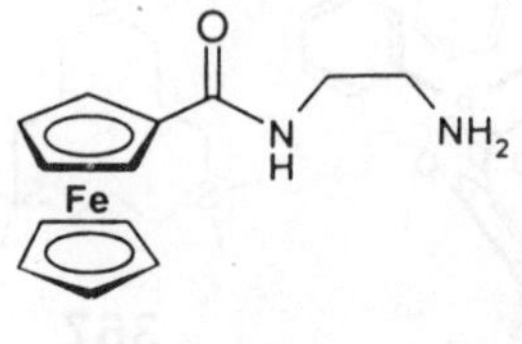

372

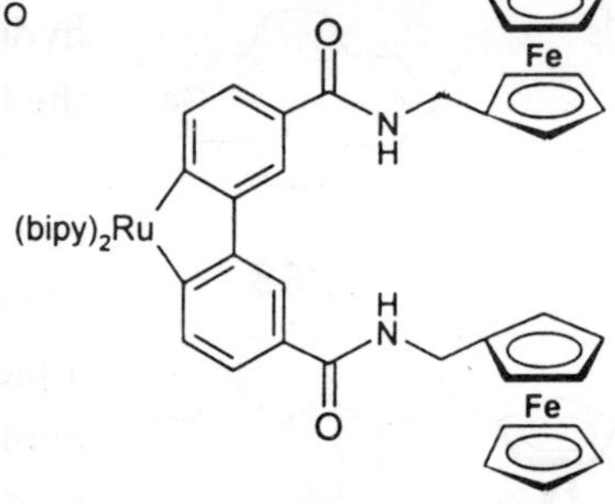

373

ferrocene to ferrocenium, thus, by switching on these interactions, stronger complexes are formed more selectively. For instance, **370** and **371** were able to detect $H_2PO_4^-$ in the presence of 10-fold excess of HSO_4^- and Cl^- anions while **372** selectively bound HSO_4^- in the presence of $H_2PO_4^-$. Such effects can be used in amperometric anion sensors. Similarly to **370** and **371**, mixed-ruthenium(II) bipyridyl-ferrocene receptor **373** sevectively binds $H_2PO_4^-$ in the presence of HSO_4^- and Cl^- [36]. However, anion complexation through second-sphere coordination is responsible for the effect observed in this case. The luminescence of the ruthenium center of **373** is quenched by ferrocene units. However, the quenching is not only removed by complexation with dihydrogen phosphate anion but it is manifested by a 20-fold increase of emission. Thus the latter receptor can be used as luminescent sensor for $H_2PO_4^-$.

7.8.3 Receptors with multiple binding sites

Receptors that simultaneously bind cations and anions (or ion- and neutral molecule) can be constructed by binding the respective fragments in one molecule. For instance, heteroditopic bis(calix[4]arene) rhenium(I) bipyridyl receptor molecules **374** are capable of simultaneous cooperative alkali metal cation and iodide anion recognition [37]. Deetz and coworkers synthesized a macrobicyclic receptor **375** which simultaneously binds an ion pair and selectively complexes dimethylsulfoxide [38]. Anion binding porphyrin **376** and cation binding calix[4]arene **377** self-assemble to yield a bifunctional receptor. The aggregation is initiated by sodium complexation by **377** then enhancing in turn the binding of SCN^- to the porphyrin unit in **376**.

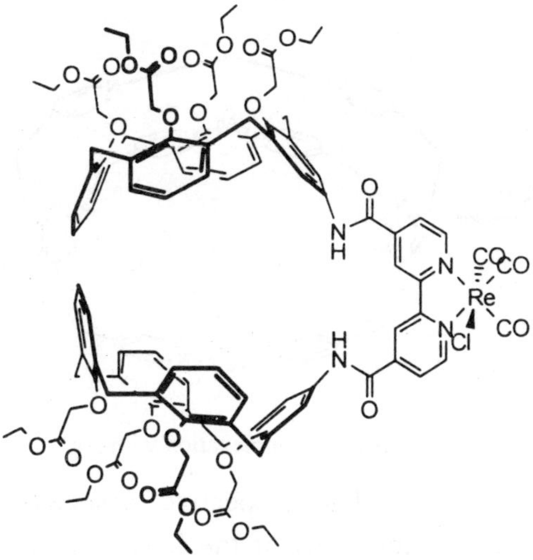

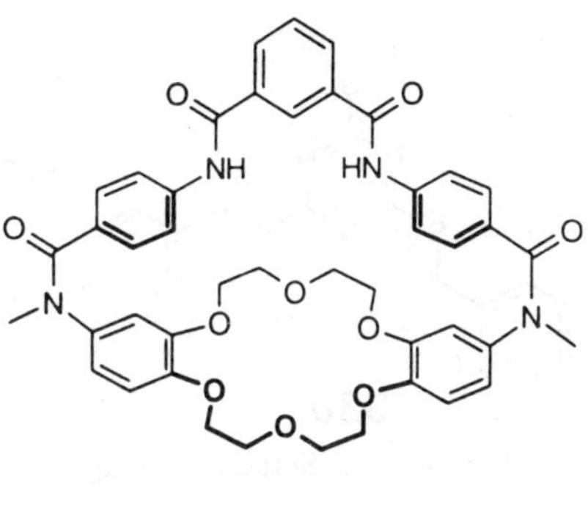

374

375

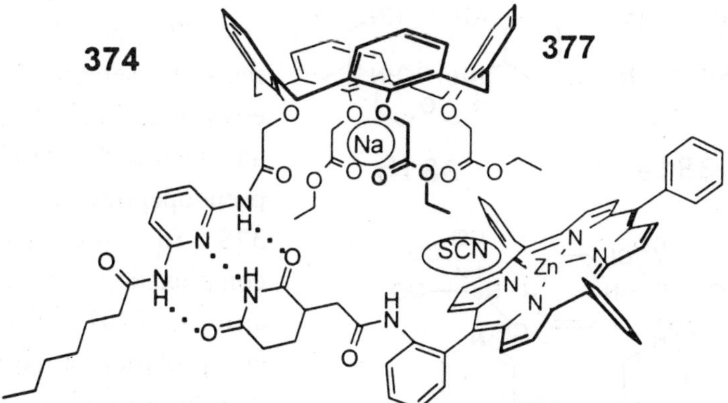

377

376

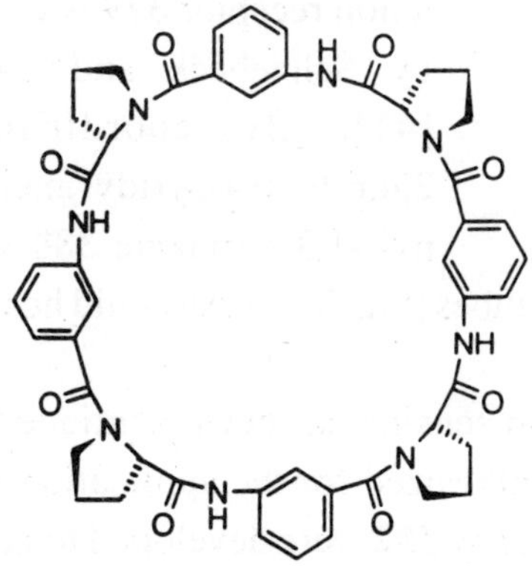

378

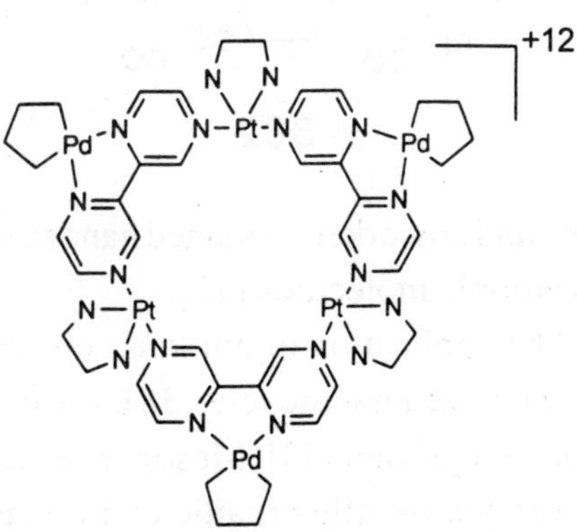

379

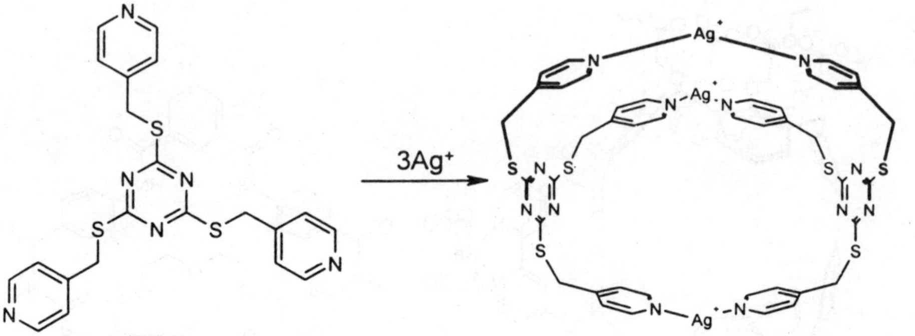

Figure 7.8.1. Self-assembling of macrobicyclic system by complexation with Ag^+.

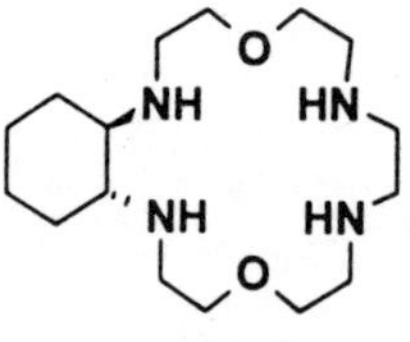

381a 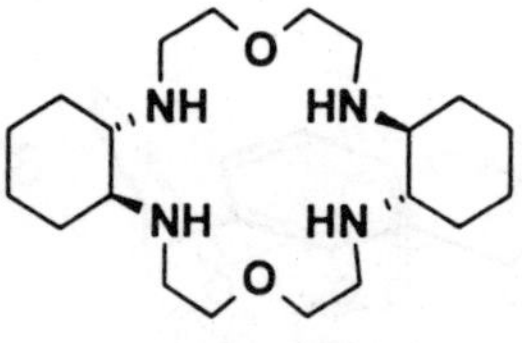 **381b**

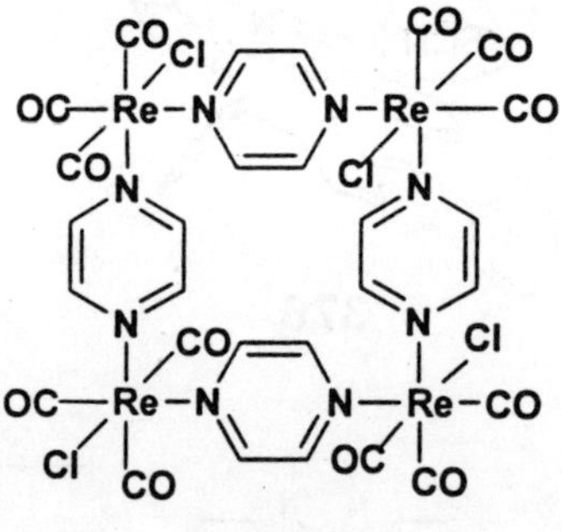

382

The respective association constants of the latter anion to free **376** is only 10 M^{-1} while for the self-assembled complex it is equal to $2.5 \cdot 10^4$ M^{-1} [39]. Kubik and Goddard obtained a cyclic pseudopeptide ditopic receptor **378** for the simultaneous complexation of cations and anions [40]. Simultaneous encapsulation of NO_3^- and PF_6^- ions by a highly charged (+12) anion receptor **379** was achieved by Schnebeck and coworkers [41]. By coordinating of 2,4,6-tris[(4-pyridyl)methylsulfanyl]-1,3,5-triazine **380** with Ag, Hong and coworkers obtained nanosized tubes (Fig.7.8.1) that could host solvent and anionic molecules [42].

The application of anion receptors in sensing has been mentioned earlier. Dioxatetraazamacrocycles **381** were synthesized for the application in chiral anion recognition [43]. Mesoporous films by **382** were developed to be used as sensors for volatile organic compounds [44]. Receptors for anions were also studied to mimic their transport through membranes [45].

References

1. (a) F. P. Schmidtchen. M. Berger. Chem. Rev.. 1997. 97. 1609; (b) M. M. G. Antonisse. D. N. Reinhoudt. Chem. Commun.. 1998, 443; (c) P. D. Beer. Acc. Chem. Res., 1998. 31. 71; (d) P. D. Beer. P. A. Gale, Angew. Chem. Int. Ed. Engl.. 2001. 40, 487; (e) The *Supramolecular Chemistry of Anions*, A. Bianchi. K. Bowman-James. E. Garcia-Espana, J. Wiley-VCH. Chichester, 1997, p. 461.
2. H. E. Simmons, C. H. Park, J. Am. Chem. Soc., 1968, 90, 2428; C. H. Park, H. E. Simmons, J. Am. Chem. Soc., 1968, 90, 2429; C. H. Park, H. E. Simmons, J. Am. Chem. Soc., 1968, 90, 2431.
3. V. I. Fetisov, A. A. Maslov, V. K. Brel', G. A. Sereda, N. S. Zefirov, Dokl. Akad. Nauk, 2000, 371; B. Hinzen, P. Seiler, F. Diederich, Helv. Chim. Acta, 1996, 79, 942.
4. L. Fabbrizzi. M. Licchelli. G. Rabaioli. A. Taglietti. Coord. Chem. Rev.. 2000. 205. 85.
5. A. I. Boldyrev. M. Gutowski, J. Simmons, Acc. Chem. Res., 1996. 29. 497.
6. H. G. Weikert, L. S. Cederbaum, J. Chem. Phys., 1993, 99, 8877; A. T. Blades, P. Kebarle, J. Am. Chem. Soc., 1994, 116, 10761.
7. H. Ohtaki, T. Radnai, Chem. Rev., 1993, 93, 1157.
8. B. Dietrich, M. Hosseini, J.-M. Lehn, R. B. Sessions, J. Am. Chem. Soc., 1981, 103. 1282.
9. R. I. Gelb, B. T. Lee, L. J. Zompa, J. Am. Chem. Soc., 1985, 107, 909.
10. J. P. Kitzinger, J.-M. Lehn, E. Kauffmann, J. L. Dye, A. I. Popov, J. Am. Chem. Soc.. 1983, 105, 7549; E. Graf, J.-M. Lehn, J. Am. Chem. Soc., 1975, 97, 5022.
11. B. Metz, J. M. Rosalky, R. Weiss, J. Chem. Soc. Chem. Commun., 1976. 533.
12. K. Ichikawa, A. Yamamoto, M. A. Hossain, Chem. Lett., 1993, 2175; M. A. Hossain, K. Ichikawa, Tetrahedron Lett., 1994, 35, 8393; F. P. Schmidtchen, Chem. Ber., 1983, 114, 597.
13. (a) U. Koert. Nachr. Chem., Tech. Lab., 1995. 43. 1302; (b) G. Mueller. J. Riede. F. P. Schmidtchen. Angew. Chem., 1988, 100, 1574.
14. V. Jubian, R. P. Dixon, A. D. Hamilton, J. Am. Chem. Soc., 1992. 114. 1120.
15. Ref. 13b.
16. P. Cudic, M. Zinic, V. Tomisic, V. Simeon, J.-P. Vignieron, J.-M. Lehn, J. Chem. Soc. Chem. Commun., 1995, 1073; Slama-Schwok, M.-P. Teulade-Fichou. J.-P. Vignieron. E. Taillandier. J.-M. Lehn. J. Am. Chem. Soc., 1995, 117. 6822.
17. J.-M. Lehn. R. Meric. J.-P. Vignieron. I. Bkouche-Waksman. C. Pascard. J. Chem. Soc. Chem. Commun., 1991. 62.
18. J. L. Sessler. A. K. Burrell. Top. Curr. Chem., 1991, 161. 177.

19. (a)H. Furuta, M. J. Cyr, J. L. Sessler, J. Am. Chem. Soc., 1991, 113, 6677; (b) M. Shionoya, H. Furuta, V. Lynch, A. Harriman, J. L. Sessler, J. Am. Chem. Soc., 1992, 114, 5714.
20. A. V. Eliseev, H.-J. Schneider, J. Am. Chem. Soc., 1994, 116, 6081; A. V. Eliseev, H.-J. Schneider, Angew. Chem. Int. Ed. Engl., 1993, 32, 1331.
21. C. Raposo, M. Almaraz, M. Martin, V. Weinrich, L. Mussons, V. Alcazar, C. Caballero, J. R. Mossan, Chem. Lett., 1995, 759.
22. P. Bühlmann, S. Nishizawa, K. P. Xiao, Y. Umezawa, Tetrahedron, 1997, 53, 1647.
23. H. Ishida, M. Suga, K. Donowaki, K, Ohkubo, J. Org. Chem., 1995, 60, 5374.
24. P. A. Gale, J. L. Sessler, V. Kral, V. Lynch, J. Am. Chem. Soc., 1996, 118, 5140.
25. J. Sheerder, J. F. J. Engbersen, A. Casnati, R. Ungaro, D. N. Reinhoudt, J. Org. Chem., 1995, 60, 6448.
26. A. P. Davis, J. F. Gilmer, J. J. Perry, Angew. Chem. Int. Ed. Engl., 1996, 35, 1312.
27. H. E. Katz, J. Org. Chem., 1985, 50, 5027.
28. X. Yang, Z. Zheng, C. B. Knobler, M. F. Hawthorne, J. Am. Chem. Soc., 1993, 115, 193.
29. S. Jacobson, R. Pizer, J. Am. Chem. Soc., 1993, 115, 11216.
30. K. Worm, F. P. Schmidtchen, A. Schier, A. Schaefer, Angew. Chem. Int. Ed. Engl., 1994, 33, 360.
31. W. Xu, J. J. Vittal, R. J. Puddephatt, J. Am. Chem. Soc., 1995, 117, 8362.
32. M. E. Jung, H. Xia, Tetrahedron Lett., 1988, 29, 297.
33. S. Aoyagi, K. Tanaka, Y. Takeuchi, J. Chem. Soc. Perkin Trans. 2, 1994, 1549.
34. D. M. Rudkevitch, Z. Brzozka, M. Palys, H. C. Visser, W. Verboom, D. N. Reinhoudt, Angew. Chem. Int. Ed. Engl., 1994, 33, 467.
35. P. D. Beer, Z. Chen, A. J. Goulden, A. R. Graydon, S. E. Stokes, J. Chem. Soc. Chem. Commun, 1993, 1834; P. D. Beer, A. R. Graydon, A. O. M. Johnson, D. K. Smith, Inorg. Chem., 1997, 36, 2112; Z. Chen, A. R. Graydon, P. D. Beer, J. Chem. Soc. Faraday Trans., 1996, 92, 97.
36. P. D. Beer, A. R. Graydon, L. R. Sutton, Polyhedron, 1996, 15, 2457.
37. P. D. Beer, J. B. Cooper, J. Chem. Soc. Chem. Commun., 1998, 129.
38. M. J. Deetz, M. Shang, B. D. Smith, J. Am. Chem. Soc., 2000, 122, 6201.
39. D. M. Rudkevich, A. N. Shivanyuk, Z. Brzozka, W. Verboom, D. N. Reinhoudt, Angew. Chem. Int. Ed. Engl., 1995, 34, 2124.
40. S. Kubik, R. Goddard, J. Org. Chem., 1999, 64, 9475.
41. R. D. Schnebeck, E. Freisinger, B. Lippert, Angew. Chem. Int. Ed. Engl., 1999, 38, 168.
42. M. Hong, Y. Zhao, W. Su, R. Cao, M. Fujita, Z. Zhou, A. S. C. Han, Angew. Chem. Int. Ed. Engl., 2000, 39, 2468.

43. B. I. Alfonso, F. Rebolledo, V. Gotor, Chem. Eur. J., 2000, 6, 3331.

44. M. H. Keefe, R. V. Slone, J. T. Hupp, K. F. Czaplewski, R. Q. Snurr, C. L. Stern, Langmuir, 2000, 16, 3964.

45. K. P. Xiao, P. Buhlmann, Y. Umezawa, Anal. Chem., 1999, 71, 1183; L. A. J. Chrisstoffels, F. de Jong, D. N. Reinhoudt, Chem. Eur. J., 2000, 6, 1376.

7.9 Porphyrin-based Hosts

The vivid interest in porphyrins **46** is triggered by both the biological significance of its analogues (photosynthetic centers, hemes responsible for the

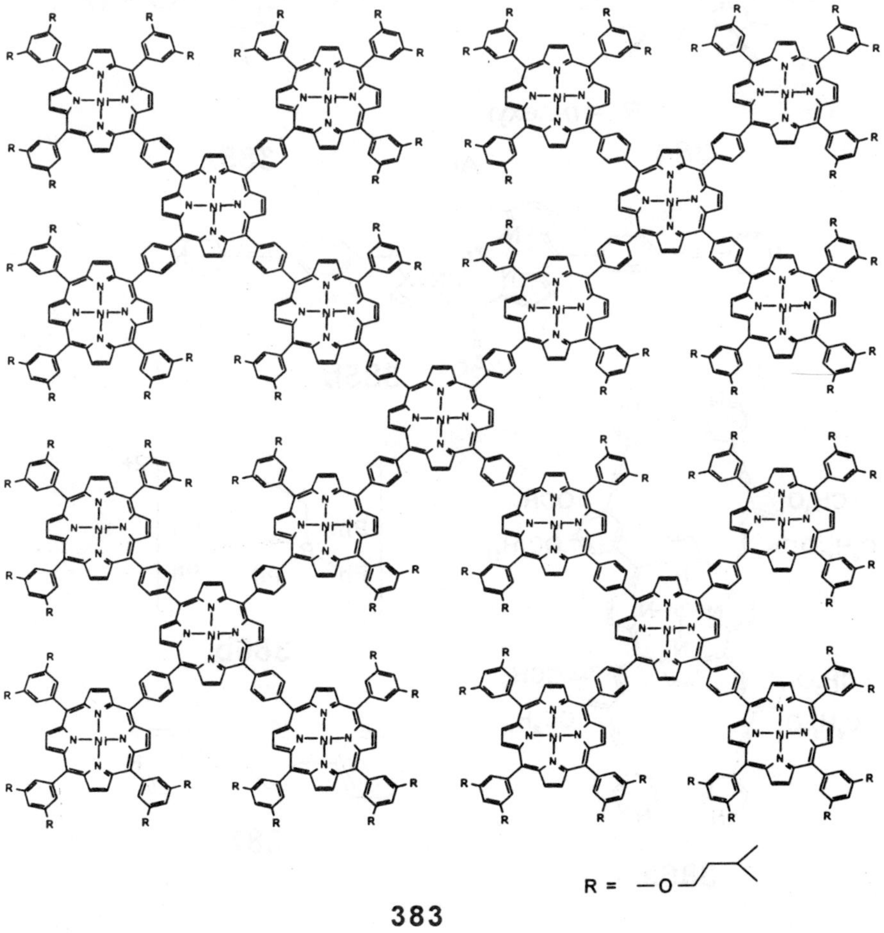

R = —O

383

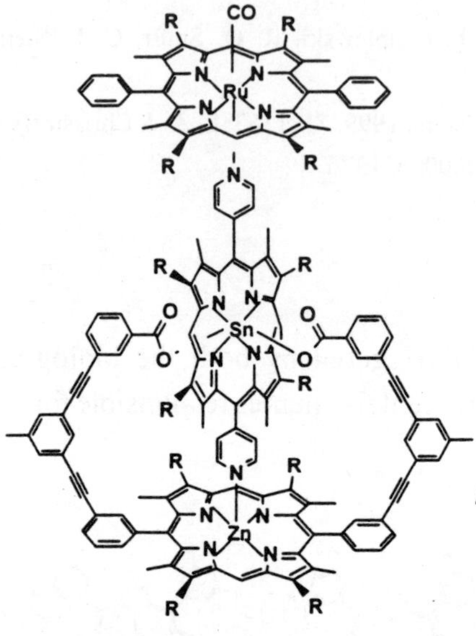

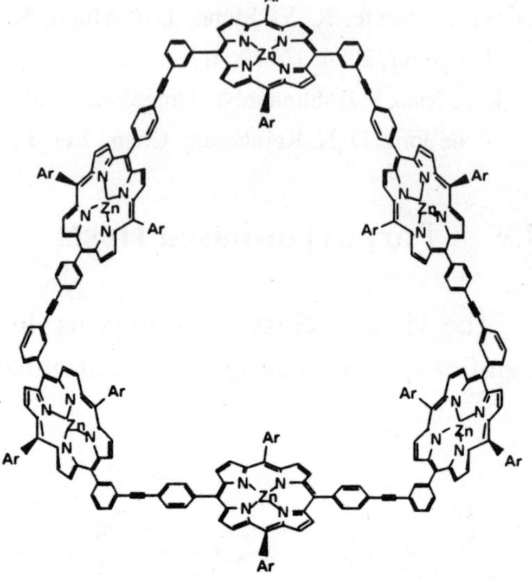

R = *n*-hexyl

384

385a

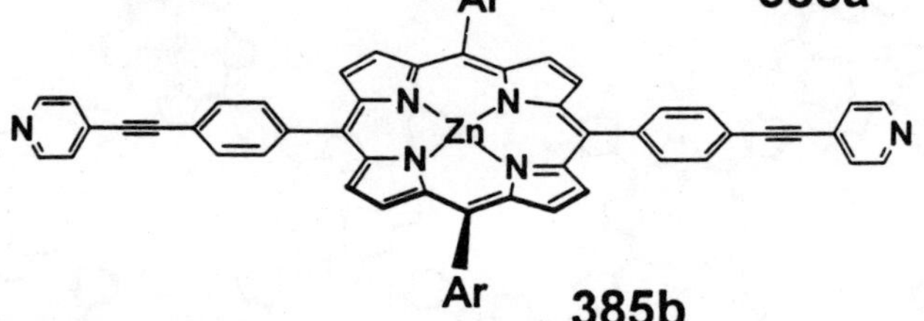

385b

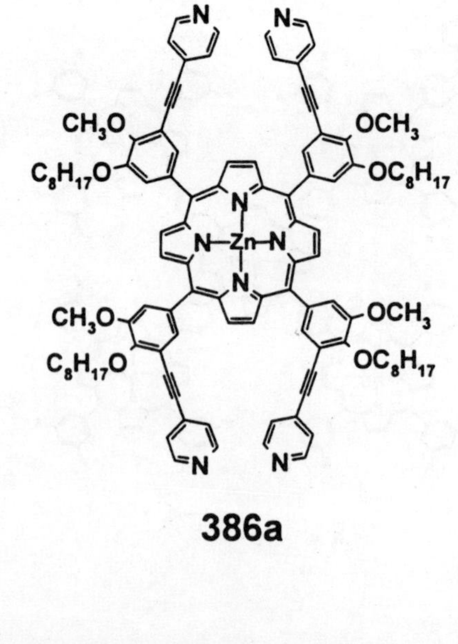

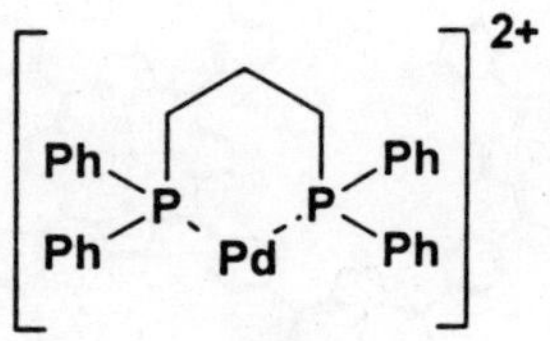

386a

386b

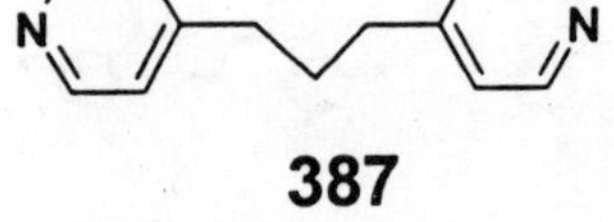

387

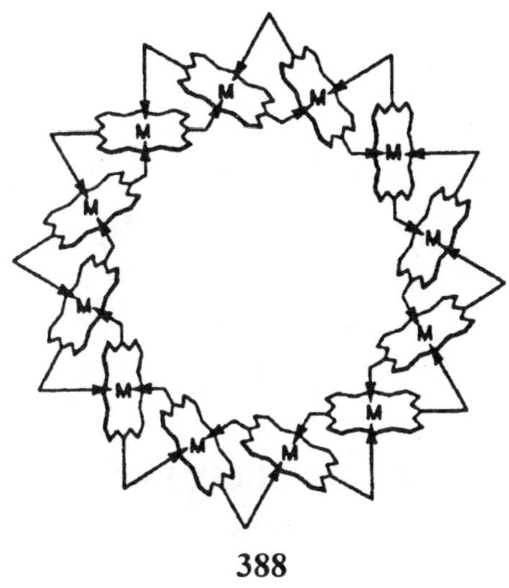

388

oxygen and CO_2 transport in living organisms [1] and by their applications as light harvesting antennae briefly presented in Section 6.3.3. Simple metal porphyrins are the subject of coordination not supramolecular chemistry but a rapid improvement in the syntheses of theses compounds, *e. g.* one-pot synthesis of **43**, obtaining a square 21-porphyrin oligomer **383** [2] capable of aggregation, **100** forming fibers, rotaxane **131** used in studies of the electron transfer, **384** involving three porphyrin moieties with different metal ions [3] or several catalysts discussed in section 6.4 justifies their discussion in this book. Similarly to fullerenes and dendrimers, porphyrins may play both the roles of a host, as in **59**, and that of a guest, as in **340**. Interestingly, in **385** (mimicking light harvesting antennae) their derivatives play these roles simultaneously [4]. Crown-ether-armed metalloporphyrins represent an unusual example of ditopic receptors, since they recognize both cations and anions [5]. Self-assembly of porphyrin derivatives may create cavities prone to accepting guest molecules. Thus **386a** forms 2:4 coordination complex with **386b** involving pyridine nitrogen atoms. This complex can host the dipyridine molecule **387** [6].

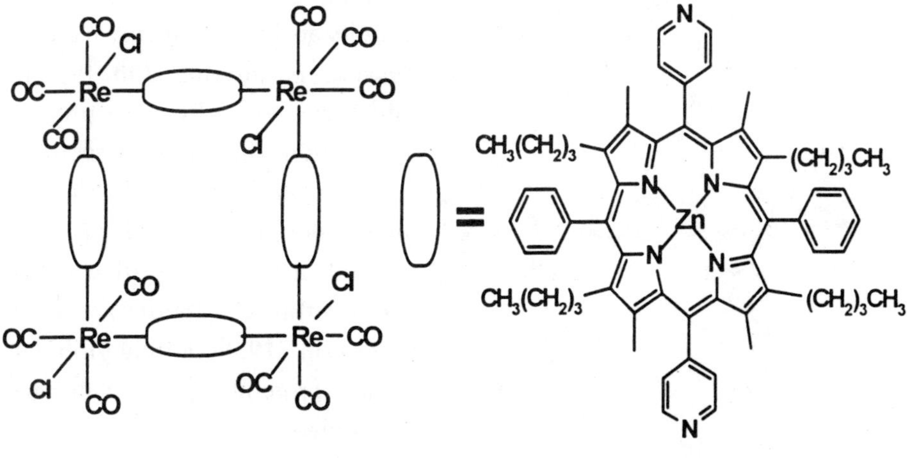

389

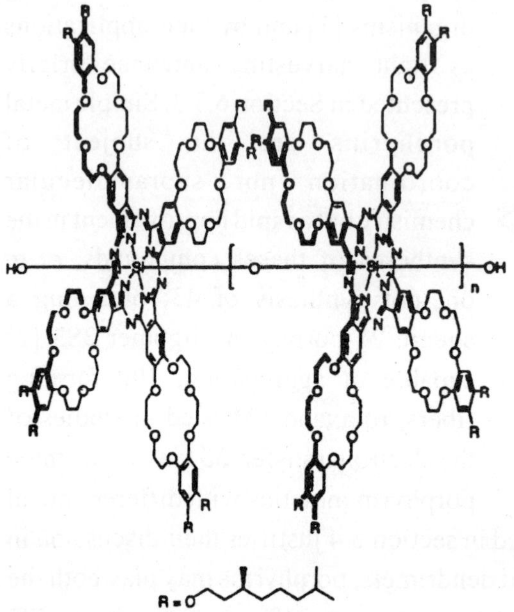

M = Zn^{2+}, Cu^{2+}, Fe^{2+}, Co^{2+}

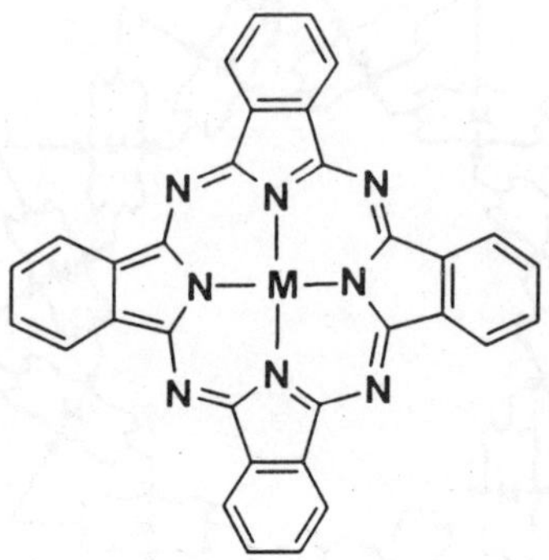

392

R = O

390

R =

391

The oligomeric porphyrin self-assembled ring **388** has a cavity which can accept guest(s) [7]. Porphyrin-based self-assembled 'molecular squares' **389** can form mesoporous thin films in which the edge of a square, thus the size of the cavity, can be adjusted by appropriate choice of substituents [8]. Fibers that form coil-coiled aggregates with distinct, tunable helicity are built from crown ethers bearing porphyrins **390** [9]. In addition to the porphyrin applications discussed in Sections 6.3.2.2 and 6.4, dendrimer metalloporphyrins **391** to be applied in catalysis [10] and the water-soluble dendritic iron porphyrin **319** modelling globular heme proteins [11] can be mentioned.

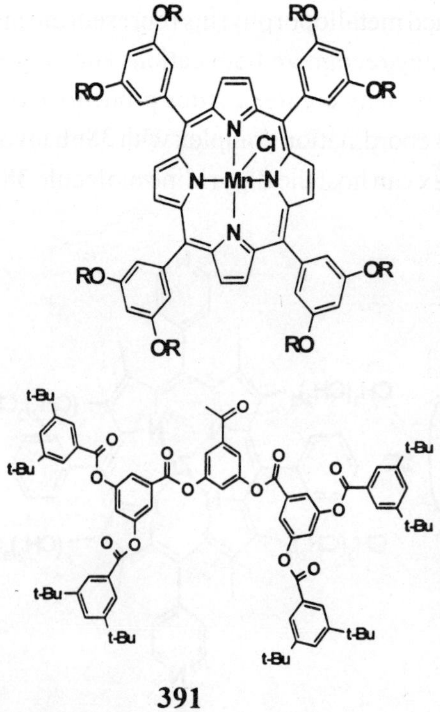

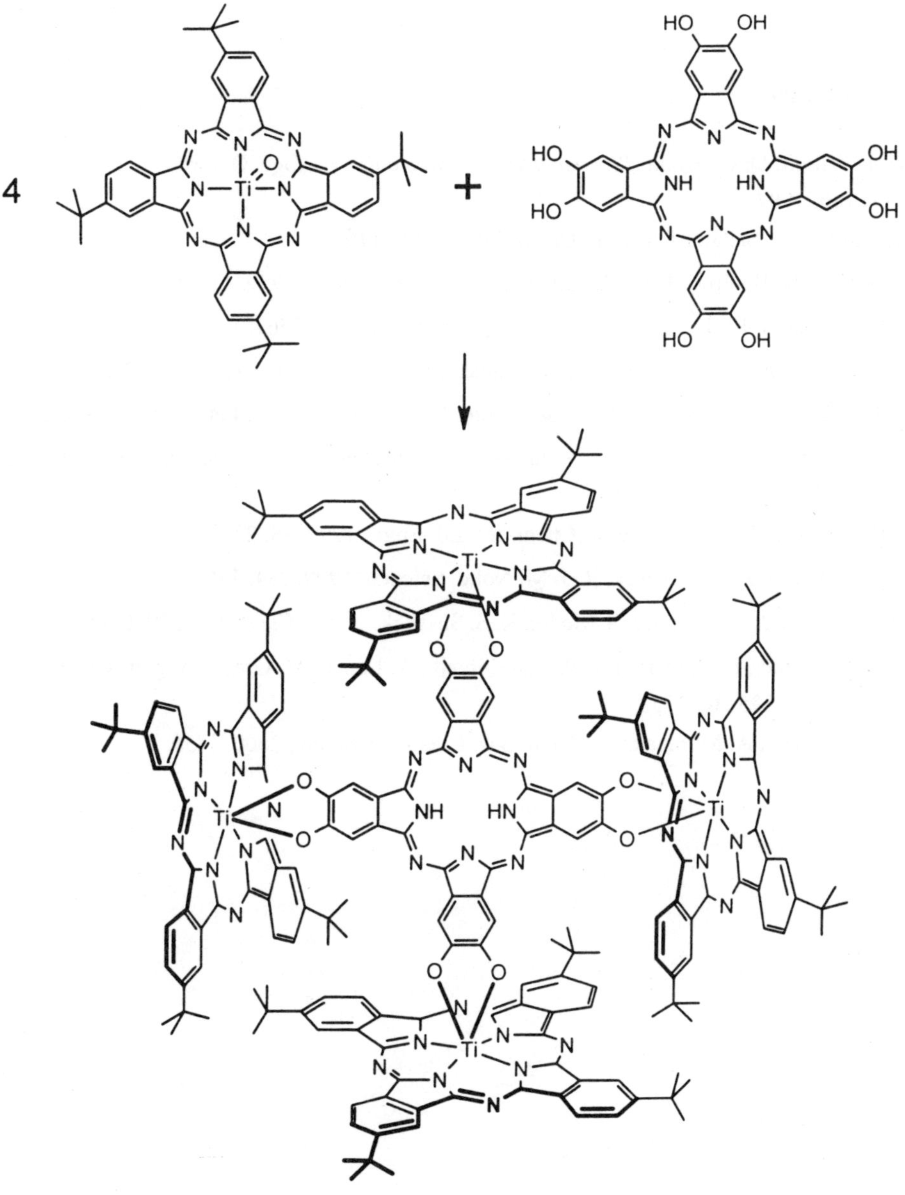

4 + →

393

Phthalocyanines **392** are porphyrin analogues embellishing Nature with exciting colours. Their pentamer **393** obtained in a one-step reaction has a cavity, thus it can play the role of host [12].

References

1. W. H. Elliott, D. C. Elliott, *Biochemistry and Molecular Biology*, Oxford University Press, Oxford, 1997.
2. K. Sugiura, Y. Sakata, T. Kawai, Chem. Lett., 1999, 1193.
3. H.-J. Kim, N. Bampos, J. K. M. Sanders, J. Am. Chem. Soc., 1999, 121, 8120.
4. A. Ambroise, J. Li, L. Yu, J. S. Lindsay, Org. Lett., 2000, 2, 2563.
5. S. Iwata, M. Suzuki, M. Shirakawa, K. Tanaka, Supramol. Chem., 1999, 11, 135.
6. A. Ikeda, M. Ayabe, S. Shinkai, S. Sakamoto, K. Yamaguchi, Org. Lett., 2000, 2, 3707.
7. R. A. Haycock, C. A. Hunter, D. A. James, U. Michelsen, L. R. Sutton, Org. Lett., 2000, 2, 2435.
8. S. Belanger, J. T. Hupp, Angew. Chem. Int. Ed. Engl., 1999, 38, 2222.
9. H. Engelkamp, S. Middlebeek, R. J. M. Nolte, Science, 1999, 284, 785.
10. P. Bhyrappa, J. K. Young, J. S. Moore, K. S. Susslick, J. Am. Chem. Soc., 1996, 118, 5708.
11. P. J. Dandliker, F. Diederich, J.-P. Gisselbrecht, A. Louati, M. Gross, Angew. Chem. Int. Ed. Engl., 1995, 34, 2752.
12. N. Kobayashi, A. Muranaka, J. Chem. Soc. Chem. Commun., 2000, 1855.

Chapter 8

OTHER EXCITING SUPRAMOLECULAR SYSTEMS

8.1 Introduction

Supramolecular chemistry has emerged about 30 years ago as a new research domain after the importance of recognition, preorganization and self-assembly in the aggregation processes was acknowledged. As described in Chapter 3, the first rules governing the inclusion of guests by macrocyclic hosts were established by Pedersen, Cram and Lehn in the late 1960s and 1970s. In spite of the established potential utility of the complexes involving the latter hosts, very few applications of macrocyclic inclusion complexes have been reported, since macrocyclization is a very inefficient kinetically controlled process. Contrary to low yields of reactions in which crown ethers like **17**, calixarenes like **18**, and most other covalently bound macrocycles are obtained, metal-directed self-assembly under thermodynamic control is frequently highly efficient [1-3]. These, often one-pot, syntheses enabling generation of large amounts of nanostructures of complicated architectures allow one to overcome the major obstacle preventing industrial use of supramolecular systems, on one hand; on the other, the apprehension that numerous intricate biological and artificial systems arise as the result of self-assembly of simple molecular subunits in a single step under equilibrium thermodynamic conditions, formed the basis of supramolecular design consisting in a controlled use of intermolecular interactions as a general building principle for the construction of supermolecular aggregates. Lehn considers chemistry as information science dealing with programmed chemical systems obtained through non-covalent interaction algorithms that operate through molecular recognition incidents based on interaction patterns of hydrogen bonding arrays, sequences of donor and acceptor, π-stacking or ions and ion coordination sites' interactions. In this language the information leading to self-assembly is contained at molecular level in the covalent structure and self-assembly is the information processing [4].

Self-assembled aggregates mimicking objects in living Nature were discussed in Chapter 4. Here, artificial superstructures obtained by a designed self-assembly on the basis of strong, directional interactions will be presented. Catenanes, knots, and other systems with distinct topological features that were introduced in Section 2.3 will be discussed first. Complex molecular architectures composed of hydrogen bonded species will be shown next. Metal directed self-assembly leading to a fascinating variety of intertwined strands (helicates), macrocycles, cages, and grids will appear next and organic zeolites will conclude the presentation. The self-assembled aggregates discussed in this Chapter are of importance not only in the new emerging field of *Material Chemistry* but also as systems mimicking the operation of biological organisms. Several such systems were presented in Chapters 4–6. It should be stressed that many self-assembled aggregates have cavities capable of guest inclusion leading to systems of higher complexity. As discussed in Section 7.3, templation by molecules or ions playing the guest role in the resulting aggregates is frequently observed [5]. The host-guest chemistry of such higher aggregates has been especially well studied in clathrate hydrates presented in Section 8.2.3 and organic zeolites discussed in Section 8.3.

References

1. P. N. W. Baxter, in *Comprehensive Supramolecular Chemistry*, J.-P. Sauvage, M. W. Hosseini, Eds., 1996, v. 9, p. 165.
2. E. C. Constable, in *Comprehensive Supramolecular Chemistry*, J.-P. Sauvage, M. W. Hosseini, Eds., 1996, v. 9, p. 213.
3. M. Fujita, Chem. Soc. Rev., 1998, 27, 417.
4. J.-M. Lehn, *Supramolecular Chemistry. Concepts and Perspectives*, VCH, Weinheim, 1995; J.-M. Lehn, Chem. Eur. J., 2000, 6, 2097.

5. D. J. Cram, J. M. Cram, *Container Molecules and Their Guests*, The Royal Society of Chemistry, Cambridge, Great Britain, 1994.

8.2 Making Use of the Preorganization Phenomenon: Topological Molecules [1]

For more than 100 years the Möbius strip, links modelling catenanes, knots, and other topological objects (discussed in Section 2.3) had been considered mathematicians' toys by chemists, and the first paper analysing the possibility of syntheses of this kind of objects was not accepted for publishing in 1960. It circulated, was widely cited as preprint, and finally appeared more than 30 years later [1a]. The synthesis of a first molecule with unusual topological properties, [2]catenane **30**, was published in the same 1960 [1b]. The syntheses of higher catenanes [2] followed soon, At first such molecules have been obtained by means of the statistical approach making use of the fact that by cyclization of long chains a few rings were formed when the starting chain was threaded through an earlier formed ring. However, obtaining more complicated higher catenanes (such as doubly intertwined catenane **33**, [3] olympiadane **7** [4], and multicatenane with bicyclic core **34** [5] as well as trefoil knot **6** [6] was made possible only by directed syntheses taking advantage of preorganization phenomenon. For this reason they are included in this book. On the other hand, it should be stressed that catenated and knotted structures are common in circular DNAs **42** [7], which even in their simplest form (the cycle consisting of two intertwined strands) model a non-trivial topological object. The AFM image of a catenated DNA is presented in Figure 8.2.1. As mentioned in Section 2.3, special enzymes topoisomerases are involved in the syntheses of such DNAs.

Figure 8.2.1. Atomic Force Microscopy image of a DNA catenane kindly provided by Prof. Akira Harada from Osaka University.

Topology is a branch of mathematics investigating relations between objects and object's properties pertinent to continuous transformations of one object into another [8]. These transformations may involve considerable deformations of the objects. However, no cutting of them or gluing their points together are allowed by the transformations. Topological singularity of such molecules as those

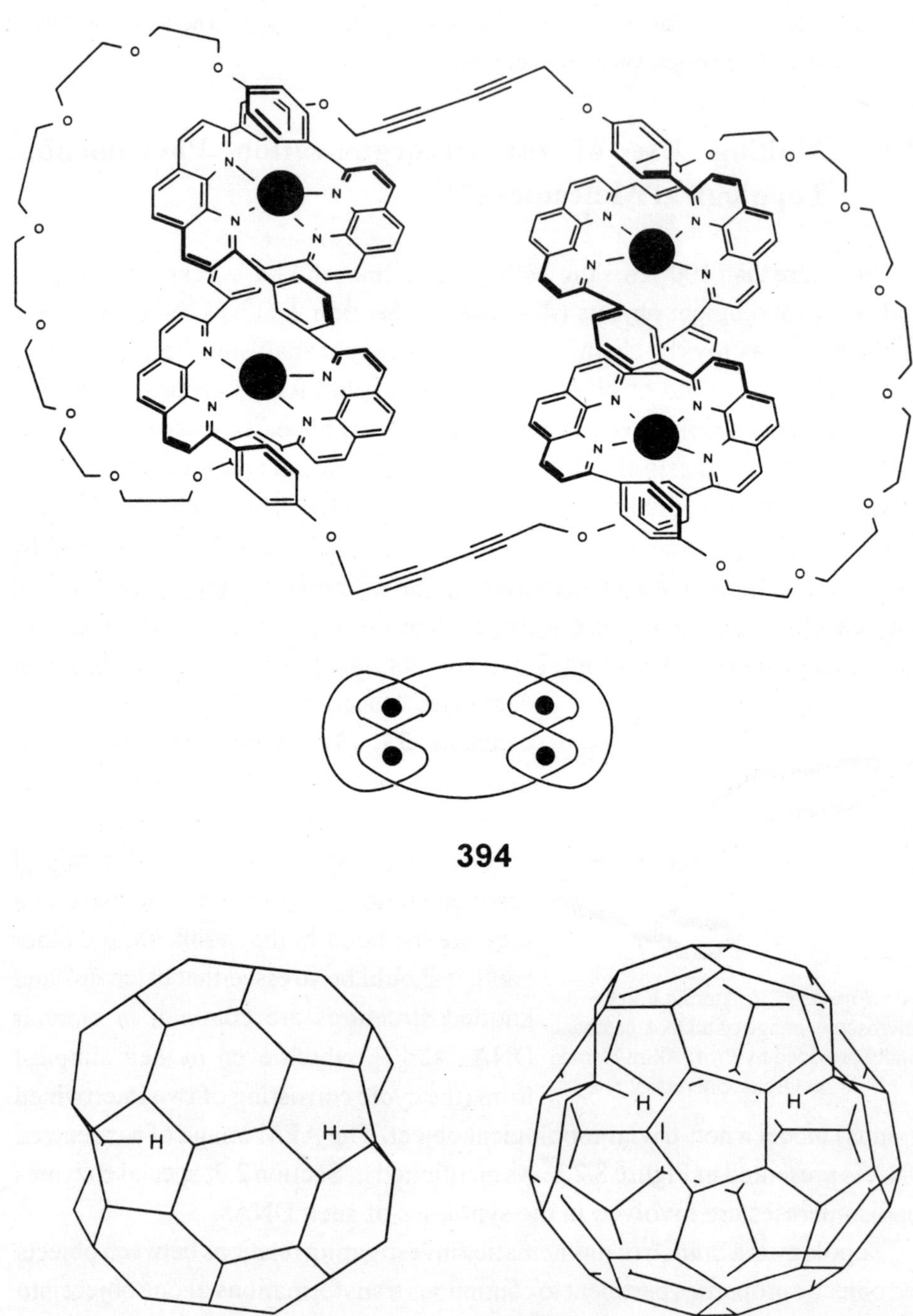

Figure 8.2.2. Two projections of $C_{60}H_{60}$ with two CH bonds pointing inside. Other hydrogen atoms omitted for clarity.

modelling links (catenanes) like **32**–**34**, the Möbius strip **31a**, **b** [1e, 9], and knots like **6** [1f], **394** [10] (double knot) was defined applying precise topological definitions allowing for considerable, unrealistic when applied to a molecule,deformations. Dumbbell systems consisting of a linear chain, threaded through a ring called rotaxanes **8,** "in" isomers of hydrogenated fullerenes (Figure 8.2.2) and endohedral fullerene complexes have only recently been included into the realm of topological molecules [11, 11] although rotaxanes have been usually discussed together with catenanes [1c, 12]. To unite **8** and the aforementioned fullerene isomers with catenanes, knots, and the molecules modelling the Möbius strip, some semiquantitative physical restrictions on the allowed deformations have to be imposed. We have recently shown [11] that it is possible to incorporate these physical restrictions into the language of topological models. The latter refer to an ideal mathematical object while molecules are physical species that cannot be distorted infinitely without bond breaking. Thus endohedral fullerene complexes [13a, b] like **307** and nested fullerenes [13c] cannot be separated into their constituent parts unless an unrealistically large extension of one ring were to enable the separation. Similarly, rotaxanes **39a** with a small central ring and big substituents cannot be separated into their parts without bond breaking. Interestingly, when the sizes of these two rotaxane components are comparable one can increase the temperature to achieve their formation or decomposition [1i]. From the point of view of topology the latter rotaxanes represent borderline case between **39a** on one hand and **39b** and **39c** on the other.

As stated above, simple catenanes were first obtained by a statistical approach. However, the successful syntheses of more complicated topological molecules would not have been possible without an enforcement of spatial orientation of reaction substrates called preorganization, introduced in Section 2.3. The latter effect can be achieved by

(1) The coordination of phenanthroline fragments with a metal atom as depicted in Figure 1.2. In this way knots like **6** [1f] and **394** [10a], a doubly-intertwined catenane **33** [3] and a mixture of [2]-, [3]-, [4]-, [5]- and [6]-catenanes **32** [14] have been synthesized by the Sauvage and Dietrich-Buchecker group.

(2) Electrostatic and π-stacking interactions between aromatic rings piled one upon another. This effect observed in the solid state of so-called *pseudorotaxanes* (that is the rotaxanes without voluminous substituents at the axle ends) [15] inspired the Stoddart group leading to the syntheses of not only numerous simple

catenanes and rotaxanes [1i, 1j] but also olympiadane **7** [4] and a molecule consisting of seven interlocked rings **395** [16].

(3) Hydrogen bonding is thought to be responsible for the formation of such molecules as the catenane **37** [17], rotaxane **396** [18] and even knot **397** [19]. And last but not least

(4) very weak but highly effective nonbonded interactions are responsible for the formation of "cyclodextrins necklace" **12** [20a] (Figure 1.5) and some other complexes. It should be stressed that the name "molecular necklace" was used by the Kim group for another kind of system **398a** involving cucurbituril "beads" **398b** [21]. Interestingly, cyclodextrins in a necklace like **12** can polymerize yielding a tubule **399** [20b].

Making use of the preorganization phenomenon also allows one to achieve

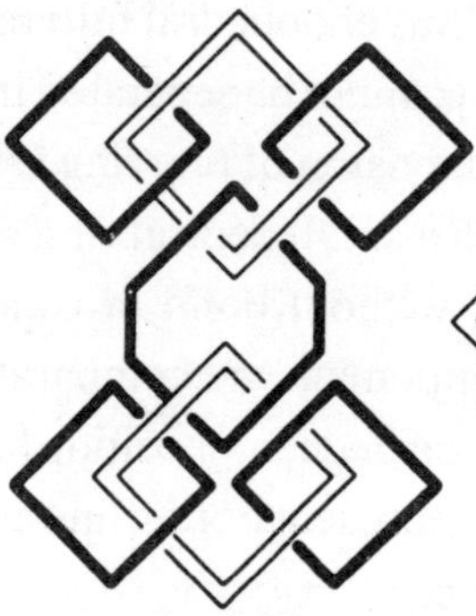

395

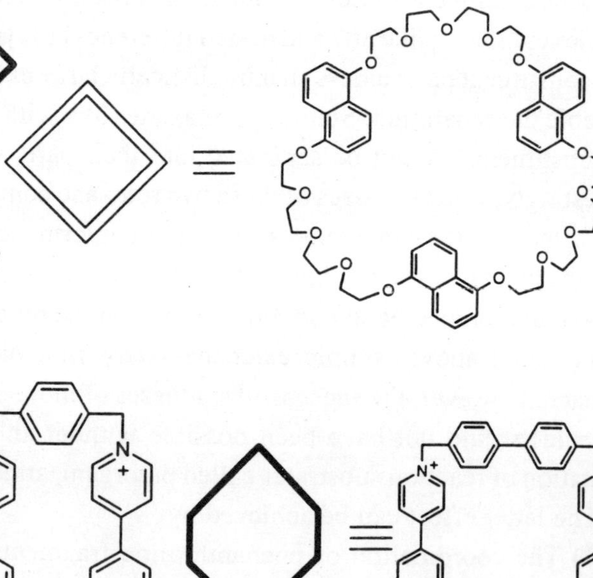

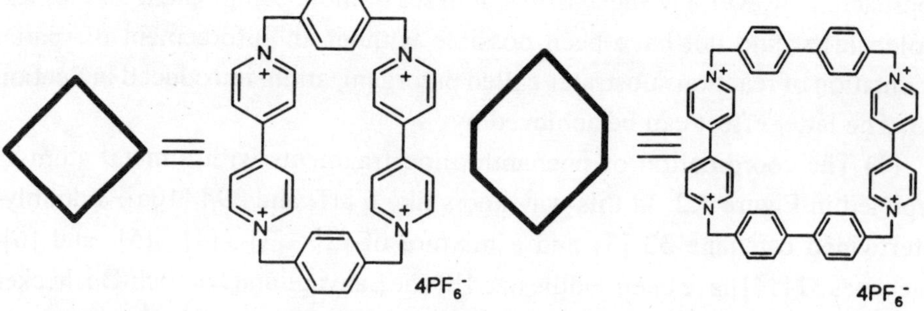

$4PF_6^-$ $4PF_6^-$

substantial yields of the products. For instance, a knot was obtained in 30% yield by the Dietrich-Buchecker and Sauvage group [22a] while some amide-based rotaxanes were synthesized with up to 41% yield by Vögtle and coworkers [22b].

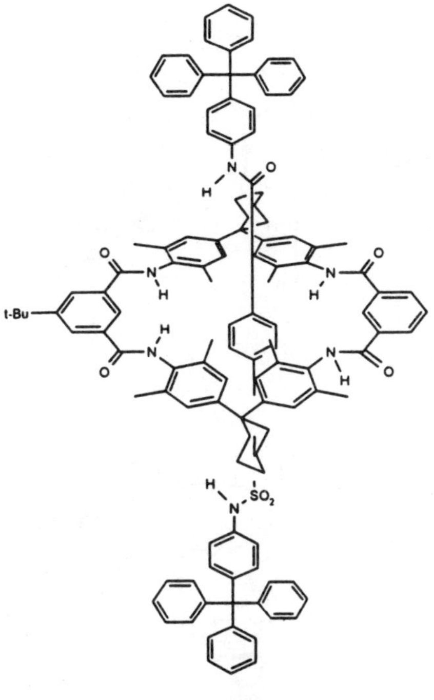

396

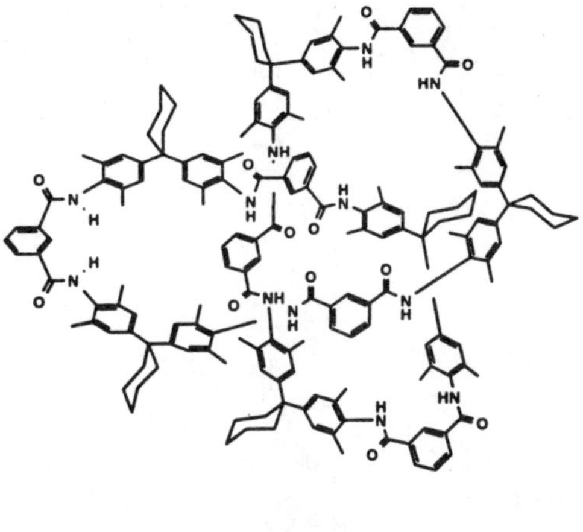

397

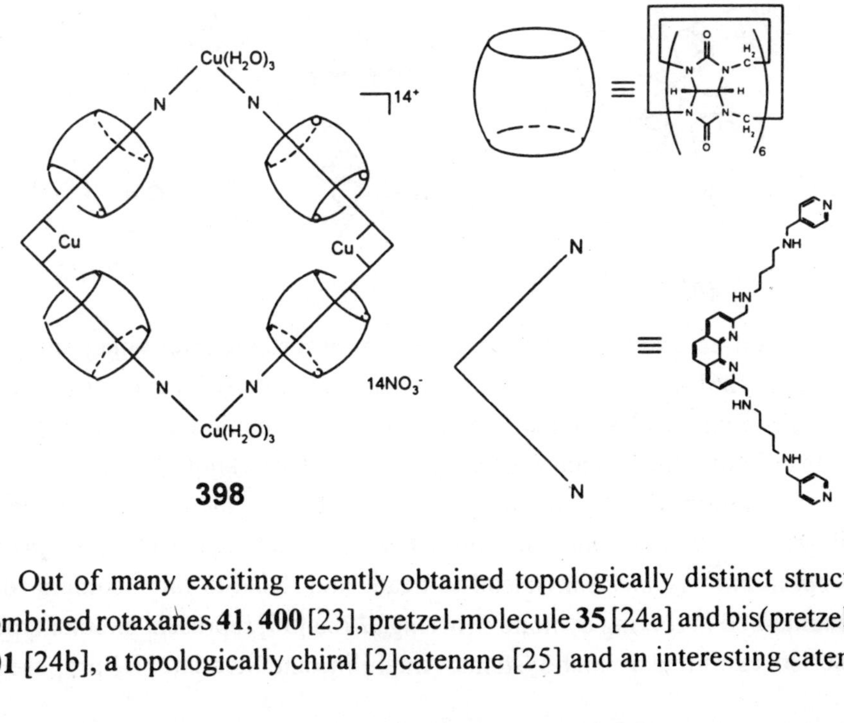

398

Out of many exciting recently obtained topologically distinct structures combined rotaxanes **41**, **400** [23], pretzel-molecule **35** [24a] and bis(pretzelane) **401** [24b], a topologically chiral [2]catenane [25] and an interesting catenated

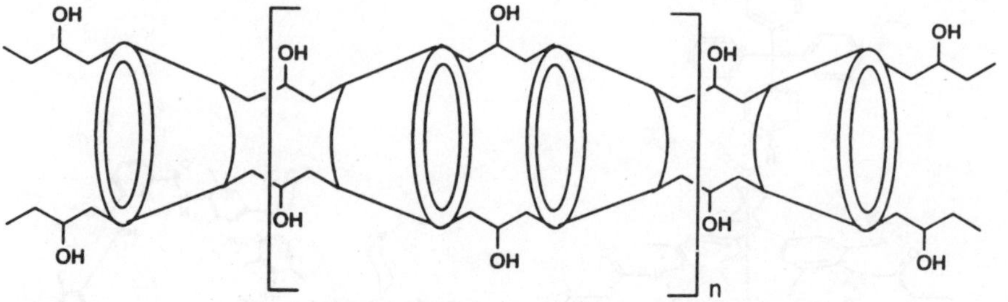

399

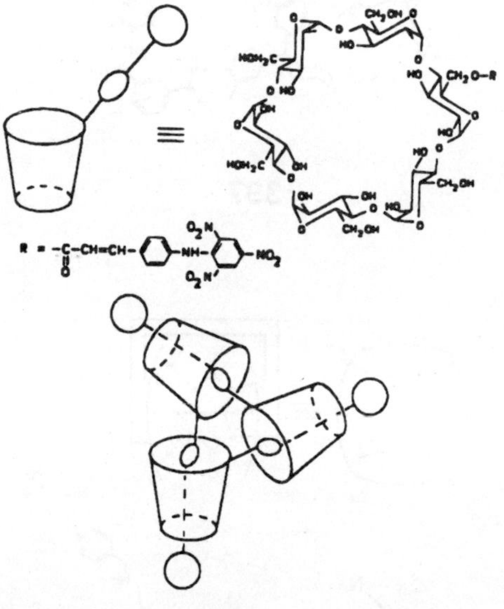

400

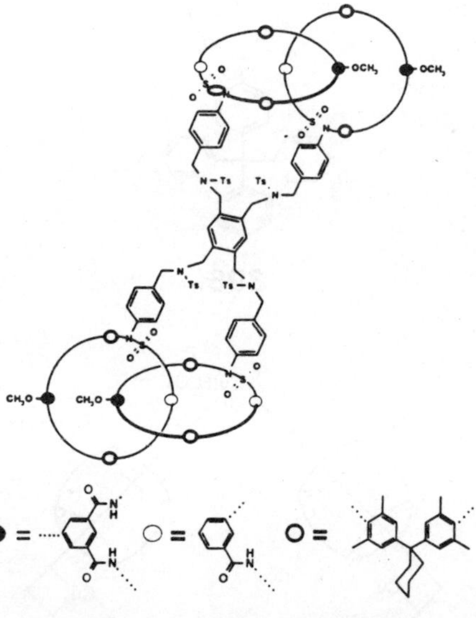

401

system that self-assembled in the solid state to form stable intertwined dimers (Figure 8.2.3) [26] should be mentioned. Rotaxanes, catenanes and some other topologically nontrivial systems [27] like **402** may exhibit *cycloisomerism* leading to chirality. Some such systems [28a] as well as a knot [28b] have been resolved to enantiomers. Two distinct positions of the ring in rotaxane **149** [29a] exemplify *translational isomerism*, while those in the catenane **152** [29b] can be called *rotational isomerism*. Head-to-head and head-to-tail isomers of [3]catenanes **403** involving cyclodextrin rings **278a** [30] represent another type of isomerism exhibited by such systems. Amabilino and Stoddart [1i] proposed

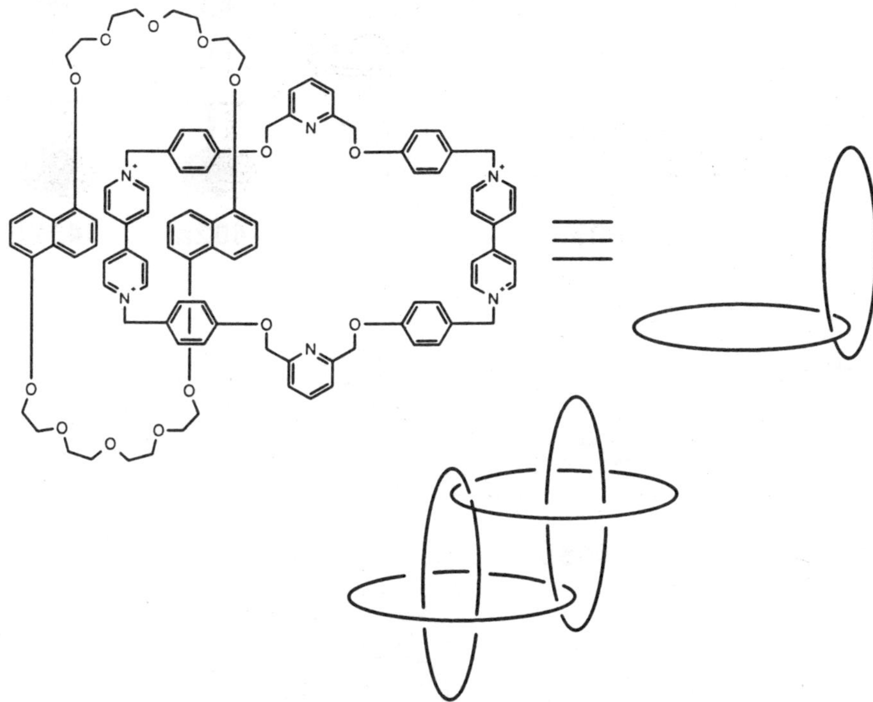

Figure 8.2.3. Schematic view of self-assembled dimer composed of two catenane.

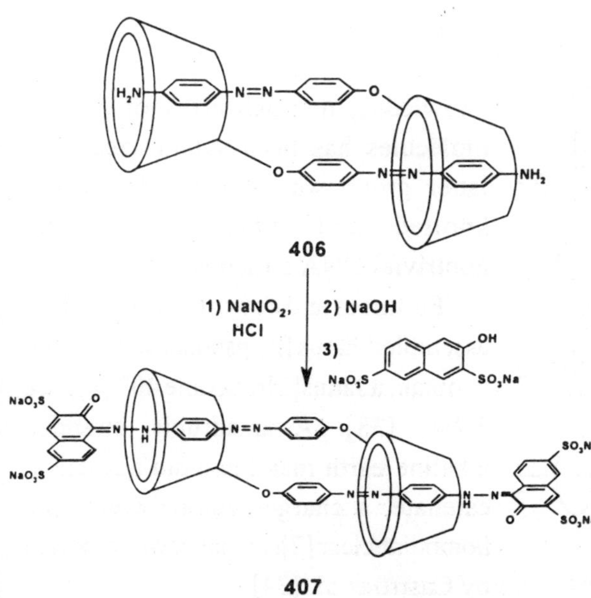

Figure 8.2.4. Rotaxane type of cyclodextrin dimer.

a Borromean ring **404** and a cyclic [6]catenane **405** as still unknown plausible synthetic targets while Dodziuk and Nowinski proposed hypothetical toroid nanotube isomers with two possible locations of an endohedral ring [11]. A carbon nanotube toroid was later reported by Martel et al. [31]. A Borromean ring system from DNA has been later obtained by the Seeman group [32a] but, to our

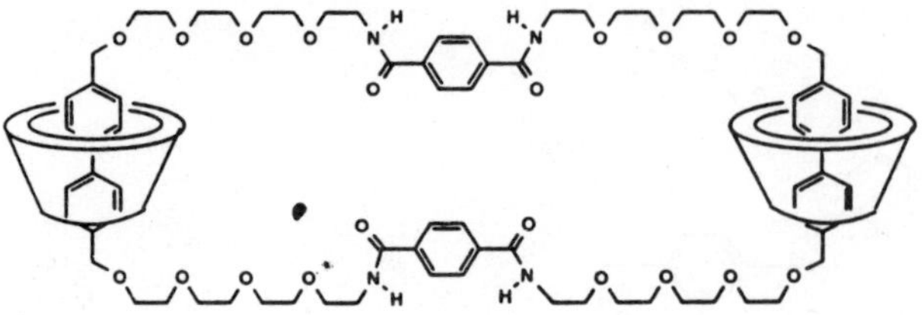

403a 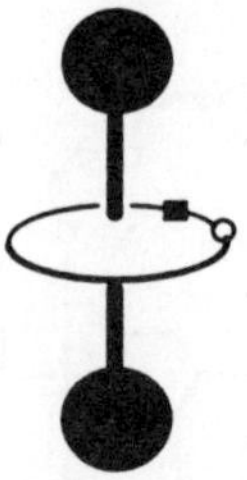 **402a** 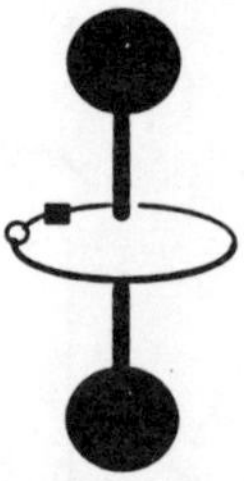 **402b**

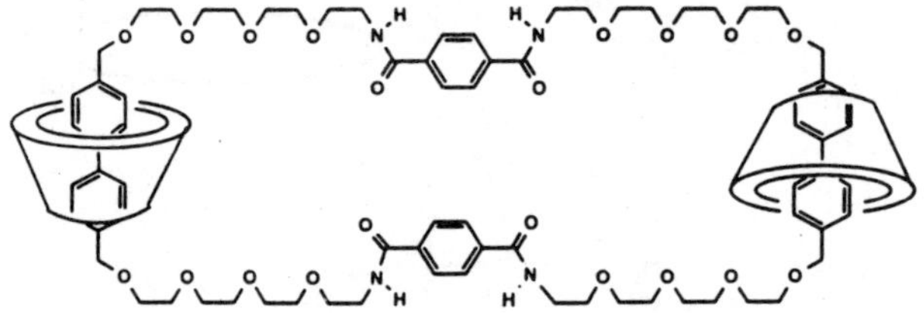

403b  **404**

405

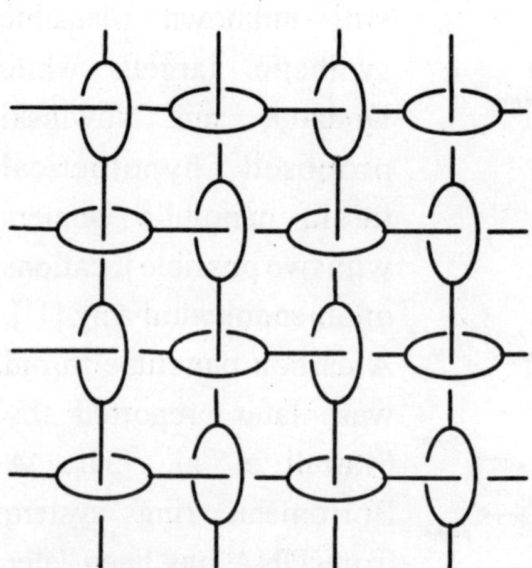

Figure 8.2.5. Rotaxane network.

knowledge, no such array of organic molecules has been synthesized. The latter group was able to obtain also knotted and other topologically nontrivial DNA structures [32b].

Fujimoto and coworkers used a self-assembled Janus [2]-pseudorotaxane **406** to obtain a Janus[2]rotaxane **407** (Figure 8.2.4) [33]. A quadruply stranded alkaline earth metal containing helical catenate, a charge-neutral heterotopic homodinuclear [2]catenane, was reported by Castro et al. [34].

Another, more complicated topological structure represents a [2]catenane formed by two interlocked nanotube rings observed by Scanning Electron Microscope by Martel and coworkers [35].

Polymers are generally not discussed in this book. However, a polymer with rotaxane structures involving cyclodextrins in side chains **408** [36], pseudorotaxane superstructures [37], doubly twisted polyrotaxane [38] as well as infinite polyrotaxane network (Figure 8.2.5) [39] can be mentioned here.

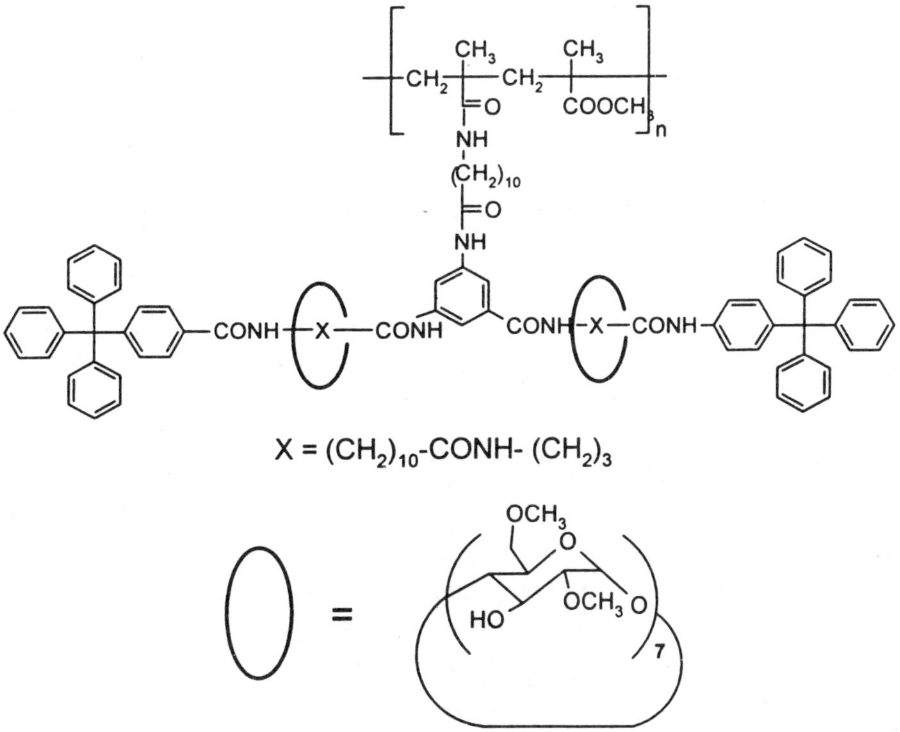

$X = (CH_2)_{10}\text{-CONH-}(CH_2)_3$

408

The synthetic skills developed in recent years has allowed researchers not only to obtain exciting structures with nontrivial topology but also to incorporate photochemically active (or other interesting) groups into topological molecules. According to Amabilino and Stoddart [1i] this enables "heading towards systems that are addressable in an electrochemical or photochemical manner". Some proposed applications of such systems have been presented in Section 6.3. The

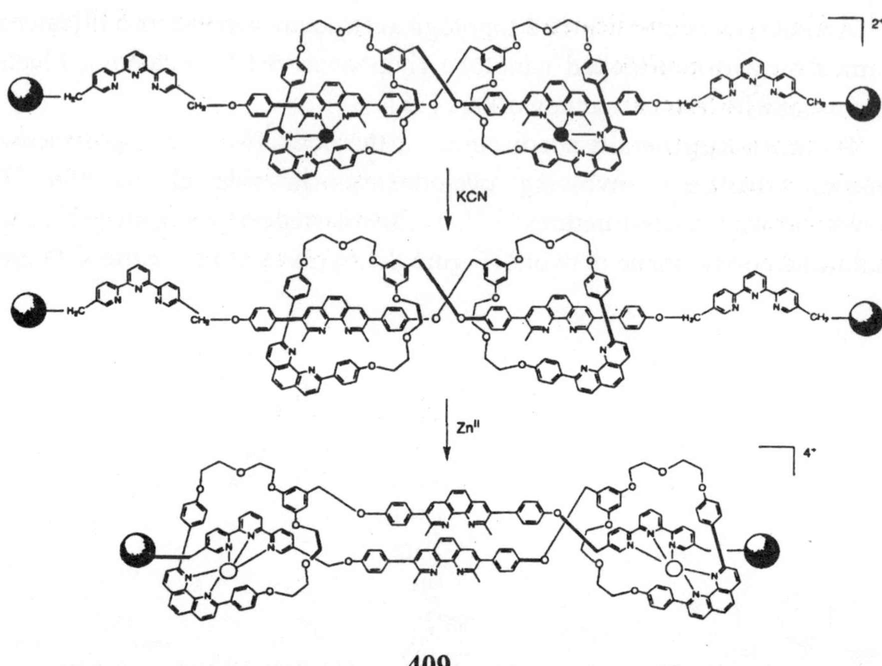

409

Figure 8.2.6. Rotaxane system mimicking contraction and stretching of muscles.

application of large amplitude motion such as those presented in Figures 6.6 and 6.7 for sensing and the construction of molecular machines and motors was discussed by Sauvage [40] and some other authors while a series of more than 50 papers entitled "Molecular meccano" by the Stoddart group was devoted to the latter application [41]. The latter group have proposed the name "molecular weaving" for the crystal structures of (pseudo)rotaxanes [42] (see also "DNA tissue" schematically presented in Figure 2.6), while the Sauvage group also developed and synthesized an interesting rotaxane system **409** (Figure 8.2.6) capable of contracting and stretching, thus mimicking muscles action [43].

References

1. (a) N. van Gulick, New J. Chem. 1993, 17, 619; (b) H. L. Frisch, E. Wasserman, J. Am. Chem. Soc., 1961, 83, 3789; E. Wasserman, J. Am. Chem. Soc., 1960, 82, 4433; (c) G. Schill, *Catenanes, Rotaxanes, and Knots*, Academic Press, New York, 1971; (d) V. I. Sokolov, Russ. Chem. Rev., 1973, 42, 452; (e) D. M. Walba, Tetrahedron, 1985, 41, 3161; (f) C. O. Dietrich-Buchecker, J.-P. Sauvage, Chem. Rev., 1987, 87, 795; (g) J.-C. Chambron, C. O. Dietrich-Buchecker, J.-P. Sauvage, Top. Curr. Chem., 1993, 165, 131;

(h) C. Liang, K. Mislow, J. Math. Chem., 1994, 15, 245; (i) D. B. Amabilino, J. F. Stoddart, Chem. Rev., 1995, 95, 2725; (j) F. M. Raymo; J. F. Stoddart, Chem. Rev., 1999, 99, 1643; (k) N. C. Seeman, J. Chen, S. M. Du, J. E. Mueller, Y. Zhang, T.-J. Fu, Y. Wang, H. Wang, S. Zhang, New J. Chem., 1993, 17, 739; (l) H. Dodziuk, K. S. Nowiński, Tetrahedron, 1998, 54, 2917.

2. Early catenane syntheses have been reviewed in Ref. 1c and 1i.

3. J.-F. Nierengarten, C. O., Dietrich-Buchecker, J.-P., Sauvage, J. Am. Chem. Soc., 1994, 116, 375.

4. D. B. Amabilino, P. R. Ashton, A. S. Reder, N. Spencer, J. F. Stoddart, Angew. Chem. Int. Ed. Engl., 1994, 33, 1286.

5. C. O. Dietrich-Buchecker, B. Fromberger, I. Lueer, J.-P. Sauvage, F. Vögtle, Angew. Chem. Int. Ed. Engl., 1993, 32, 1434.

6. C. O. Dietrich-Buchecker, J.-P. Sauvage, Angew. Chem. Int. Ed. Engl., 1989, 28, 189.

7. E. M. Shekhtman, S. A. Wasserman, N. Cozzarelli, M. J. Solomon, New J. Chem., 1993, 17, 757.

8. C. Rourke, B. Sanderson, *Introduction to Piecewise-linear Topology*, Ergeb. der Math. u. Ihrer Grenz. 69, Springer Verlag, 1972.

9. D. M. Walba, R. M. Richards, R. C. Haltiwanger, J. Am. Chem. Soc., 1982, 104, 3219.

10. R. F. Carina, C. O. Dietrich-Buchecker, J.-P. Sauvage, J. Am. Chem. Soc., 1996, 118, 9110.

11 H. Dodziuk, K. Nowiński, Chem. Phys. Lett., 1996, 249, 406.

12. The issue of New Journal of Chemistry, 1993, 17, (10 - 11).

13. (a) H. Shinohara, in *Fullerenes, Chemistry, Physics and Technology*, K. M. Kadish, R. S. Rueff, Eds., Chapter 8; S. Nagase, K. Kobayashi, T. Akasaka, T. Wakahara, *ibid.*, Chapter 9; J. H. Weaves, Acc. Chem. Res., 1992, 25, 143; (b) H. Dodziuk, G. Dolgonos, O. Lukin, Carbon, in press; (c) Nested fullerenes are giant fullerenes having one (or more) smaller fullerenes inside: S. Ijima, Nature, 1991, 354, 56; H. Dodziuk, G. Dolgonos, O. Lukin, Chem. Phys. Lett., 2000, 329, 351.

14. F. Bitsch, C. O. Dietrich-Buchecker, A.-K. Khemiss, J.-P. Sauvage, A. Van Dorsselaer, J. Am. Chem. Soc., 1991, 113, 4023.

15. J.-Y. Ortholand, A. M. Z. Slavin, N. Spencer, J. F. Stoddart, D. J. Williams, Angew. Chem. Int. Ed. Engl., 1989, 28, 1394.

16. D. B. Amabilino, S. E. Ashton, J. Y. Boyd, S. Lee, N. Menzer, J. F. Stoddart, D. J. Williams, Angew. Chem. Int. Ed. Engl., 1997, 36, 2070.

17. A. G. Johnston, D. A. Leigh, R. J. Pritchard, M. D. Deegan, Angew. Chem. Int. Ed. Engl., 1995, 34, 1209; A. G. Johnston, D. A. Leigh, L. Nezhat, J. P. Smart, M. D. Deegan, M. D., *ibid.*, 1212.

18. T. Dünnwald, R. Jäger, F. Vögtle, Chem. Eur. J., 1997, 3, 2043.

19. O. Safarowsky, M. Nieger, R. Frohlich, F. Vögtle, Angew. Chem. Int. Ed. Engl., 2000, 39, 1616.

20. (a) A. Harada, J. Li, M. Kamachi, Nature, 1992, 356, 325; J. F. Stoddart, Angew. Chem. Int. Ed. Engl., 1992, 31 846; (b) A. Harada, J. Li, M. Kamachi, Nature, 1993, 364, 516.

21. S. G. Roh, K. M. Park, G. J. Park, S. Sakamoto, K. Yamaguchi, K. Kim, Angew. Chem. Int. Ed. Engl., 1999, 38, 638.

22. (a) C. O. Dietrich-Buchecker, J.-P. Sauvage, A. De Cian, J. Fisher, J. Chem. Soc. Chem. Commun., 1994, 2231; (b) F. Vögtle, R. Jäger, M. Händel, S. Ottens-Hildebrandt, W. Schmidt, Synthesis, 1996, 353.

23. T. Hoshino, M. Miyauchi, Y. Kawaguchi, H. Yamaguchi, A. Harada, J. Am. Chem. Soc., 2000, 122, 9876.

24. (a) C. Reuter, A. Mohry, A. Sobanski, F. Vögtle, Chem. Eur. J., 2000, 6, 1674; (b) A. Mohry, H. Schwierz, F. Vögtle, Synthesis, 1999, 1753.

25. D. K. Mitchell, J.-P. Sauvage, Angew. Chem. Int. Ed. Engl., 1988, 27, 930.

26. M. B. Cabezon, J. Cao, F. M. Raymo, J. F. Stoddart, A. J. P. White, D. J. Williams, Angew. Chem. Int. Ed. Engl., 2000, 39, 148.

27. For instance, in topology a cycle is a trivial knot.

28. (a) G. Rapenne, C. O. Dietrich-Buchecker, J.-P. Sauvage, J. Am. Chem. Soc., 1996, 118, 10932; (b) C. Yamamoto, Y. Okamoto, T. Schmidt, R. Jäger, F. Vögtle, J. Am. Chem. Soc., 1997, 119, 10541.

29. (a). R. A. Bissell, E. Cordova, A. E. Kaifer, J. F. Stoddart, Nature, 1994, 369, 133; (b) J.-P. Sauvage, Bull. Pol. Acad. Sc., 1998, 46, 289.

30. D. Armsbach, P. R. Ashton, R. Ballardini, V. Balzani, A. Godi, C. P. Moore, L. Prodi, N. Spencer, J. F. Stoddart, M. S. Tolley, T. J. Wear, D. J. Williams, Chem. Eur. J., 1995, 1, 33.

31. R. Martel, H. R. Shea, P. Avouris, Nature, 1999, 398, 299.

32. (a) C. D. Mao, W. Q. Sun, N. C. Seeman, Nature, 1997, 386, 137; (b) J. E. Mueller, S. M. Du, N. C. Seeman, J. Am. Chem. Soc., 1991, 113, 6306; S. M. Du, N. C. Seeman, J. Am. Chem. Soc., 1992, 114, 9652.

33. T. Fujimoto, Y. Sakata, T. Kaneda, J. Chem. Soc. Chem. Commun., 2000, 2143.

34. S. L. Castro, O. Just, W. S. Rees, Jr., Angew. Chem. Int. Ed. Engl., 2000, 39, 933.

35. R. Martel, H. R. Shea, P. Avouris, J. Phys. Chem. B, 1999, 103, 7552

36. M. Born, H. Ritter, Angew. Chem. Int. Ed. Engl., 1995, 34, 309.

37. S. J. Cantrill, A. R. Pease, J. F. Stoddart, J. Chem. Soc. Dalton Trans., 2000, 3715.

38. M. C. Jimenez, C. Dietrich-Buchecker, J.-P. Sauvage, A. De Cian, Angew. Chem. Int. Ed. Engl., 2000, 39, 3284.

39. B. F. Hoskins, R. Robson, D. A. Slizys, J. Am. Chem. Soc., 1997, 119, 2952.

40. J.-P. Sauvage. Acc. Chem. Res., 1998, 31, 611.

41. P. R. Ashton, V. Baldoni, V. Balzani, C. G. Claessens, A. Credi, H. D. A. Hoffmann, F. M. Raymo, J. F. Stoddart, M. Venturi, A. J. P. White, D. J. Williams, Eur. J. Org. Chem., 2000, 7, 1121.

42. P. R. Ashton, A. C. Collins, M. C. T. Fyfe, S. Menzer, J. F. Stoddart, D. J. Williams, Angew. Chem. Int. Ed. Engl., 1997, 36, 735; P. R. Ashton, A. C. Collins, M. C. T. Fyfe, P. T. Glink, S. Menzer, J. F. Stoddart, D. J. Williams, *ibid.*, 1997, 36, 59.

43. M. Consuelo, C. Dietrich-Buchecker, J.-P. Sauvage, Angew. Chem., 2000, 112, 3284.

8.3 Multiple Hydrogen-bonded Systems

8.3.1 Rosettes, tapes (ribbons), fibers and two-dimensional networks

Due to its strength, directionality, and biological relevance, hydrogen bonding is one of the most important interactions leading to self-assembly, thus to the formation of supramolecular aggregates. Simple hydrogen-bonded aggregates like the dimer in micelle shown in Figure 4.8, **111** – **115** base pairs, the aggregates **118**

410

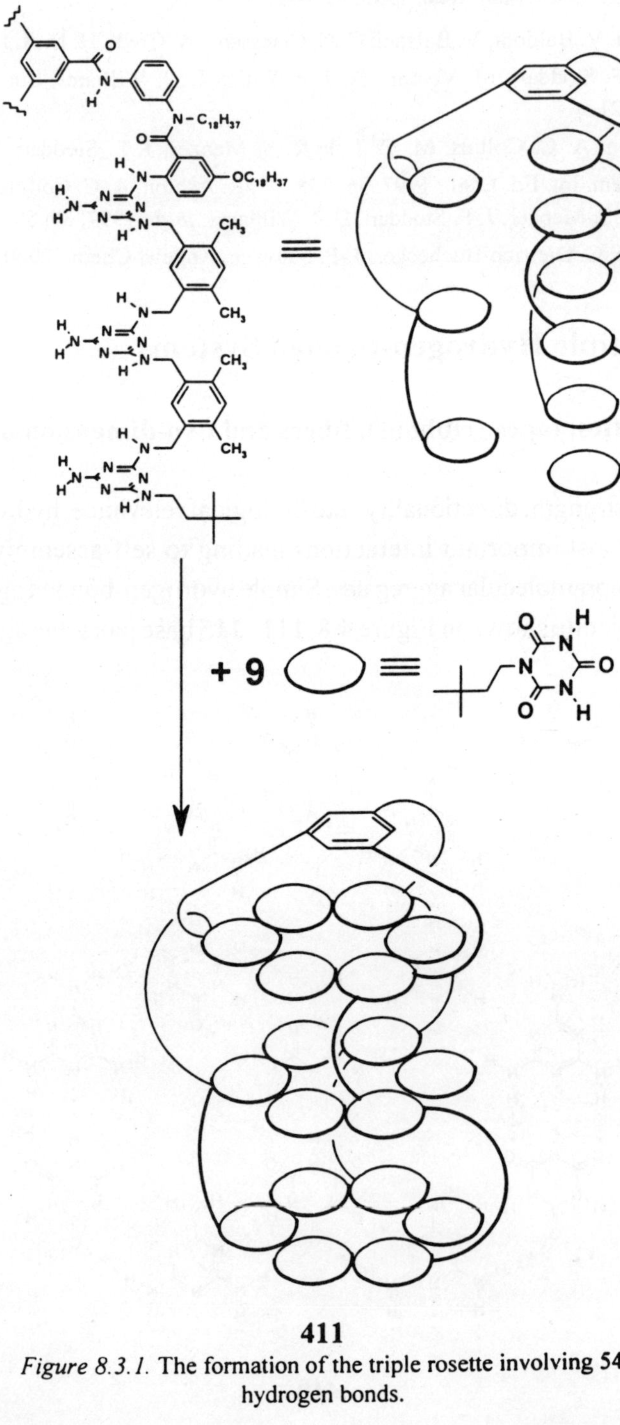

411

Figure 8.3.1. The formation of the triple rosette involving 54 hydrogen bonds.

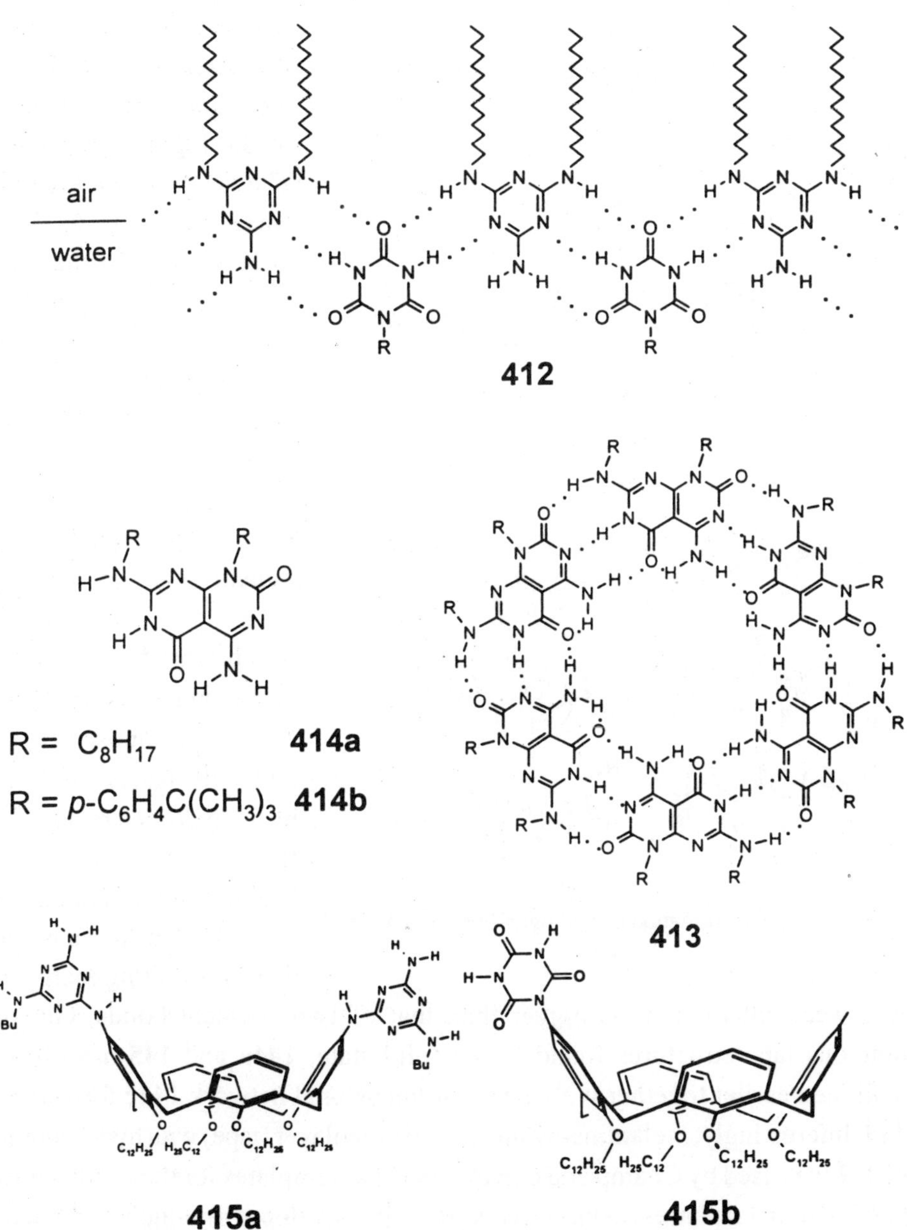

(Figures 5.2 and 6.2), the complex involving hydrogen-bonded barbiturate that should serve as sensor (Figure 6.9) and **234** have been presented earlier. The Whitesides rosette **3a** [1], tapes **3b** [2] and **109**, crinkled tape **3c** [1] (Figure 1.1) formed by melamine **1** and cyanuric acid **2** also shown earlier as well as

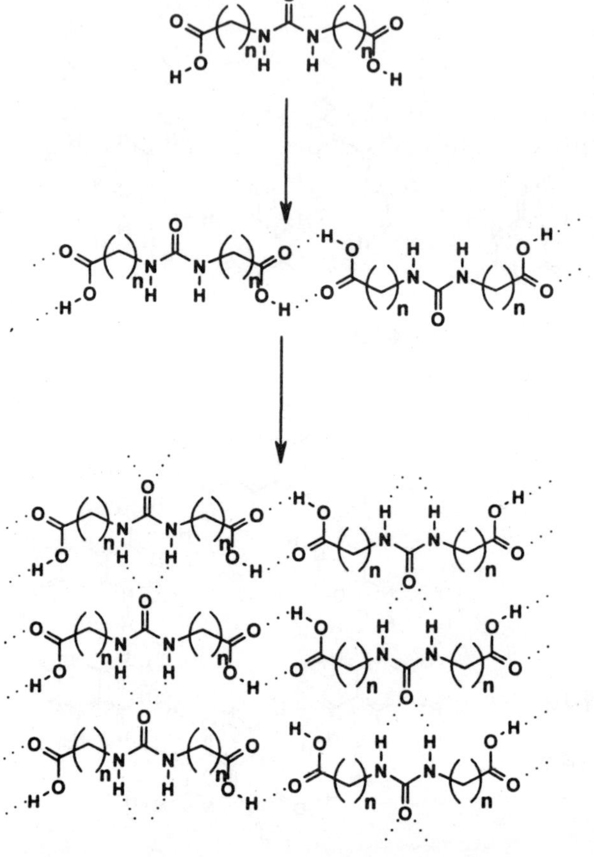

Figure 8.3.2. The formation of hydrogen-bonded tapes and layers.

hydrogen-bonded motifs discussed in Section 6.2 are a good representation of the higher aggregates. The rosette is formed only when R and R1 are different from H otherwise a two-dimentional infinite net **410** is formed. The net observation by Wang and coworkers [3] inspired Whitesides to the study of not only **3a**-**3c** but also of more sophisticated three-dimentional structures. One of the most complicated of such aggregates **411** (Figure 8.3.1) is composed of 10 molecules and is stabilized by 54 hydrogen bonds [4]. If conservatively 4 kcal/mol are assumed for the energy of such a bond, then the resulting energy of these weak interactions is bigger than that of two covalent bonds. Out of numerous tape structures found in the solid state, **144a** and **145** have been mentioned earlier together with the two-dimensional network **144c** formed by **144d**. Interestingly, melamine-cyanuric acid monolayer (tape) with hydrophobic tail **412** was used by Champ and coworkers [5] as templates for the synthesis of modified mineral phases. Lehn and coworkers [6] developed a multiply hydrogen-bonded rosette **413** analogous to **3a** formed from molecules **414a** and **414b**. By substituting calixarene with either cyanuric acid or melamine **415a**, **b** the Reinhoudt group obtained bis-rosette structures [7] with a considerable cavity which, at least in principle, is capable of including large aromatic molecules. The Lauher group studied several hydrogen bonded two-dimentional networks formed

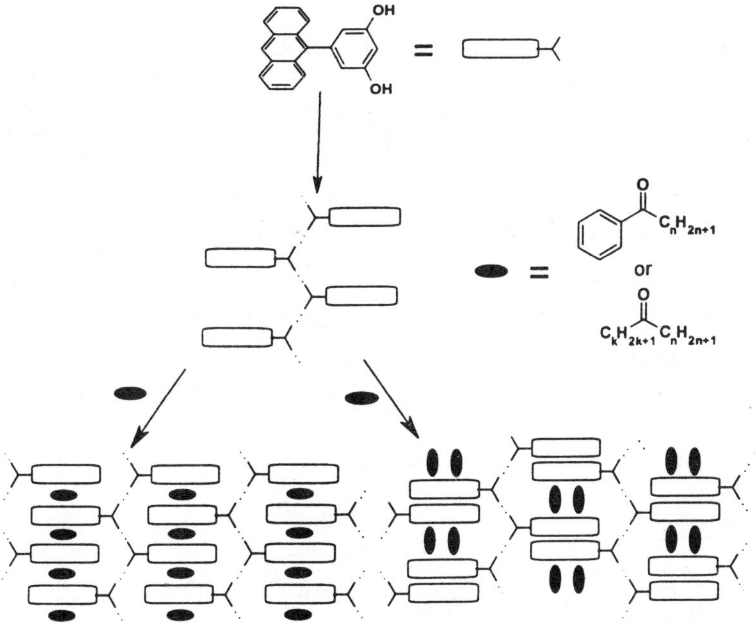

Figure 8.3.3. Guest-dependent formation of various 2D hydrogen bonded networks.

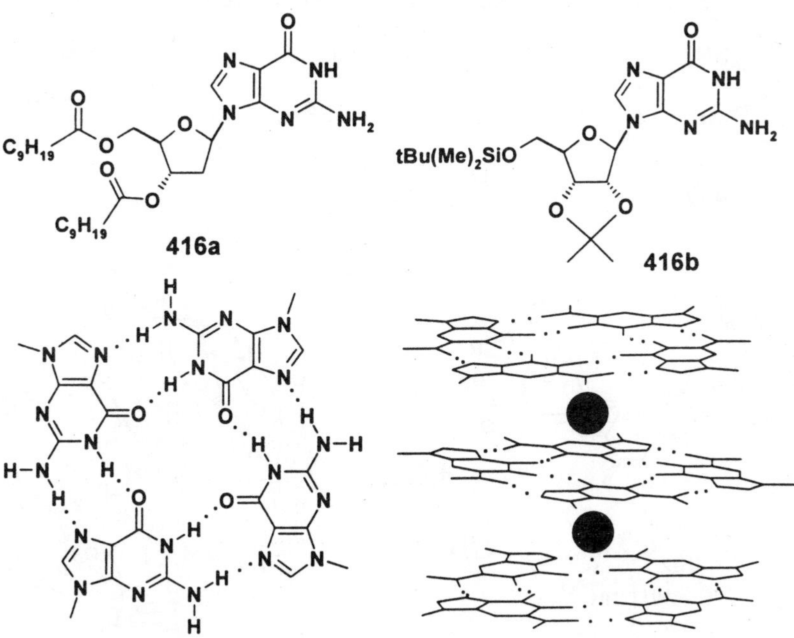

Figure 8.3.4. Quadruplets of guanosine derivatives (left) forming a channel with alkaline metal ions.

by two types of hydrogen bonds [8]. The first one typical of carboxylic acid dimers resulting from the interactions of the groups situated at the ends of linear molecules produces tapes, while the urea fragment is resposible for the second type of hydrogen bonds between the tapes yielding the network (Figure 8.3.2). Aoyama and coworkers adopted a different approach to the problem of two-dimensional networks formation (Figure 8.3.3) since their hydrogen-bonded tapes self-assemble due to the stacking interactions and additional guest intercalation [9].

The fibers formed from cyclic octapeptides **94** and **95** by hydrogen bonds perpendicular to the average ring planes were discussed in Section 4.2.4. The presence of the hydrophobic side chains in the latter molecule enabled Ghadiri

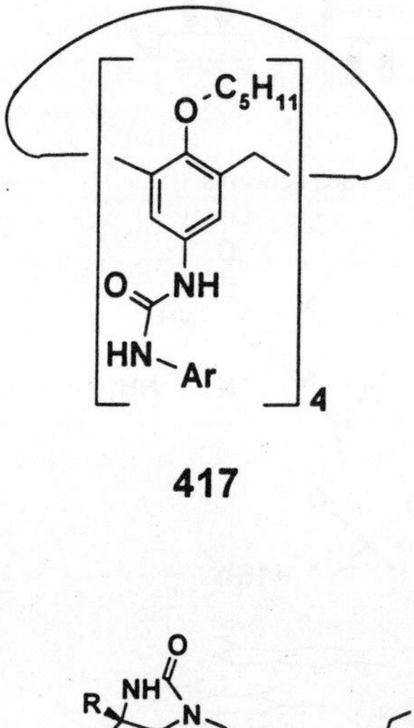

417

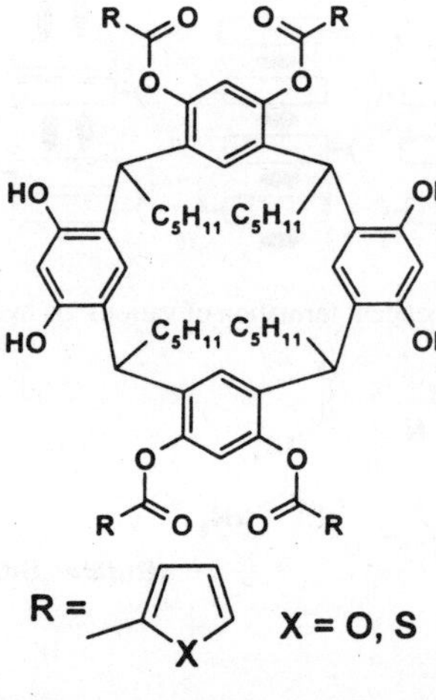

418a

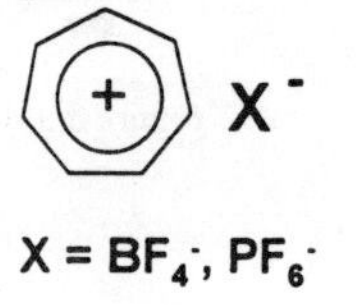

418b

$X = BF_4^-, PF_6^-$

R = ... X = O, S

2

+ CH

CH =

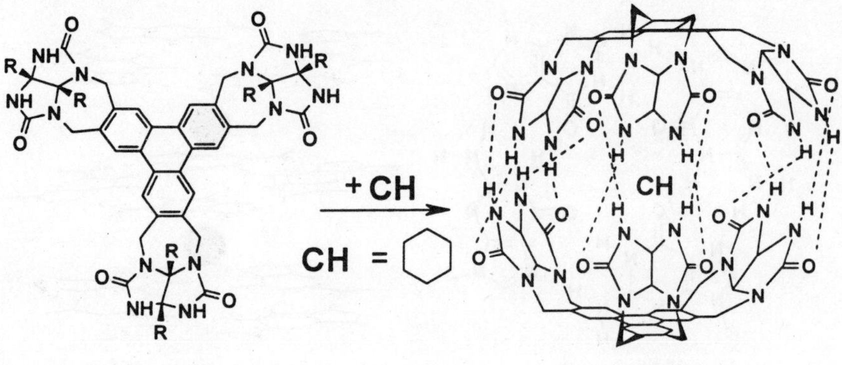

419

Figure 8.3.5. The formation of a capsule with cyclohexane template.

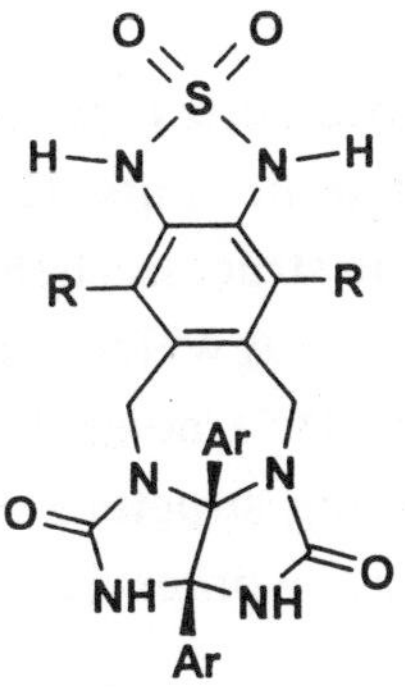

420

and coworkers the incorporation of these fibers into a membrane. Another type of highly ordered structure of importance in biomimetic studies is formed by lipophilic guanosine derivatives **416** (Figure 8.3.4). They form hydrogen-bonded tetramers which self-assemble in the presence of alkali metal anions in organic solvents to form tubular polymers [10].

8.3.2 Hydrogen-bonded capsules and other higher architectures

Most hydrogen-bonded capsules are dimers formed by calixarenes like **417** [11], resorcarenes like **418a** [12], cyclodextrins like **13** [13] and Rebek tennis ball compound **81b** built of two hydrogen-bonded **81a** units [14]. In many cases, selectively included guests stabilize the structures. Thus, the **418a** dimer is stabilized by the included tropylium cation **418b**. As shown by the selectivity of inclusion of camphor enantiomers **28** by **13**, the capsules can exhibit enantioselectivity [13]. They can also catalyze chemical reactions [15]. Although hydrogen-bonded capsules are relatively 'soft' structures, they can influence conformational mobility of the included guest. For instance, the capsule **419** (Figure 8.3.5) restricts the included cyclohexane molecule freezing its inversion. The latter is characterized by a free energy of activation at coalescence (ΔG^*) of 10.55 ± 0.05 kcal/mol for the included species, while the corresponding value for

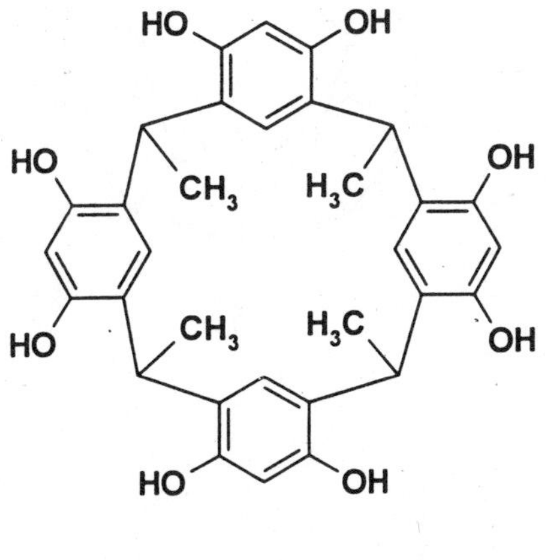

421

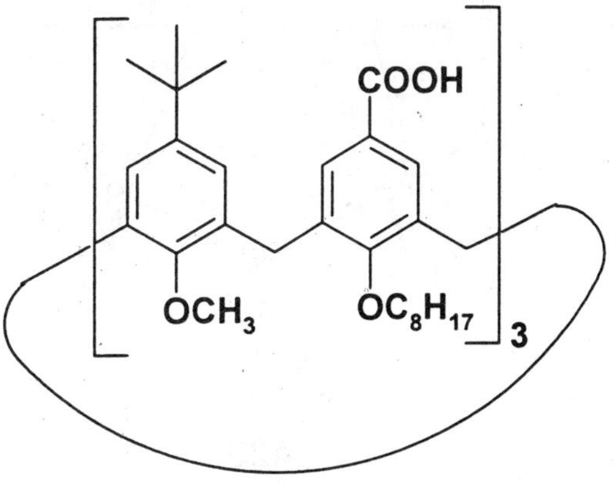

422

the free one was about 0.3 kcal/mol lower [6].

Tetrameric capsules like that formed by **420** [7] and the hexameric capsule formed by **421** [18] have been also reported. Six resorc[4]arene units **421** crystallize with eight water molecules, yielding a spherical superstructure, held together by 60 hydrogen bonds, with a voluminous cavity of 1,375 Å which can host guest molecules [19], while resorc[6]arene **422** self-assembles to a molecular cube [20]. A macrocyclic organometallic hydrogen-bonded ring structure in which 24 nickel atoms are involved was reported by Dearden and coworkers [21].

8.3.3 Clathrate hydrates of gases

As follows from his laboratory notes, the first discovered clathrate hydrate (of chlorine) was observed, but not recognized, by Davy in 1810. Then Cl_2, Br_2, SO_2, CO_2, CH_3Cl, CH_4, C_2H_6, and numerous other gases were shown to form clathrate hydrates [22, 23]. Contrary to inorganic stoichiometric hydrates, those involving hydrocarbons are both non-stoichiometric and crystalline. In addition, gas hydrate composition was found to depend on temperature, pressure, and some

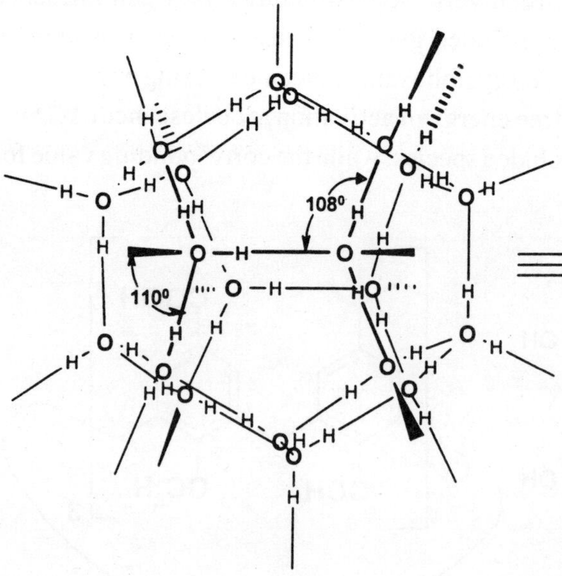

423

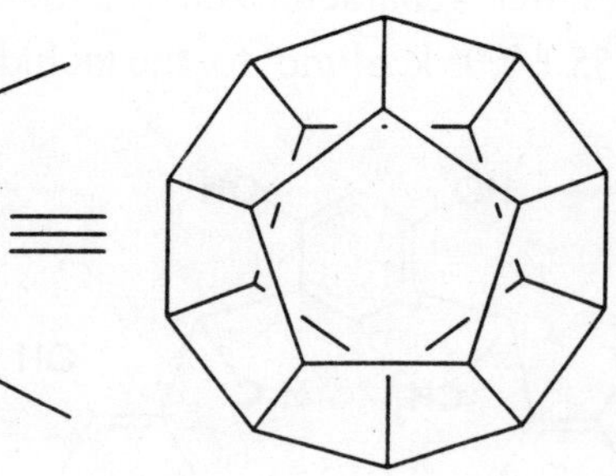

5^{12}

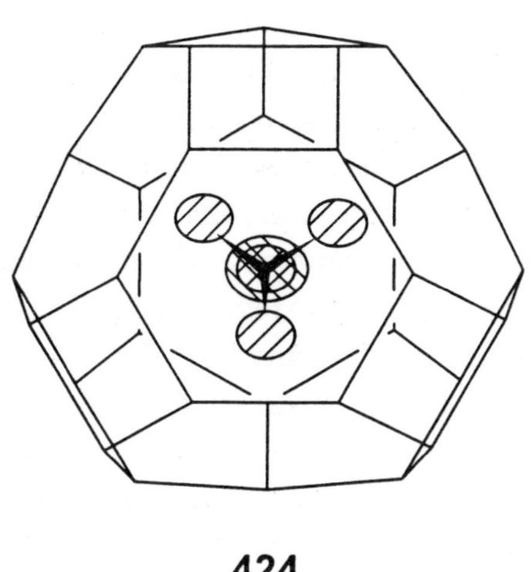

424

other factors. Hydrates of natural gases present certain industrial hazards, since they plug pipelines so causing considerable losses. They are also probably the cause of some atmospheric disasters. On the other hand, they are expected to offer a huge amount of methane gas - a many times richer energy source than the existing supply of carbon, oil, and gas together.

Dodecahedrane **423** of I_h symmetry like C_{60} (the notation 5^{12} shows that it consists of twelve five-membered rings) is one of the most frequently formed

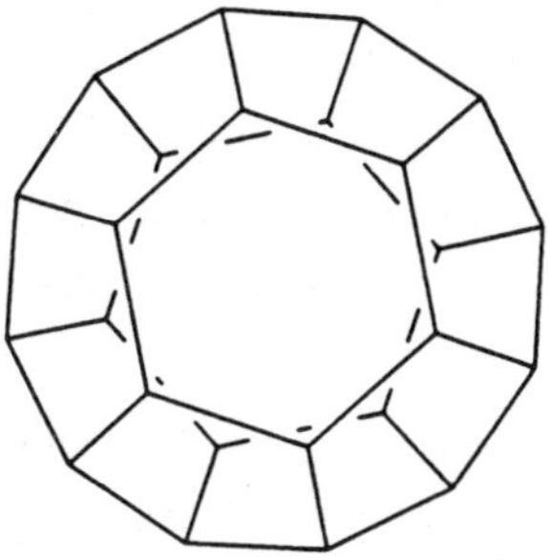

$5^{12}6^2$ 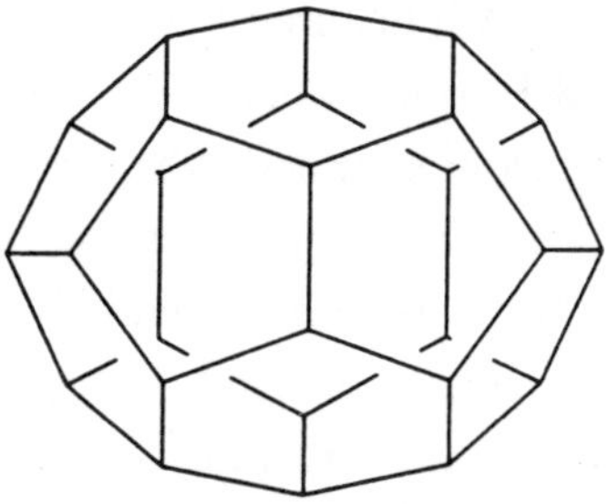 $5^{12}6^3$ 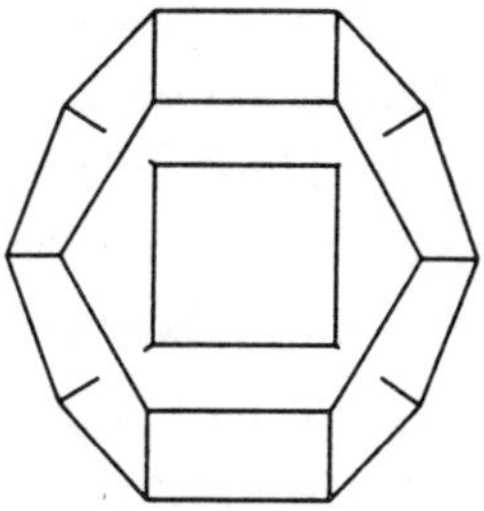 $4^35^66^3$

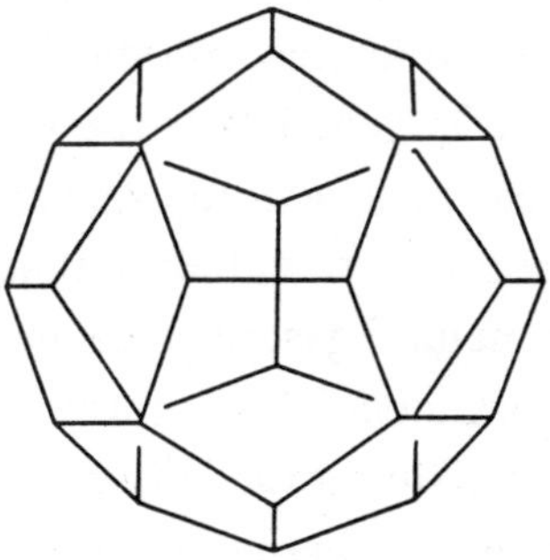

$5^{12}6^4$ 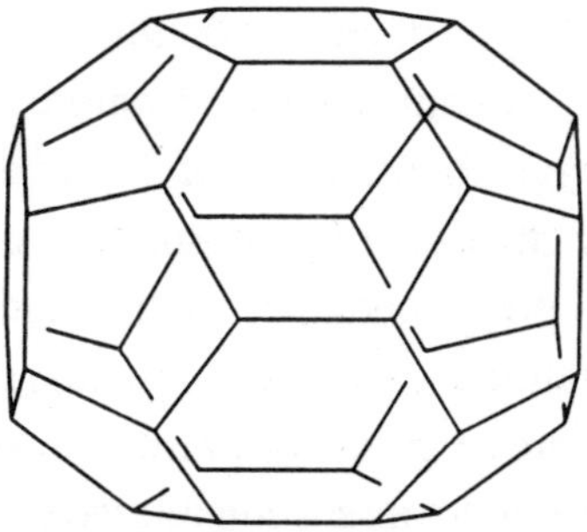 $5^{12}6^8$

Figure 8.3.6. Examples of hydrate cages.

water clusters held together by hydrogen bonds. Few other cages representing the clusters are shown in Figure 8.3.6. The clusters are dynamic entities since the hydrogen bonds holding the cage together are so weak that they are constantly broken and reconstructed. Interestingly and understandably [24], the cage's stability is increased considerably by the included guest. For instance, *t*-butylamine stabilizes $4^3 5^9 6^2 7^3$ **424**. Having C_5 symmetry axes, **423** cannot fill the space, thus hydrate structures are characterized by numerous voids. Only four types of clathrates crystal structures have been found upto today by using X-ray analysis and molecular modeling. Cubic structures CS-I and CS-II are the most popular. The former one consists of two types of voids formed by 46 water molecules in a crystallographic cell. It can include cyclopropane in the voids of the first type and methane or xenon in the second one. A cell of the CS-II crystal structure is built by 136 water molecules also forming two types of voids (capable of including SF_6, tetrahydrofuran or CCl_4 in the first type, Ar or Kr in the second). Less frequently found are the tetragonal structure TS-I (of 172 H_2O molecules in elementary cell forming three type of voids that can include Br_2) and the hexagonal structure HS-III (built out of 34 H_2O having three type of voids capable of hosting methylcyclohexane and H_2S). The domain of hydrate clathrate studies is relatively young and vividly expanding. Thus a discovery of other structures of clathrate hydrate cannot be excluded.

Due to its tetrahedral arrangement and the possibility of hydrogen bond formation, water is known to build various different structures which are close in energy. Therefore a change in experimental conditions (concentration, pressure, temperature) can lead to different clathrate structures for the same guest. For instance, all four types of structures have been found for $N^+(t\text{-Bu})_4Br$ [25]. In general, small guests seem to stabilize better the structures with smaller voids, while larger guests are better accommodated by the larger voids. However, to achieve an effective stabilization of the latter structures an additional small guest must occupy at least part of smaller voids forming so-called double hydrates. As said before, methane, ethane, and other natural gases can be and are, included into clathrate cages (and/or voids) stabilizing them. They form huge deposits both on-shore (*e. g.*, in permafrost in Alaska and Siberia) and off-shore on the ocean bed. The deposits are considered to be a potential alternative energy resource since most of them contain methane. Estimated to be as large as 10^{16} m^3, the hydrate resources of hydrocarbon gases are thought to be at least twice as large as the combined fossil fuel reserve [26]. It should be stressed, however, that

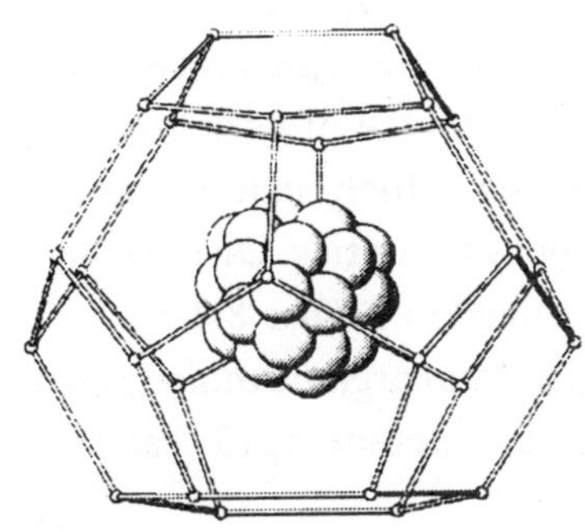

Figure 8.3.7. A hydrate cage with highly disordered Br_2 molecule. The figure kindly provided bu Prof. Ripmeester.

a rapid pace of discovery of clathrate hydrates deposits constantly changes these proportions disfavouring traditional fuel resources. Many serious technical problems have to be solved prior to the commercial use of methane from gas clathrates. The avoidance of the formation of clathrate hydrates is also of considerable practical interest today since they can block gas transmission lines and create problems in deep ocean drilling, and so forth. The formation of the clathrate plugging that damages pipelines presents a serious problem for gas companies, since their removal increases the production costs by up to 20% [27]. Over the long term, the deposits of clathrate hydrates are influenced by the Earth's climate and, in turn, their decomposition has influenced, does influence and will influence the future climate [28] thanks to the "greenhouse effect" of methane that is considerably stronger than that of carbon dioxide. The existence of carbon dioxide hydrates was also suggested on Mars [29], in Saturn's rings [30] and in comets [31].

As mentioned earlier, clathrate hydrates are often in a very fragile equilibrium, depending, among others, on pressure and temperature. Thus some eruptions of underwater vulcanoes in the Caspian sea [32] and the sea bordering Panama [33] are thought to be caused by the destruction of clathrate hydrates. Interestingly, the "secret of the Bermuda triangle" was also explained by the same phenomenon [34]. Namely, huge amounts of gases released by the decomposition of clathrates coming to the water's surface could result in very big waves in which a ship could sink in relatively short time. When released into the atmosphere the gases could also produce tremendous winds in the air high above the water's surface, which could be the reason for the disappearance of aircraft in this region.

Geological studies have recently proved that at least four times during the past 60,000 years there were massive releases of methane from hydrates [35].

As mentioned before, clathrates are highly dynamic structures. Not only are hydrogen bonds forming the clathrate cages constantly disrupted and reformed, but also guest molecules can exhibit considerable mobility. The X-ray structure of bromine hydrate reported by the Ripmeester group illustrates this point [36].

Their single crystal diffraction study of 16 different crystals of distinct composition (Br_2*8.62H_2O to Br_2*10.68H_2O) and morphologies revealed the existence of only one structure with a considerable degree of variation in the occupancy of large voids by bromine molecules which can assume 41 crystallographically independent sites in three types of cavities. Disordering of one Br_2 molecule in one of these cavities is visualized in Figure 8.3.7.

Clathrate hydrates can be formed not only with neutral organic molecules but also with alkylammonium salts with small anionic counteranions [37] and with guests involving polymeric anions [38].

References

1. J. A. Zerkowski, G. M. Whitesides, J. Am. Chem. Soc., 1992, 114, 5473; J. P. Mathias, E. E. Simanek, J. A. Zerkowski, C. T. Seto, G. M. Whitesides, J. Am. Chem. Soc., 1992, 114, 5473.
2. J. A. Zerkowski, C. T. Seto, D. A. Wierda, G. M. Whitesides, J. Am. Chem. Soc., 1990, 112, 9025.
3. Y. Wang, B. Wei, Q. Wang, J. Crystallogr. Spectrosc. Res., 1990, 79.
4. J. P. Mathias, E. E. Simanek, C. T. Seto, J. A. Zerkowski, G. M. Whitesides, Angew. Chem. Int. Ed. Engl., 1993, 32, 1766.
5. S. Champ, J. A. Dickinson, P. S. Fallon, B. R. Heywood, M. Mascal, Angew. Chem. Int. Ed. Engl., 2000, 39, 2716.
6. A. Marsh, M. Silvestri, J.-M. Lehn, J. Chem. Soc. Chem. Commun., 1996, 1527.
7. R. H. Vreekamp, J. P. M. Duynhoven, M. Hubert, W. Verboom, D. N. Reinhoudt, Angew. Chem. Int. Ed. Engl., 1996, 35, 1215.
8. S. Coe, J. J. Kane, T. L. Nguyen, L. M. Toledo, E. Wininger, F. W. Fowler, J. W. Lauher, J. Am. Chem. Soc., 1997, 119, 86.
9. K. Endo, T. Ezuhara, M. Koyanagi, H. Masuda, Y. Aoyama, J. Am. Chem. Soc., 1997, 119, 499.
10. E. Mezzina, P. Marani, R. Itri, S. Masiero, S. Pieraccini, G. P. Spada, F. Spinozzi, J. T. Davies, G. Gottarelli, Chem. Eur. J., 2001, 7, 388.
11. M. O. Vysotsky, I. Thondorf, V. Böhmer, Angew. Chem. Int. Ed. Engl., 2000, 39, 1264.
12. A. Shivanyuk, E. F. Paulus, V. Böhmer, Angew. Chem. Int. Ed. Engl., 1999, 38, 2906.
13. H. Dodziuk, A. Ejchart, O. Lukin, M. O. Vysotsky, J. Org. Chem., 1999, 64, 1503.

14. R. Wyler, J. de Mendoza, J. Rebek, Jr., Angew. Chem. Int. Ed. Engl., 1993, 32, 1699; M. M. Conn, J. Rebek, Jr., Chem. Rev., 1997, 97, 1647.
15. J. Kang, J. Santamaria, G. Hilmersson, J. Rebek, Jr., J. Am. Chem. Soc., 1998, 120, 7389.
16. B. M. O'Leary, R. M. Grotzfeld, J. Rebek, Jr., J. Am. Chem. Soc., 1997, 119, 11701.
17. F. Hof, C. Nuckolls, S. L. Craig, M. Tomas, J. Rebek, Jr., J. Am. Chem. Soc., 2000, 122, 10991.
18. T. Gerkensmeier, W. Iwanek, C. Agena, R. Frohlich, S. Kotila, C. Nather, J. Mattay, Eur. J. Org. Chem., 1999, 9, 257.
19. L. R. MacGillivray, J. L. Atwood, Nature, 1997, 389, 469.
20. A. Arduini, L. Domiano, L. Ogliosi, A. Pochini, A. Secchi, R. Ungaro, J. Org. Chem., 1997, 62, 7866.
21. A. L. Dearden, S. Parsons, R. E. P. Winpenny, Angew. Chem. Int. Ed. Engl., 2001, 40, 151.
22. E. D. Sloan, *Clathrate Hydrates of Natural Gases*, Marcel Dekker, New York, 1997.
23. V. Rodionova, D. V. Soldatov, Yu. A. Dyadin, Khimya v Interesakh Ustoichivogo Razvitiya, (Chemistry in Favour of Sustainable Development, Russ.), 1998, 6, 49.
24. In agreement with the old philosophic rule 'Horror vacui', Nature abhors a vacuum.
25. Yu. A. Dyadin, I. V. Bondariuk, L. S. Aladko, Zh. Strukt. Khim., 1995, 36, 1088.
26. Yu. F. Makogon, paper presented at the Third Chemical Congress of North America. Toronto, Canada, June 5-10, 1988; K. A. Kvenvolden, Chem. Geol., 1988, 71, 41.
27. S. Sh. Byk, Yu. F. Makogon, V. I. Fomina, *Gazovye Gidraty*, (Gas Hydrates, Russ.), Khimiya, Moskva, 1980.
28. K. A. Kvenvolden, Global Biogeochem. Cycles, 1988, 2, 221; E. G. Nisbet, Can. J. Earth Sci. 1990, 27, 148; E. G. Nisbet, J. Geophys. Res., 1992, 97 D12, 12854; L. C. Sloan, J. C. G. Walker, T. C. Moore Jr., D. K. Rea, J. C. Zachos, Nature, 1992, 357, 320; see also Ref. 2 and references cited therein.
29. S. L. Miller, Marine Sciences, 1974, 3, 151.
30. K. D. Pang, C. C. Voga, J. W. Rhoads, J. M. Ajello, Proc. 14th Lunar and Planet. Sci. Conf., Houston, 1983, 14, 592.
31. A. H. Delsemme, S. L. Miller, Plant. Space Sci., 1970, 18, 709.
32. G. N. Ginzburg, R. A. Gusseinov, A. A. Dadashev, G. A. Ivanova, S. A. Kazantsev, V. A. Solov'ev, Ye. V. Telepnev, R. E. Askeri-Nasirov, A. D. Yesikov, V. I. Mal'tseva, Yu. G. Mashirov, I. Yu. Shabaeva, Izv. AN SSSR. Ser. Geol., 1992, 7, 5.
33. K. A. Kvenvolden, Ann. New York Acad. Sci., 1994, 715, 232.
34. E. D. Sloan, Ann. New York Acad. Sci., 1994, 715, 1; P. Englezos, *ibid.*, 1994, 715, 75; Canadian Chem. News, 1990, May, 5.
35. J. P. Kennett, Cannariato, I. L. Hendy, R. J. Behl, Science, 2000, 288, 128.

36. K. A. Udachin, G. D. Enright, C. I. Ratcliffe, J. A. Ripmeester, J. Am. Chem. Soc., 1997, 119, 11486.
37. G. A. Jeffrey, in *Inclusion Compounds*, J. L. Atwood, D. D. Nicol, J. E. D. Davis, Eds., Academic Press, New York, 1984, v. 1, p. 757.
38. H. Nakayama, Bull. Chem. Soc. Japan, 1987, 60, 2319.

8.4 Organic Zeolites

Zeolites are a class of microporous inorganic crystals consisting of tetrahedrally arranged aluminosilicates which are able to catalyse chemical reactions by making use of a reversible binding of guest molecules in their cavities [1, 2]. The design, creation, and study of mesoporous compounds constitute one of the most important branches of crystal engineering discussed in Sect. 6. 2. Organic zeolites, which sometimes bear the name pillar compounds, could become much more versatile than their inorganic counterparts.

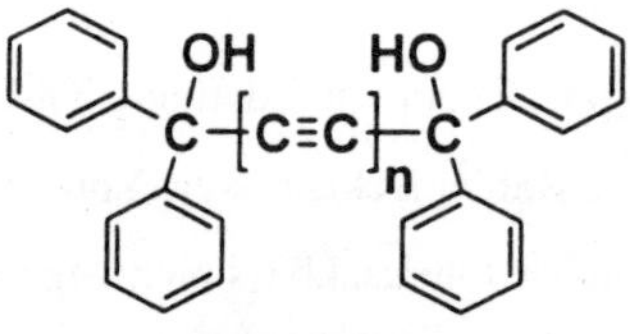

Porosity may be generated, for instance, either by using awkwardly shaped molecules like **425** or by creating a net with large cages or channels (avoiding interpenetration) due to hydrogen bonds or coordination. In this way three- (3D) or two-dimensional (2D) nets are created which can host smaller molecules. Few

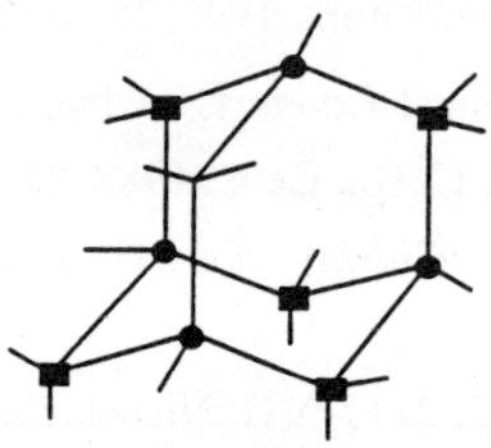

425

426

a
● = ■ = OH_2
●–■ = O-H...O

b
● = ■ = (HOOC)(COOH)(COOH)(COOH)-adamantane
●–■ = –C(=O)O-H··O=C(O-H··O)– (O··H-O / O-H··O)

c
● = $C[C_6H_4-CN]_4$
■ = Cu^+
●–■ = C–C_6H_4–CN···Cu^+

d
● = hexamethylenetetramine (N_4 cage)
■ = CBr_4
●–■ = N···Br

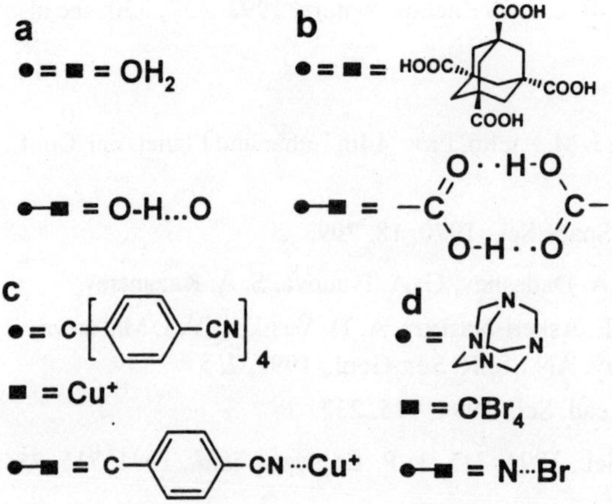

Figure 8.4.1. Examples of moieties forming porous adamantane type of structures.

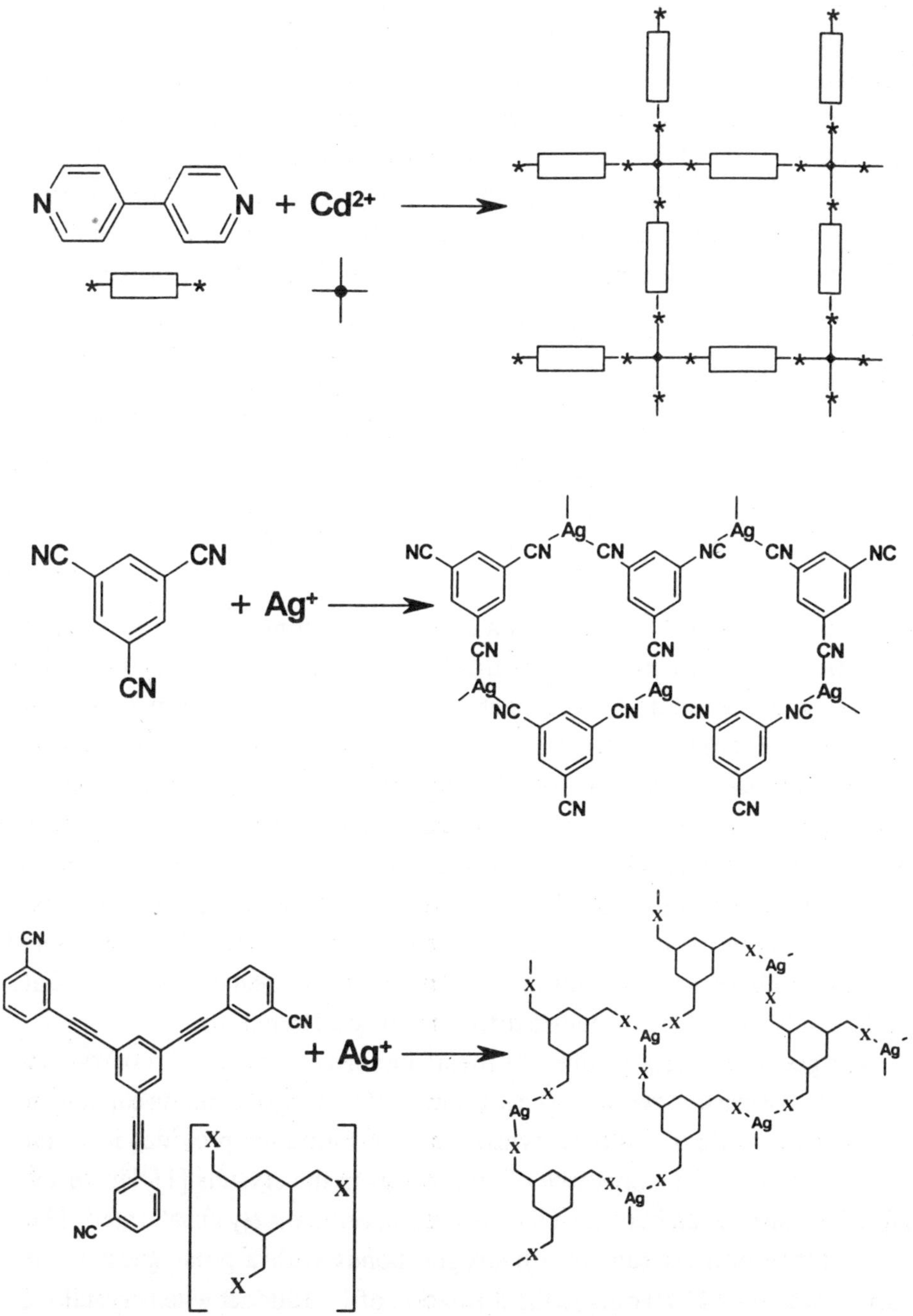

Figure 8.4.2. The formation of a zeolite type of 2D networks by complexation with Ag^{+} or Cd^{2+} ions.

examples of 3D diamond-like lattices and layered 2D nets are shown in Figure 8.4.1 and 8.4.2. The porosity of such materials may be characterized either by the surface area of the material (which may exceed 1000 m^2/g [3]) or by the volume of cavities or channels (occupying in favourable cases up to 70% of the total volume [1]). The classical zeolites also called molecular sieves are aluminosilicate materials with a pore size less than 15 Å, while recently developed novel inorganic zeolites have well-defined larger pores of 15 - 100 Å [4]. On the basis of pore sizes, microporous (<20 Å), mesoporous (20 - 500 Å)

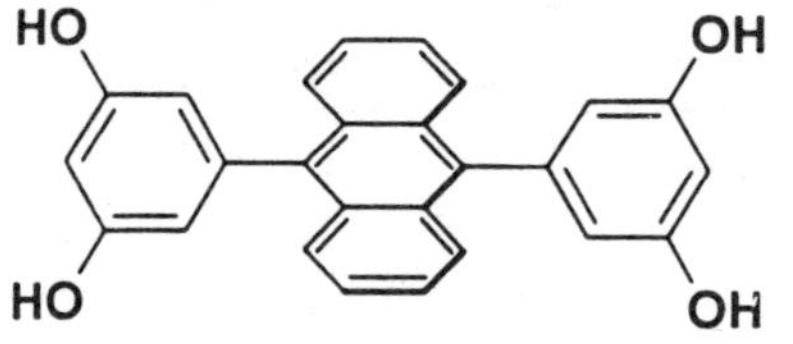

427 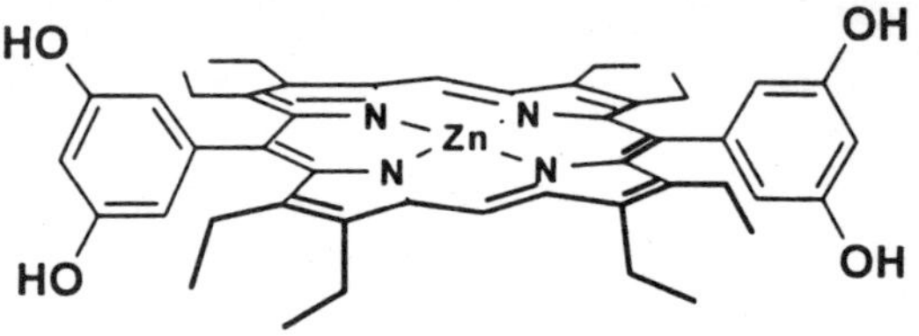 **428**

and macroporous (> 5000 Å) materials are distinguished [5]. Zeolites applications are based on their capability of selective guest binding. It should be stressed once more that the pore sizes (> 20 Å) and void volumes (> 50%) of organic zeolites are larger than those of inorganic ones. As discussed in Section 8.3.3 on clathrate hydrates, a guest can stabilize the host cage. When the guest is removed the hydrogen-bonded network can either be preserved or collapse to a more dense structure. However, the latter systems may be flexible enough to readsorb the guest in favourable conditions and restore the porous structure. Thus two main problems in the studies of organic zeolites are the formation of large but stable permanent voids in the absence of any guest in organic crystals, and the study of molecular guests diffusion in solid materials.

Host-guest interactions are of crucial importance for the properties and applications of organic zeolites. The saturation of hydrogen bonding capacity for many hosts capable of multiple hydrogen bond formation provides for robust networks in spite of the presence of large voids in the crystals [1]. However, a simple hydrogen bond between OH groups is coordinatively unsaturated. Thus, two such free protons can form hydrogen bonds with a polar guest for the diresorcinol host **427** leading to the formation of 1 : 2 adduct when crystallized from ethyl acetate. Interestingly, when crystallization is carried out from a mixture of ethyl acetate and benzene a ternary adduct **427**:2(ester):2(benzene) is

429  **430**

431

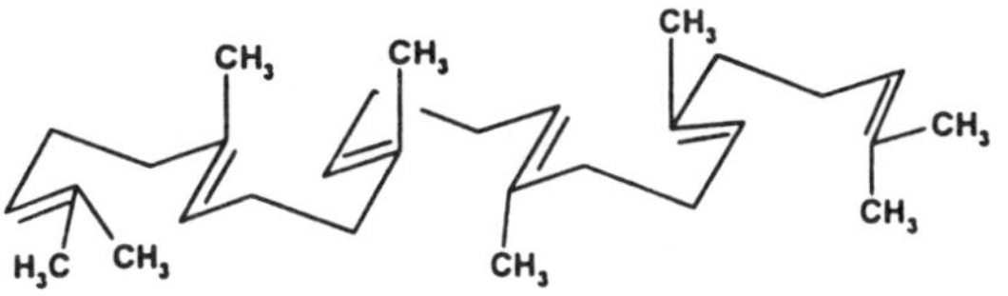

432

created [6]. This means that the volume control of the inclusion is very strict in this case, allowing only for the entering of nonpolar guests into the cavities partly filled with the polar ester molecules. The enclosure of a polar and nonpolar guests within a cavity of **427** provides a basis for application of this organic zeolite as a solid catalyst for bimolecular Diels-Alder reactions [7]. Larger cavities are formed in the net of analogous diresorcinol derivative of Zn(II)-porphyrin **428** allowing in addition for greater flexibility of design making use of cooperation of coordination and hydrogen bonding since the tetrahydrofuran guests have been found to exhibit both hydrogen bonding and coordination interactions with the host [8]. Similarly, in most crystal structures of metalloorganic solids studied, the metal centres are coordinatively saturated and there is no guest-metal interaction unless guest/ligands exchange occurs. In the absence of strong host (metal)-guest interaction, in large cavities or channels there are many guest molecules of the solvent that are highly disordered. Thus, by high porosity of the crystal adduct of **426c** with $x(C_6H_5NO_2)$ ($x \geq 7.7$) two thirds of its volume is essentially liquid [9]!

Although several highly enantioselective [10], regioselective [11], or functional group-selective [12] solid state complexations have been reported, the unambiguous description of such a selectivity is not clear since the crystal packing forces are of a magnitude comparable to those responsible for the complex formation. However, it is obvious that pore size and shape are important factors determining, amongst others, the selectivity. Their control is an important target in the design of organic zeolites. Sometimes the size and shape of the pores may be systematically varied by modifying the host. However, these tactics do not always work since an apparently minor structural alteration can significantly change the crystal structure. This is not the case for a family of chiral bicyclic

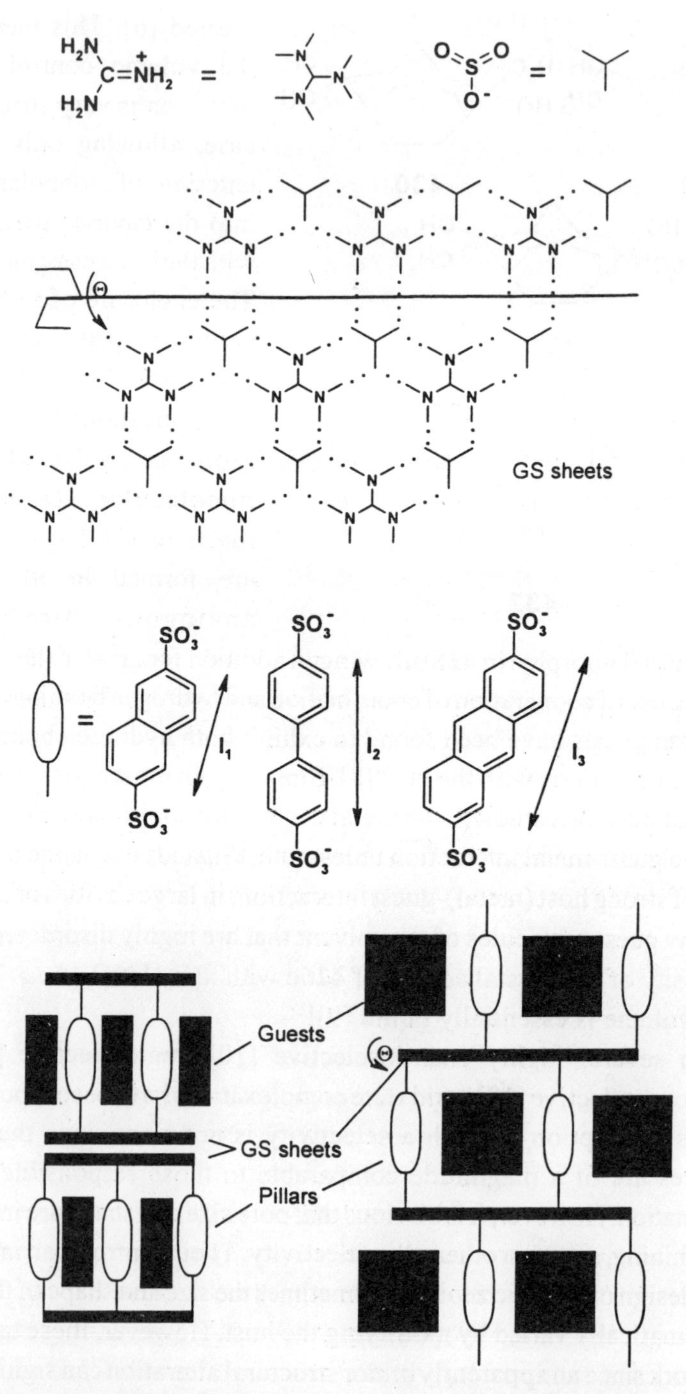

Figure 8.4.3. Guest-dependent association leading to a 2D network.

diols such as **429–431**, since the basic spiral motif is preserved in the family [13] while the canal cross-sectional areas are changed from 0 to 35 Å. Thus guest molecules of varying sizes such as ferrocene $Fe(C_5H_5)_2$ and squalene **432** can be incorporated into the tuned intrahelix channels. Another procedure enabling pore expansion was schematically presented in Figure 4.9 (Section 4.2.3). Holman and Ward were able to control and fine-tune the cavity sizes by creating a layered structure in which the interlayer distance (*i. e.* the pillars length) was changed (Figure 8.4.3) [14]. Interestingly, the layered structure is not rigid. It can partly adapt itself to the guest by accordion-like puckering of the layers and rotational freedom of the pillars.

There is also rapid development in the domain of standard silica-based zeolites. Their versatility can be extended by imprinting. For instance, Davis and Katz [15] recently successfully carried out imprinting and obtained a silica framework with pore walls anchoring three aminopropyl groups in cavities. Another achievement was reported by Ramamurthy, Schefer and coworkers [16]. The latter authors were able to obtain 90% diastereomeric excess of a product of the photochemical reaction in a commercially available zeolite containing chiral tropolone ether **433** in its pores.

Numerous applications of organic zeolites have been proposed. In the future they could serve as catalytic sites and miniature reaction chambers as well as storage compartments and stationary chromatographic phases to be used for the separation of components from mixtures or purification. Similarly to microporous oxides, organic zeolites could probably also be applied as novel functional materials with unique electronic, optical, or mechanical properties. Since 1950 such materials have made possible cheaper and lead-free gasoline, low cost synthetic fibers, and plastics with various versatile applications such as insulated windows, automobile air-conditioning, and air brakes on trucks, to mention but a few [18]. Zeolites are also used to improve energy efficiency, reduce automobile exhaust and other emissions and to clean up hazardous waste (*e. g.*, leakage of radioactive substances by the Three Mile Island nuclear power plant). They have also enhanced environmental protection by facilitating the replacement of ozone-depleting chlorofluorocarbons by new refrigerants. An interesting application

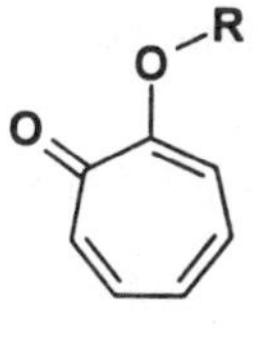

R = CH_3, C_2H_5, $CH(CH_3)_2$

433

may arise from nanowires interconnection pores in silica aerogels [18] and from tunable size- and shape-selective microcavities in nanoporous channels [19].

References

1. Y. Aoyama, Top. Curr. Chem., 1998, 198, 131.
2. S. L. Suib, Chem. Rev., 1993, 93, 803.
3. J. Y. Ying, C. P. Mehnert, M. S. Wong, Angew. Chem. Int. Ed. Engl., 1999, 38, 57.
4. C. T. Kresge, M. E. E. Leonowicz, W. J. Roth, J. C. Vartuli, J. S. Beck, Nature, 1992, 359, 710; J. S. Beck, J. C. Vartuli, W. J. Roth, M. E. E. Leonowicz, C. T. Kresge, K. D. Schmitt, C. T.-W. Chu, D. H. Olsen, E. W. Sheppard, S. B. McCullen, J. B. Higgins, J. L. Schlenker, J. Am. Chem. Soc., 1992, 114, 10834.
5. P. Behrens, Adv. Mater., 1993, 5, 127.
6. K. Kobayashi, K. Endo, Y. Aoyama, H. Masuda, Tetrahedron Lett., 1993, 34, 7929.
7. K. Endo, T. Koike, T. Sawaki, O. Hayashida, H. Masuda,Y. Aoyama, J. Am. Chem. Soc., 1997, 119, 4117.
8. K. Kobayashi, M. Koyanagi, K. Endo, H. Masuda, Y. Aoyama, Chem. Eur. J., 1998, 4, 417.
9. B. F. Hoskins, R. Robson, J. Am. Chem. Soc., 1990, 112, 8719.
10. F. Toda, K. Tanaka, I. Miyahara, S. Atsutsu, K. Hirotsu, J. Chem. Soc. Chem. Commun., 1994, 1795; P. P. Korkas, E. Weber, M. Czugler, G. Naray-Szabo, *ibid.*, 1995, 2229.
11. M. Fujita, Y. J. Kwon, S. Washizu, K. Ogura, J. Am. Chem. Soc., 1994, 116, 1151.
12. C. E. Mario, R. Bishop, D. C. Craig, A. O'Brien, M. L. Scudder, J. Chem. Soc. Chem. Commun., 1994, 2513.
13. A. T. Ung, D. Gizachew, R. Bishop, M. L. Scudder, I. G. Dance, D. C. Craig, J. Am. Chem. Soc., 1995, 117, 8745.
14. K. T. Holman, M. D. Ward, Angew. Chem. Int. Ed. Engl., 2000, 39, 1653.
15. M. E. Davis, A. Katz, Nature, 2000, 403, 286.
16. A. Joy, V. Ramamurthy, Chem. Eur. J., 2000, 6, 1287; W. G. Dauben, K. Koch, S. L. Smith, O. L. Chapman, J. Am. Chem. Soc., 1963, 85, 2615; A. Joy, J. R. Sheffer, V. Ramamurty, Org. Lett., 2000, 2, 119.
17. J. D. Sherman, Proc. Natl. Acad. Sci. USA, 1999, 96, 3471.
18. J. V. Ryan, A. D. Berry, M. L. Anderson, J. W. Long, R. M. Stroud, V. M. Cepak, V. M. Browning, D. R. Rolison, C. I. Merzbacher, Nature, 2000, 406, 169.
19. Y. Shin, J. Liu, L.-Q. Wang, Z. Nie, W. D. Samuels, G. E. Fryxell, G. J. Exarhos, Angew. Chem. Int. Ed. Engl., 2000, 39, 2702.

8.5 Metal-directed Self-assembly of Complex Supramolecular Architecture: Chains, Racks, Ladders, Grids, Macrocycles, Cages, Nanotubes and Self-intertwining Strands (Helicates)

8.5.1 Chains, racks, ladders, grids, macrocycles and cages [1]

434

Metal coordination is a strong directional interaction capable of inducing an abundance of various supramolecular architectures. Today oligomeric linearly arranged *cis*-platinum ions held together by ligands are probably the most important chain systems **434** due to antitumor activity of $PtCl_2(NH_3)_2$ [2]. Ladders, grids, macrocycles, cages, helicates, or even cones are formed, depending on the nature of cation(s) and ligand(s) which, according to the terminology developed by Lehn [1], bear information in their structures. The recognition process leading to the formation of

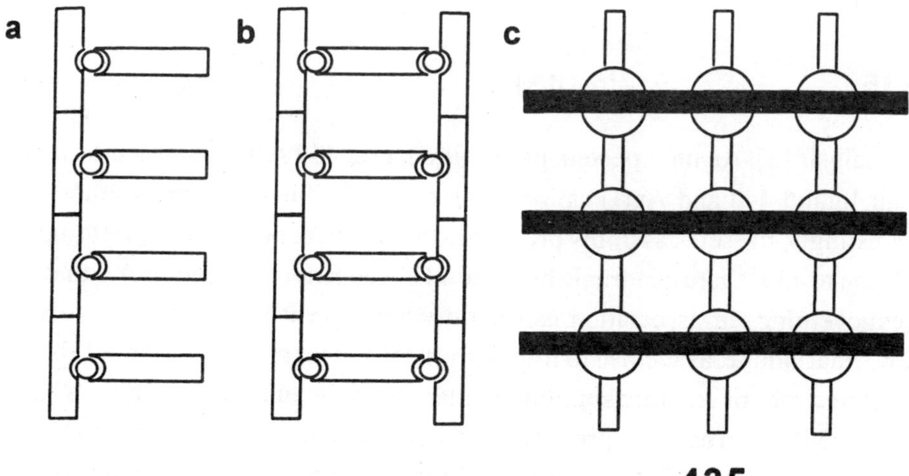

Figure 8.5.1. Symbolic view of a rack (left), ladder (middle), and grid (right).

supermolecule is 'the program' reading this information and utilizing it in the aggregation process. An understanding of this process allows one to design supramolecular systems with specific features which could find applications. For instance, racks (Figure 8.5.1) were obtained when the bipyridine unit **8** was bound to **436** via Cu(I) or another metal ion of tetrahedral coordination, while an analogous reaction involving **436** and **437** with the same type of ion yielded a ladder. A grid **435** was designed using the same reasoning by making use of six ligands **438** and nine Ag(I) cations [1b]. The latter structure was proved by X-

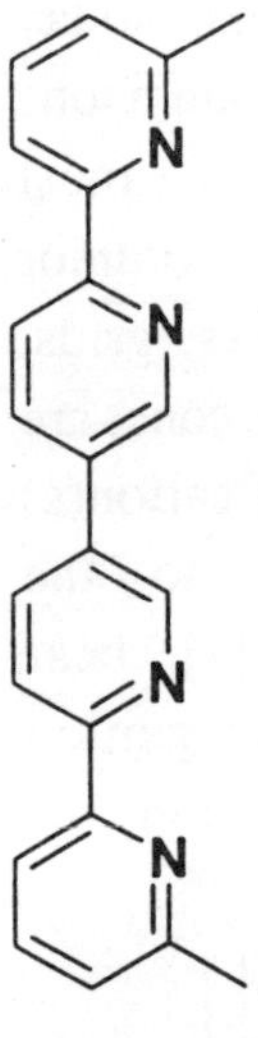

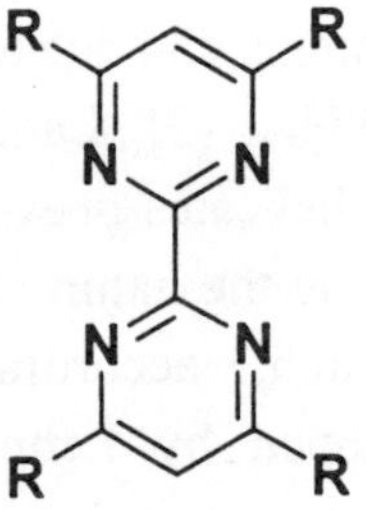

436 **437** **438** **439**

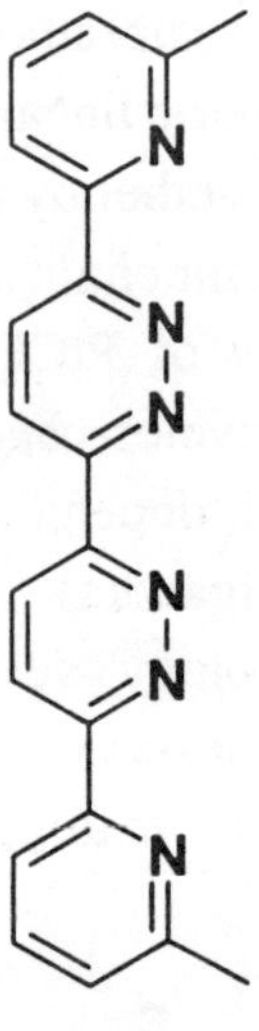

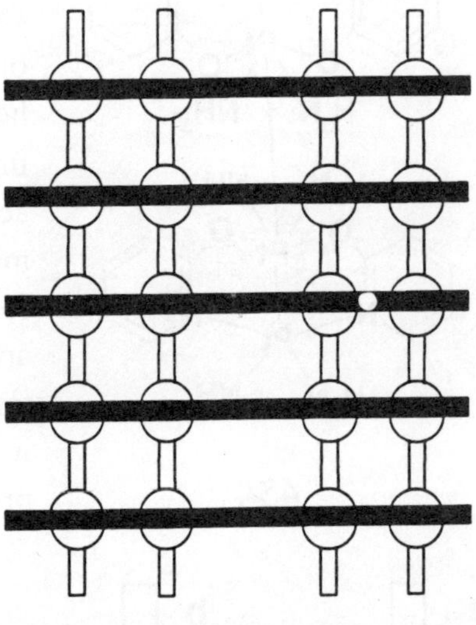

ray analysis [3]. An unexpected, partly filled, grid **439** self-assembled from the linear ligand **440** and Ag(I) cations was established by the same method [4]. Interestingly, the self-assembly process in this case was not specific, yielding, in addition to **439**, a rare quadruple helicate **441**. The latter example shows that the rules governing self-association, especially when solvent molecules are involved, are not that simple, as seemed to follow from the first examples presented above.

Other types of exciting supramolecular motifs include square **382**, **389** and circular **388** macrocycles presented earlier. Similar to the former ones are molecular rectangles [5]. By using chiral strands **442a** and six Ag(I) ions the von Zelewsky group succeeded in obtaining the single-stranded **442b** macrocycle of defined chirality [6]. Hexaruthenium macrocycle **443** built, also of six,

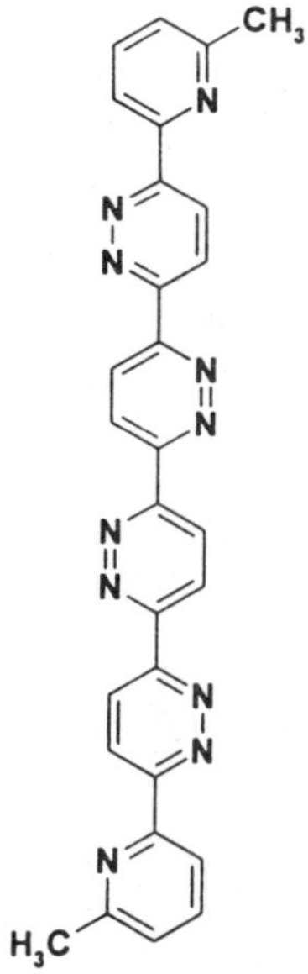

10+

440

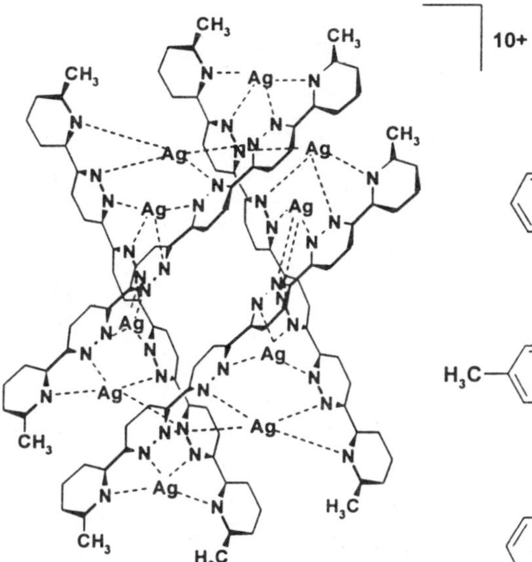

441

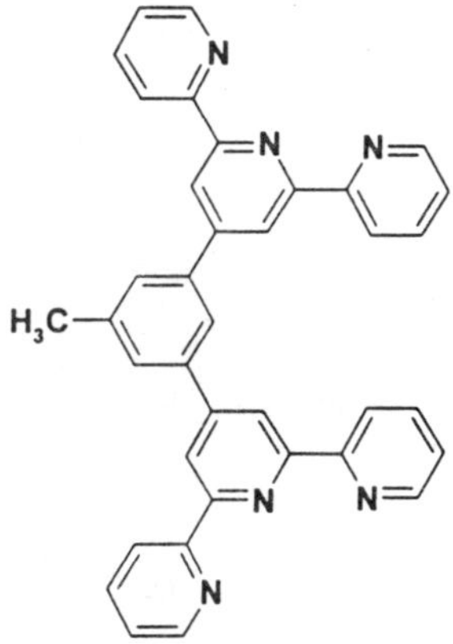

444

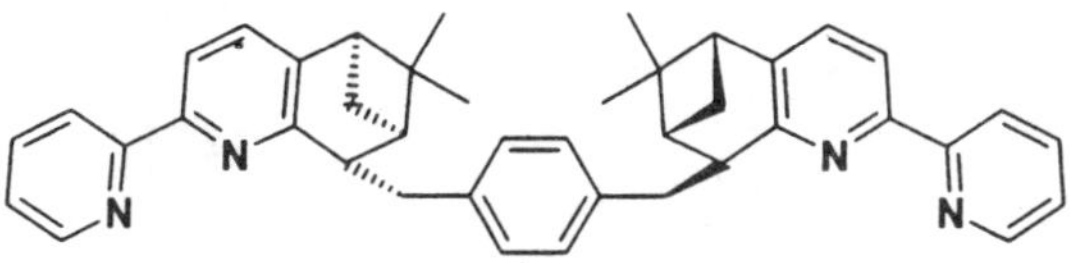

442a

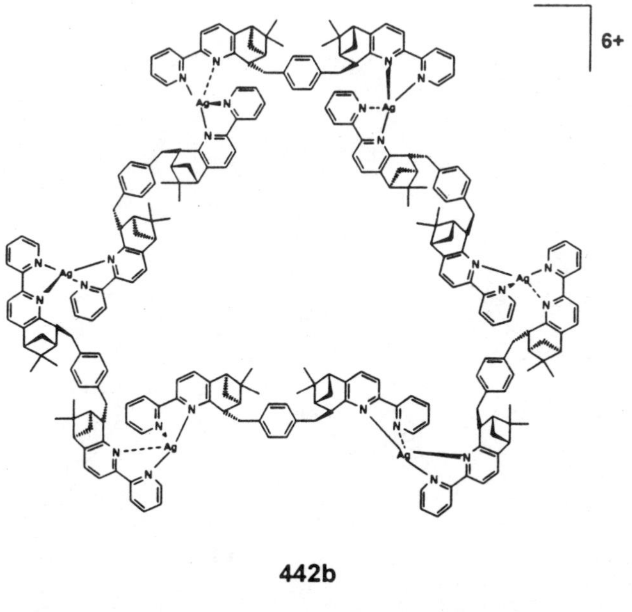

6+

442b

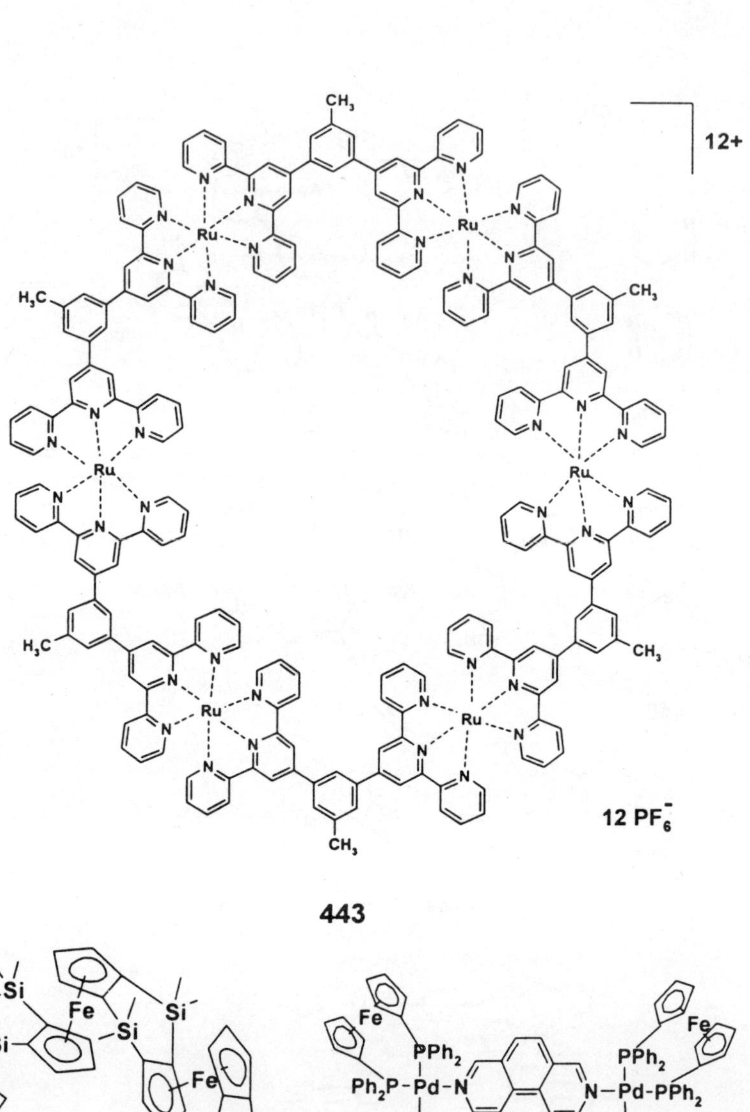

12+

12 PF_6^-

443

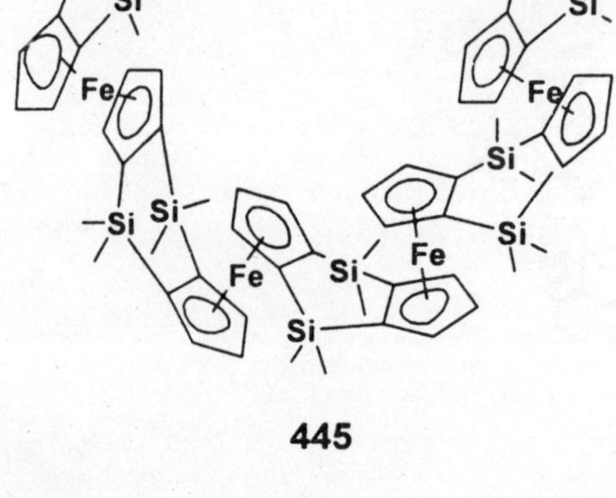

445

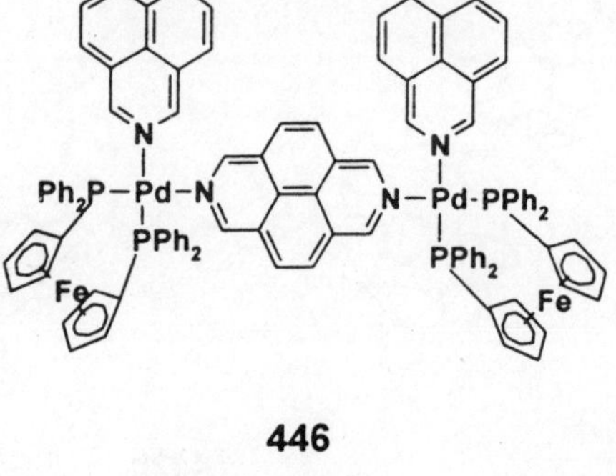

446

bis(terpyridyl) monomers **444** was developed in the hope to obtain a material for a prospective use [7] similarly to the ruthenium-based dendrimer **321** which has interesting luminescent and redox-active properties. The quite unusual circle **445** is composed of seven doubly bridged ferrocene units [8]. **388** mentioned above is self-assembled of approximately 12 cobalt porphyrins bearing similar pyridine substituents [9]. A series of square macrocycles with various aromatic fragments as edges was reviewed by Stang and Olenyuk [10a]. They include, among others, ferrocene **446** [10b], crown ether [10c], or calixarene [10c] units at vertices. More complicated aggregation which cannot not be rationalized in simple terms has taken place to produce the macrocyclic dimer involving a phenanthroline derivative **447**, sugar-boronic acid **448** and Cu(I) cations [11].

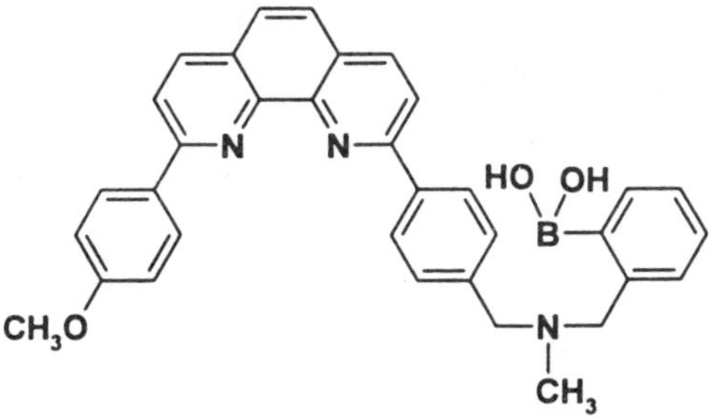

447

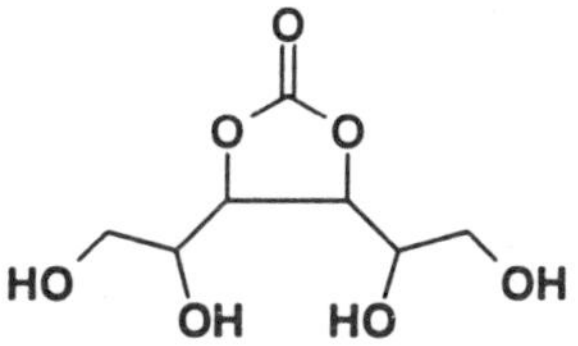

448

A few supramolecular cages self-assembled by means of metal coordination have been presented earlier in this book. They include **26b**, capable of the inclusion of tetramethylammonium cation, the self-assembled cage built of **386a** and **386b** with **387** guest, as well as **264** and **269** in which short-lived species **267** and **268**, respectively, have been stabilized. A nanometer-sized hexahedral capsule (Figure 8.5.2) self-assembled from 18 Pd ions and six 1,3,5-tris(3,5-pyrimidyl)benzene ligands was obtained by Takeda and coworkers [13]. The tetranuclear Fe(III) complex **449a** with a hydrophobic cavity lined with the six phenyl groups (obtained by reacting deprotonated acetylacetonate derivative like **449b** with the metal ions) was described as adamantane-like but it resembles a tetrahedral structure [14]. A supramolecular nanometer-sized cube with O_h symmetry was obtained by Hong, Cao, Chan and coworkers using metal coordination [15]. Today the biggest abiological self-assembled supermolecule is a dodecahedron (Figure 8.5.3) with the general formula $C_{2900}H_{2300}N_{60}P_{120}S_{60}O_{200}F_{180}Pt_{60}$ and a molecular weight of 61,955 [16]. The system, stable for weeks as a solid in the refrigerator or for more than a day

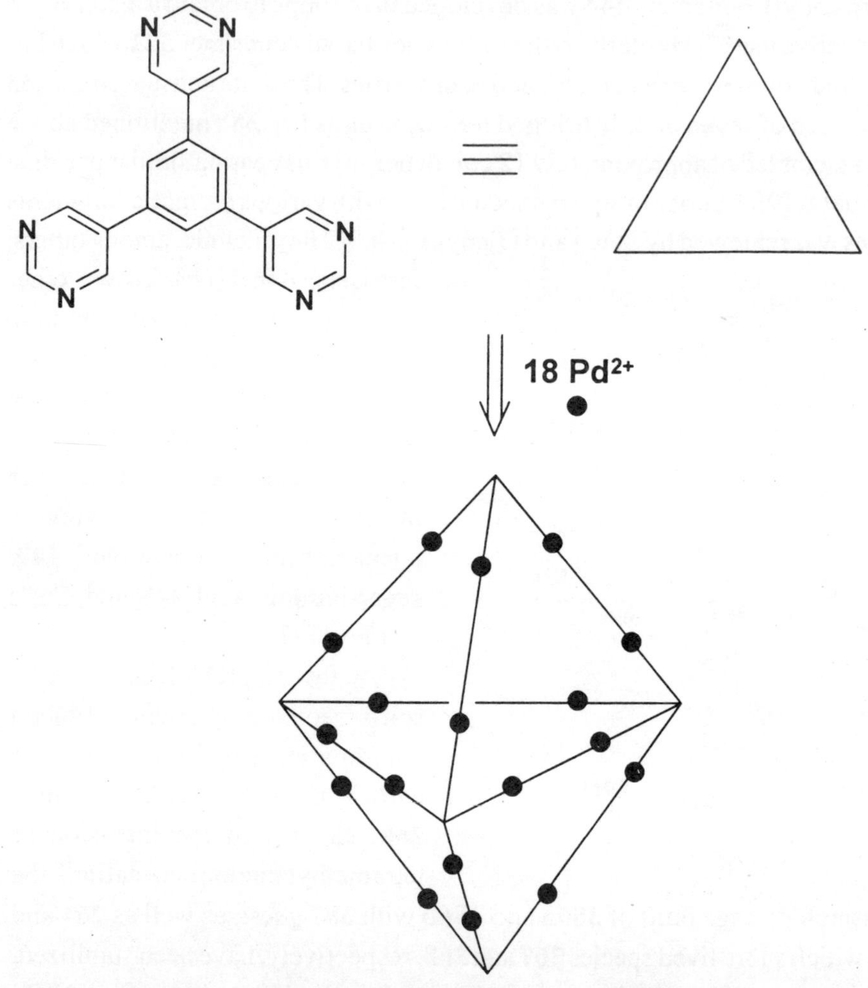

Figure 8.5.2. Formation of a hexahedral complex involving 18 Pd^{2+} ions.

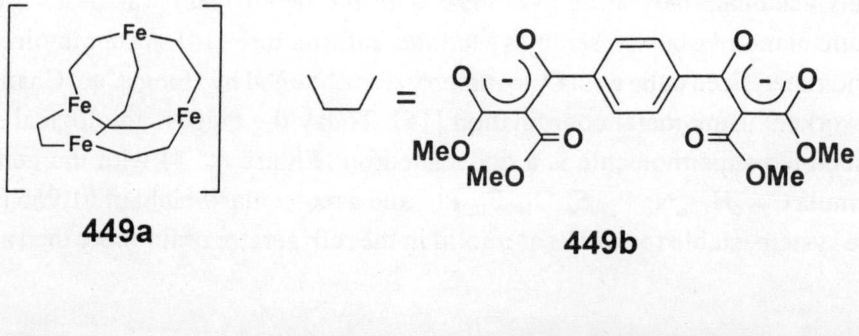

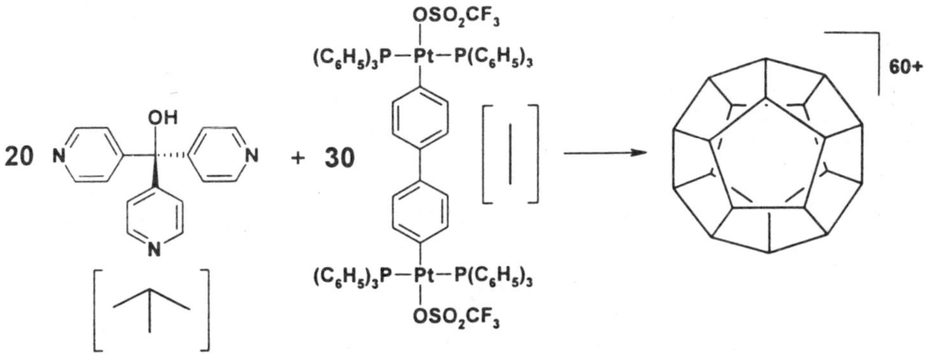

Figure 8.5.3. Self-assembly of 50 molecules yielding huge structure with dodecahedrane architecture.

in solution at room temperature, may serve as a host for several large molecules even as big as fullerenes.

Still another type of a self-assembled structures, an open cone and a coordination nanotube, was obtained by Fujita's group [17]. The process of self-assembly in the latter case is guest controlled, yielding either a cone or a tetrahedron.

Metal-directed self-assembled cyclic complexes can exhibit remarkable selectivity of neutral organic molecules complexed in their voids. The macrocycle **450** [18] exhibits association constant of 2680 $L{\cdot}mol^{-1}$ with *p*-dimethoxybenzene while the corresponding constant for *p*-dinitrobenzene is equal to only 30 $L{\cdot}mol^{-1}$.

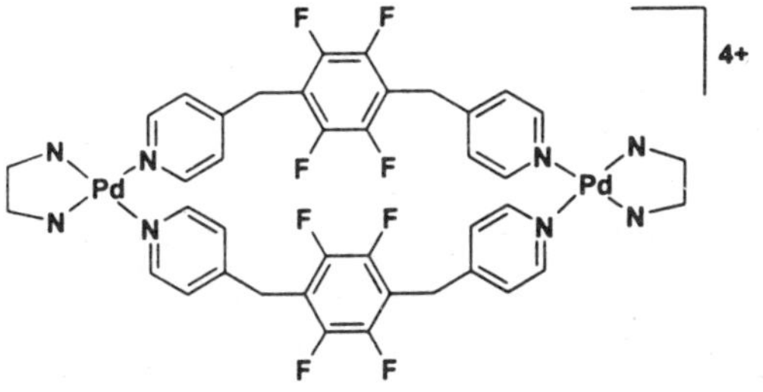

450

8.5.2 Helicates

Helical structures are typical of proteins, nucleic acids and oligosaccharides. Therefore self-assembling helicates and helical fibers have focused the attention of several research groups [19]. Most of the efforts in this field were carried out by the Lehn group. Thus the helicates presented below were mainly obtained by them. The double helicate **9** (Figure 1.3) has been mentioned in Chapter 1. By studying a series of complexes involving oligoethers **451** of different lengths, the

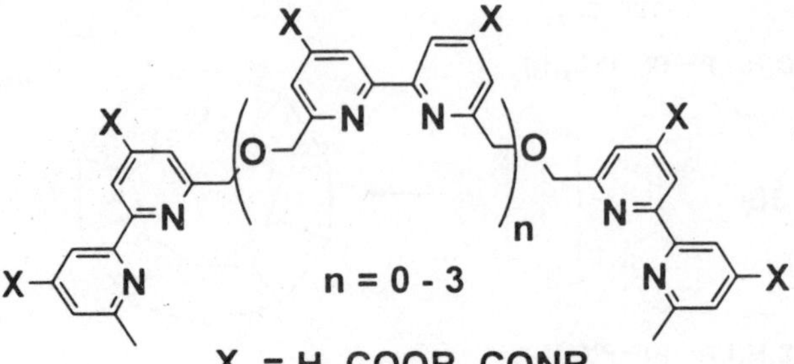

n = 0 - 3

X = H, COOR, $CONR_2$

451

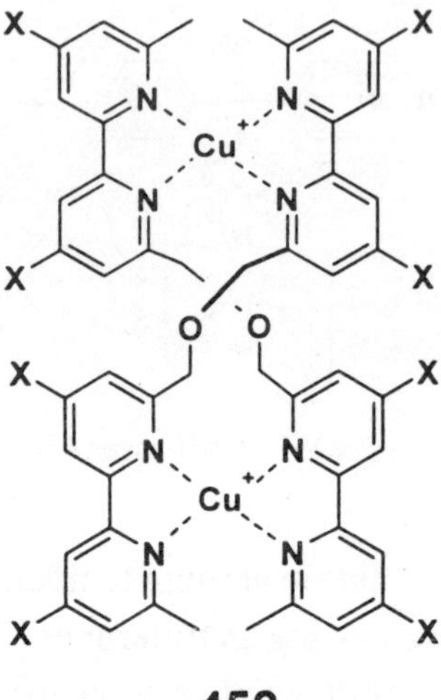

452

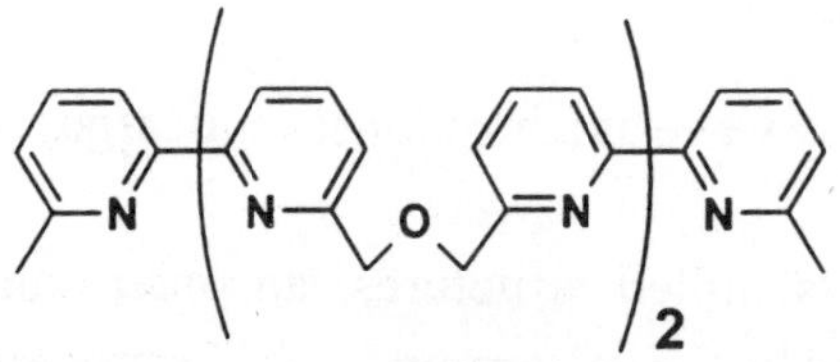

453a

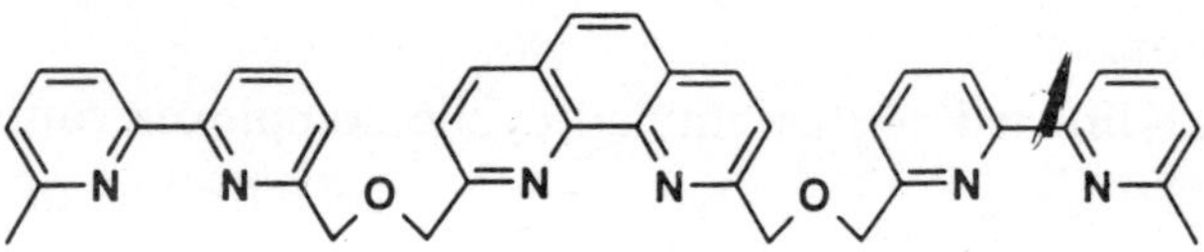

453b

selectivity (in particular, self-recognition) and cooperativity of the helicates **452** formation have been established by the Lehn group in analogy to the simple reasoning enabling them to design hosts for the linear, tetrahedral and spherical guests briefly presented in Section 3.1. The double helix formation resulting from the tetrahedral-like coordination put on by each Cu(I) cation involved, on the one hand, and from the design of the ligand, on the other, was formulated by Lehn in terms of the recognition process constituting "the algorithm" and the molecular steric "program". Understanding the rules governing the helicates' formation leads to a rich variety of such structures including double-, triple- [20] and quadruple- **441** helicates [4]. Attractive molecules promising interesting applications are helicates with pendant nucleosides that bind to nucleic acids [21].

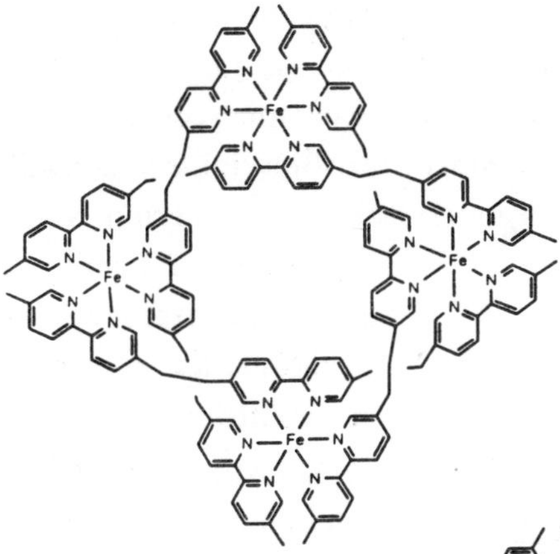

454 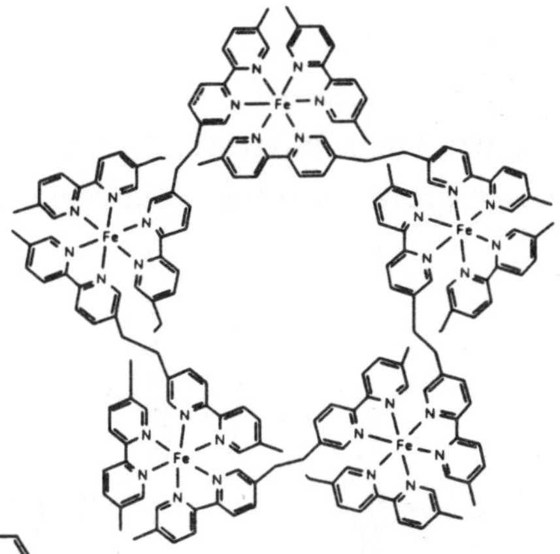 **455**

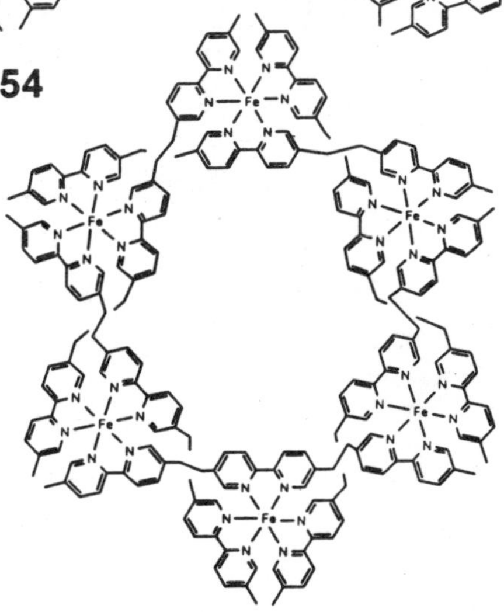

456

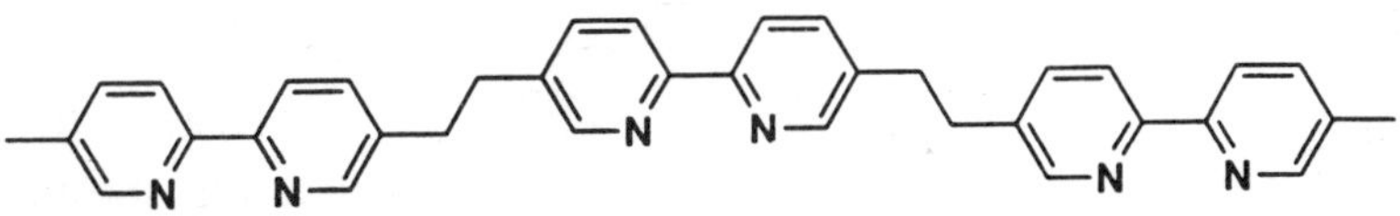

457

As mentioned above, oligoethers **451** exhibited self-recognition forming the doubly-stranded helicates consisting of two identical fragments. However, a mixed, so-called heteroleptic, helicate built of two different strands **453a** and **453b** of a similar length but different constitution was reported by Cohen's group

[22]. Heteronuclear helicates involving Co^{2+} and Ag^{+} cations are also known [23]. The beautiful circular tetra- **454** [24], penta- **455** [25] and hexa-nuclear **456** [24] helical structures are built of the ligands each of which coordinates with three metal ions. Interestingly, the self-assembly of trisbipyridine ligand **457** to afford the penta- or hexanuclear complex is governed by the counterion. Namely, the use of $FeCl_2$ salt yields the smaller macrocycle, with a chloride anion proved to be situated in the cavity by X-ray analysis. On the other hand, applying $Fe(BF_4)_2$, $FeSO_4$, or $FeSiF_6$ salts provides the larger complex, which probably carries one counterion in its centre. By extending the strands, the gorgeous dodecanuclear helicate **458** [Cu_{12}(**458a**)$_4^{12+}$] has been obtained [26]. Under certain experimental conditions not only macrocyclic **454**–**456** and **458** are obtained, but also grid-type structures like **435** or **439** depending on the number and the type of ions used.

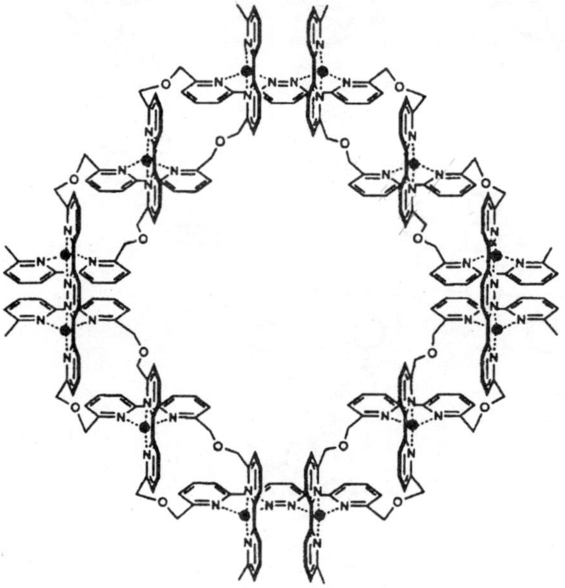

458

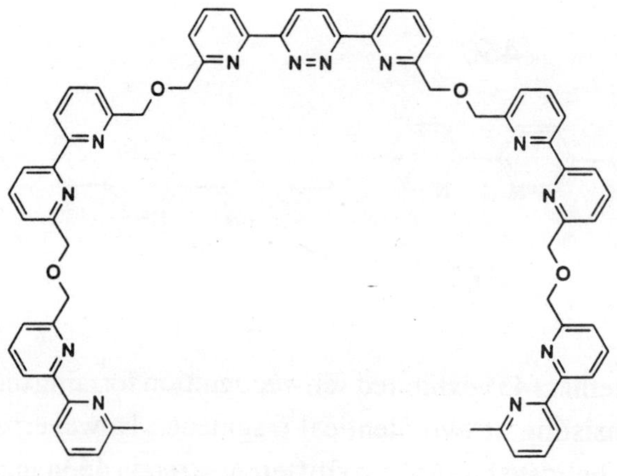

458a

Helical intertwined structures should exhibit chirality. However, the helicates obtained from strands **451** are racemic mixtures of the right- and left-handed double helices. Jodry and Lacour resolved a dinuclear triple helicate **459**

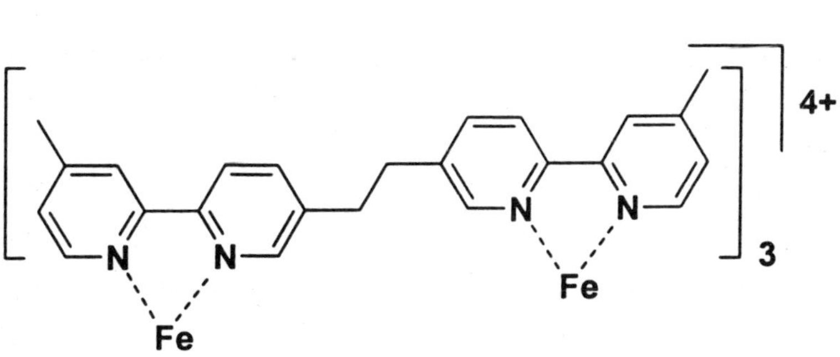

459

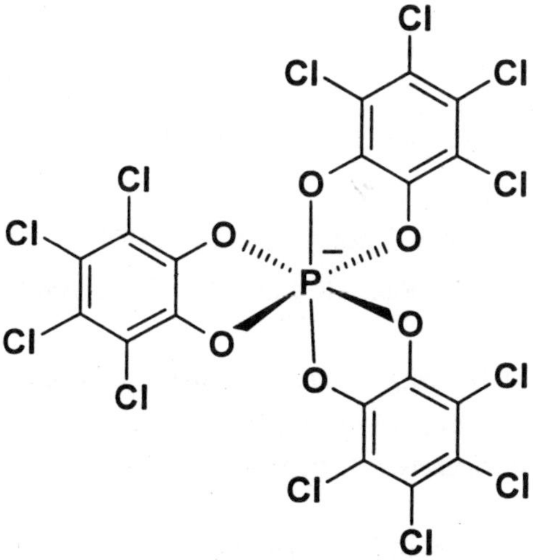

460

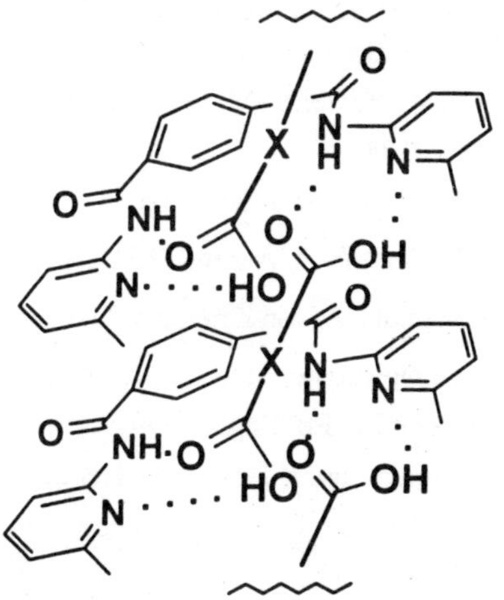

$X = (CH_2)_3$

461

by asymmetric extraction/precipitation with TRISPHAT anions **460** as resolving agents [27].

Lehn and coworkers also discussed the principles of design of self-assembling helical structures on the basis of conformational control. Depending on the solvent used, they were able to obtain helicate fibers and bundles possessing extended molecular channels characterized by hollow cores of limited diameter of ca. 8 Å. This finding could form the basis of functional materials for multichannel ion active transport [28].

Metal coordination is the main building factor used in helicate synthesis. Few examples utilizing hydrogen bonding with purpose in mind include among others polymeric **461** synthesized in the Hamilton group [29], quadruple helices **104** forming in water and self assembling **102** with **103** discussed in Section 2.5.

References

1. (a) J.-M. Lehn, *Supramolecular Chemistry. Concepts and Perspectives*, VCH, Weinheim, 1995; (b) p. 157.
2. J. K. Barton, H. N. Rabinovitz, D. J. Szalda, S. J. Lippard, J. Am. Chem. Soc., 1977, 99, 2827; J. K. Barton, D. J. Szalda, H. N. Rabinovitz, J. V. Waszczak, S. J. Lippard, J. Am. Chem. Soc., 1979, 101, 1434.
3. P. N. W. Baxter, J.-M. Lehn, J. Fischer, M.-T. Youinou, Angew. Chem. Int. Ed. Engl., 1994, 33, 2284.
4. P. N. W. Baxter, J.-M. Lehn, G. Baum, D. Fenske, Chem. Eur. J., 2000, 6, 4510.
5. A. K. Duhme, Z. Anorg. Chem., 1998, 624, 1922.
6. O. Mamula, A. von Zelewsky, G. Bernardinelli, Angew. Chem. Int. Ed. Engl., 1998, 37, 290.
7. G. R. Newkome, T. J. Cho, C. N. Moorefield, G. R. Baker, R. Cush, P. S. Russo, Angew. Chem. Int. Ed. Engl., 1999, 38, 3717.
8. B. Grossmann, J. Heinze, E. Herdtweck, F. H. Kohler, H. Noth, H. Schwenk, M. Spiegler, W. Wachter, B. Weber, Angew. Chem. Int. Ed. Engl., 1997, 36, 387.
9. R. A. Haycock, C. A. Hunter, D. A. James, U. Michelsen, L. R. Sutton, Org. Lett., 2000, 2, 2435.
10. (a) P. J. Stang, B. Olenyuk, Acc. Chem. Res., 1997, 30, 502; (b) P. J. Stang, B. Olenyuk, J. Fan, M. A. Arif, Organometallics, 1996, 15, 904; (c) P. J. Stang, D. H. Cao, K. Chen, G. M. Gray, D. C. Muddiman, R. D. Smith, J. Am. Chem. Soc., 1997, 119, 5163.
11. M. Yamamoto, M. Takeuchi, S. Shinkai, Tetrahedron Lett., 2000, 41, 3137.
12. A. Ikeda, M. Ayabe, S. Shinkai, S. Sakamoto, K. Yamaguchi, Org. Lett., 2000, 2, 3707.
13. N. Takeda, K. Umemoto, K. Yamaguchi, M. Fujita, Nature, 1999, 398, 794.
14. R. W. Saalfrank, B. Horner, D. Stalke, J. Salbeck, Angew. Chem. Int. Ed. Engl., 1993, 32, 1179.
15. M. Hong, Y. Zhao, W. Su, R. Cao, M. Fujita, Z. Zhou, A. S. C. Chan, J. Am. Chem. Soc., 2000, 122, 4819.
16. B. Olenyuk, M. D. Levin, J. A. Whiteford, J. E. Shield, P. J. Stang, J. Am. Chem. Soc., 1999, 121, 10434.
17. (a) M. Aoyagi, K. Biradha, M. Fujita, J. Am. Chem. Soc., 2000, 122, 7150; (b) K. Umemoto, K. Yamaguchi, M. Fujita, J. Am. Chem. Soc., 2000, 122, 7150.

18. M. Fujita, S. Nagao, M. Iida, K. Ogata, K. Ogura, J. Am. Chem. Soc., 1993, 115, 1574.

19. C. Piguet, G. Bernardinelli, G. Hopfgartner, Chem. Rev., 1997, 97, 2005; A. E. Rowan, R. J. M. Nolte, Angew. Chem. Int. Ed. Engl., 1998, 37, 63.

20. S. Rigault, C. Piguet, G. Bernardinelli, G. Hopfgartner, Angew. Chem. Int. Ed. Engl., 1998, 37, 169.

21. U. Koert, M. M. Harding, J.-M. Lehn, Nature, 1990, 346, 339; B. Schoentjes, J.-M. Lehn, Helv. Chim. Acta, 1995, 78, 1.

22. M. Greenwald, D. Wessely, E. Katz, I. Willner, Y. Cohen, J. Org. Chem., 2000, 65, 1050.

23. E. C. Constable, V. J. Walker, J. Chem. Soc. Chem. Commun., 1992, 884; E. C. Constable, A. J. Edwards, R. Raithby, V. J. Walker, Angew. Chem. Int. Ed. Engl., 1993, 32, 1465.

24. B. Hasenknopf, J.-M. Lehn, N. Boumediene, A. Dupont-Gervais, A. van Dorsselaer, B. Kneisel, D. Fenske, J. Am. Chem. Soc., 1997, 119, 10956.

25. B. Hasenknopf, J.-M. Lehn, J.-M. Baum, B. O. Kneisel, D. Fenske, Angew. Chem. Int. Ed. Engl., 1996, 35, 1838.

26. D. P. Funeriu, J.-M. Lehn, J.-M. Baum, D. Fenske, Chem. Eur. J., 1997, 3, 99.

27. J. J. Jodry, J. Lacour, Chem. Eur. J., 2000, 6, 4297.

28. L. A. Cuccia, J.-M. Lehn, J.-C. Homo, M. Schmutz, Angew. Chem. Int. Ed. Engl., 2000, 39, 233.

29. S. J. Geib, C. Vicent, E. Fan, A. D. Hamilton, Angew. Chem. Int. Ed. Engl., 1993, 32, 119.

Chapter 9

THE PROSPECTS OF FUTURE DEVELOPMENT OF SUPRAMOLECULAR CHEMISTRY

In spite of its lack of definition, supramolecular chemistry is booming and only a few examples of research in this rapidly developing area could be presented in this book. The field is expanding so fast, it has widened to such an extent that, in the author's opinion, it has ripened to be split into three areas differing by the objects studied and experimental techniques used. The first one would deal with small aggregates, that is, mainly with inclusion complexes. *Inclusion* or *host-guest chemistry*, is an accepted name for the greatest part of it. Another well-defined area of supramolecular chemistry is *crystal engineering*, devoted to the design and studies of crystals with desired properties built of two- or more components. In between these two fields is a domain which could be dubbed *aggregate chemistry*. This area lying between supramolecular chemistry of small aggregates and those crystal studies dealing with huge aggregates would be devoted to the studies of aggregates of intermediate size: mono-, bi-, and multi-layers, micelles, vesicles, fibers and similar mesoscopic systems. It should be stressed that, while small aggregate chemistry deals with complexes having a well-defined mass, the number of subunits forming layers, micelles, and other objects falling into this category is not defined. In this respect small aggregates resemble simple molecules while the latter objects and supramolecular crystals are analogues of polymers.

On the basis of the present achievements in supramolecular chemistry, some trends in its future development can be foreseen. The prospects for supramolecular chemistry are bright. Situated between chemistry, physics, biology, and technology, it should enhance our understanding of the operation of living organisms at the molecular level, on the one hand, and form a basis for the creation of not only smaller (ideally unimolecular) but environmentally friendly and more efficient devices, on the other. The trend to obtaining bigger self-assembled species is obvious. Beautiful huge cages, such as those discussed in Section 8.5, mark one direction of development. Physicochemical studies leading to a better understanding of the driving forces of self-assembly, in particular for inclusion complex formation, is another. Such studies combined with theoretical methods will allow one to predict in the future the recognition ability of a given host. They will build a sounder basis for the rational design of complexes exhibiting desired properties. In the terminology developed by Lehn [1], we will be able to read the information encoded in a molecular structure, using another structure to direct the formation of supramolecular aggregates.

A better understanding of complicated recognition processes, such as that of hexakinase enzyme in which the host cavity is created in the guest presence [2], will be achieved.

The creation of aggregates of higher complexity, such as the self-assembled channels which are subsequently built into a layer (such as those consisting of the alternating *D*- and *L*-amino acids discussed in Section 4.2.4), is another booming area that will expand rapidly. Such channels enable studying transport through membrane. The latter example brings us to the recently founded field of *biomimetic supramolecular chemistry*. Mimicking a spider's web, enzyme activity (*e. g.* by cyclodextrins discussed in Section 3.5) or the smelling sensitivity of dogs' noses will not only deepen our knowledge of the living world but also help by the creation of smaller and much more efficient devices. Some of these developments were discussed at "NanoSpace 2000. Advancing the Human Frontier" conference organized by the National Aeronautics & Space Administration (NASA) [3]. The application of antibodies for fullerenes and nanotubes selection and formation of composite nanosized implant materials (exhibiting much better properties than the existing ones) presented there are representative examples of this development [4], As discussed in Section 7.5, carbon nanotubes are indispensable in the drive to miniaturize computers, sensors and other devices. Today the available amounts of high-quality uniform nanotubes are very small. Massive efforts to obtain larger amounts of pure fullerenes and, especially, nanotubes should be mentioned here. The considerable funds supporting the studies of their formation allow one to expect that a considerable breakthrough will change this situation soon. This, in turn, will enable industrial production of electronic devices built of elements consisting of one molecule or one aggregate. The first transistor based on a single nanotube discussed in Section 6.3.2.2 (still large with all its connecting elements) and logical gates, such as the one proposed by Balzani, Stoddart *et al.* [5], exemplify such devices. A tiny fuel cell smaller, but similar, to those that power NASA space shuttles [6] and a palm-sized chemical analysis device [7] are significantly larger than the monomolecular devices mentioned above, but they represent a technological trend in the same direction. In addition to such devices, supramolecular chemistry will not only form the foundations for the development of novel catalysts and drugs, but also for new methods of drug administration as well as numerous other new supramolecular materials. Alkalides and electrides discussed in Section 7.1.3 and stable cyclobutadiene **4** in the hemicarcerand **5** cage discussed in Section 7.3 show how dramatically can change the properties of a molecule or ion when it is involved in a supramolecular complex. The latter example points to a fruitful cooperation between supramolecular chemistry, on the one hand and theoretical and synthetic chemistry of unusual hydrocarbons, on the other, which would enable capturing and studying short-lived species in supramolecular complexes.

Such a development would be impossible without the generation of new experimental techniques extending the possibilities of studies of supramolecular systems. One such new technique is the vibrational (or infrared-visible) spectroscopy sum frequency generation [8] enabling one to look at just the outer monolayer, thus providing more information than more standard methods like contact angle and surface tension measurements [9].

The question "Will chemistry survive in XXI century?" posed at The 215th Meeting of the American Chemical Society in Dallas in March 1998 [10] was answered positively. But it was stressed that chemistry will undergo a certain evolution. Rapidly developing exciting domain of supramolecular chemistry will certainly occupy an important position in this development.

References

1. J.-M. Lehn. Angew. Chem. Int. Ed. Engl., 1990, 29, 1304.

2. R. B. Prince. S. A. Barnes, J. S. Moore, J. Am. Chem. Soc., 2000, 122, 2758.

3. R. Dagani, Chem. Eng. News, 2000, Feb. 28, p. 36.

4. R. Dagani, Chem. Eng. News, 2000, Feb. 28, p. 39.

5. A. Credi, V. Balzani, S. J. Langford, J. F. Stoddart, J. Am. Chem. Soc., 1997, 119, 2679.

6. T. Campbell, http://chemweb.com/alchem/2000/news/nw_000505_fuel.html

7. T. Campbell, http://chemweb.com/alchem/2000/news of 20 April 2000.

8. G. A. Somorjai, G. Rupprecht, J. Chem. Phys. B, 1999, 103, 1623.

9. A. M. Rouhi, Chem. Eng. News, 2000, June 26, 33.

10. N. S. Getty, J. Chem. Educ., 1998, 75, 665.

Index